Statistical Tests

Statistical Tests

An introduction with MINITAB commentary

G. P. BEAUMONT

J. D. KNOWLES

PRENTICE HALL

London   New York   Toronto   Sydney   Tokyo   Singapore
Madrid   Mexico City   Munich

First published 1996 by
Prentice Hall International (UK) Limited
Campus 400, Maylands Avenue
Hemel Hempstead
Hertfordshire, HP2 7EZ
A division of
Simon & Schuster International Group

Typeset in 10/12pt Times
by Mathematical Composition Setters Ltd., Salisbury, UK

Printed and bound in Great Britain by
Hartnolls Limited, Bodmin, Cornwall

Library of Congress Cataloging-in-Publication Data

Beaumont, G. P.
    Statistical tests : an introduction with Minitab commentary / G.P.
    Beaumont, J.D. Knowles.
        p.   cm.
    Includes index.
    ISBN 0-13-842576-0 (pbk. : alk. paper)
    1. Statistical hypothesis testing.   2. Minitab.   I. Knowles, J.
    D.   II. title.
    QA277.B385   1996
    519.5'6–dc20   95-30474
                                                                        CIP

British Library Cataloguing in Publication Data

A catalogue record for this book is available from
the British Library

ISBN 0-13-842576-0

1 2 3 4 5 00 99 98 97 96

Contents

Preface

This book is intended to provide a second course in statistics, concentrating on statistical tests but omitting those topics such as regression and analysis of variance which commonly appear in a parallel course. We discuss a range of statistical procedures, adding a theoretical presentation to a running commentary on the numerical aspects as presented by the statistical package MINITAB. It is not necessary to have access to MINITAB – the presentation of the material is independent of the package.

Introductory courses vary considerably both in scope and difficulty. In the first chapter we provide brief outlines of the background knowledge we assume – recording the basic results of probability and establishing the notation (Section 1.1), explaining the MINITAB commands PDF, CDF, INVCDF and RANDOM (Section 1.2), reviewing those standard distributions which play a key role in the main text (Section 1.3), and in the final section (Section 1.4) discussing the sampling distributions. The topic of sampling from a finite population – frequently omitted from a first course – finds no natural place in the text but is addressed in Appendix 2.

The first three chapters of the main text discuss the basic ideas for the topics of data analysis (Chapter 2), estimation of parameters (Chapter 3) and hypothesis testing (Chapter 4). In hypothesis testing, our prevailing attitude is to report the p-value of a statistic, rather than stating its position relative to a particular critical value, and we provide more discussion of two-sided tests than is usual. Chapters 5–8 provide extensive cover for certain distribution-free tests which lend themselves to a detailed discussion of their properties and we include an introduction to multiple testing – an important subject not often discussed at this level. The final chapter (Chapter 9) provides a formal basis for those tests which were treated earlier in an informal way. Each section throughout these chapters concludes with a variety of problems and complete solutions are provided at the end of the chapter.

The mathematical knowledge required rarely exceeds that to be found in first courses in calculus, together with such techniques of counting and manipulation that are common in a treatment of probability and the distribution of random variables. This self-imposed restriction has occasionally led to a choice of method which is not the most sharp or elegant, and to the omission of limiting results for large samples. Otherwise, an attempt has been made to derive the properties of the statistics employed – including an extended derivation for the chi-squared, t and F densities in Appendix 1.

In the use of MINITAB, we assume only that the reader can enter data and perform the basic arithmetical operations, with the other commands explained as they arise in the text. Here, we use the command line method of MINITAB version 8 (and earlier) and leave readers having access to MINITAB for Windows free to enter their instructions using the mouse operations. The few discrepancies between the results delivered by MINITAB and those anticipated on theoretical grounds appear mostly to arise from our use of perhaps unreasonably small samples when illustrating the theory – in such cases, MINITAB is likely to provide a warning message. By relating the MINITAB procedures to the theory, we hope to discourage any temptation to use the package mechanically.

Acknowledgements

We acknowledge, with gratitude, permission to quote tabulated values for the standard distributions given in *The New Cambridge Elementary Statistical Tables* by D. V. Lindley and W. F. Scott, and help rendered by the MINITAB Author Assistance Program.

Particular assistance was provided by Professors Brian Connolly and Ken Bowen. Brian has provided freshly computed tables for the distribution functions of several statistics used in non-parametric tests, and, as editor, was most helpful and encouraging in the early stages of the project. Ken has amazed us with his stamina as he read most of the many drafts of the text – his green biro was dishearteningly active! We are pleased to record our thanks.

Several charming ladies battled to convert our miserable handwriting into type – in particular, Mrs B. Alderman has borne the brunt of preparing a large part of the manuscript. To them our thanks are tinged with sympathy. Finally, we record assistance with some late amendments and the diagrams from our colleague Dr D. L. Yates.

Note

Geoffrey Beaumont passed away shortly after the manuscript was delivered to the publisher. This book, Geoffrey's fifth, was conceived to mark his retirement and draw upon his long experience of teaching statistics at the undergraduate level. Credit for any merits the book may be deemed to possess is his, responsibility for any blemishes rests with me (J.D.K.).

Chapter One

Preliminaries

1.1 Probability and distributions

We collect here the basic probability notation and formulae we use throughout the book.

1.1.1 Probability spaces

The starting point in a discussion of probability is to identify the sample space S of all possible outcomes, and the assignment of probability to collections of outcomes, called events and regarded as subsets of S. Basic properties are

P(i) $\Pr(\phi) = 0, \Pr(S) = 1$
P(ii) $0 \leqslant \Pr(E) \leqslant 1$ for all events E
P(iii) $\Pr(E \cup F) = \Pr(E) + \Pr(F)$ for mutually exclusive (disjoint) events E and F
P(iv) $\Pr(E \cup F) = \Pr(E) + \Pr(F) - \Pr(E \cap F)$ for any two events E and F
P(v) $\Pr(E^c) = 1 - \Pr(E)$ for any event E where E^c = complement of E = *not* E
P(vi) E and F are independent events if and only if $\Pr(E \cap F) = \Pr(E)\,\Pr(F)$.

1.1.2 Random variables

Many problems are more conveniently described in terms of a random variable rather than identifying the sample space. In fact, a random variable X is simply a function on the sample space, giving, to each element of S, the numerical value of interest in the problem at hand.

For each random variable X, the probability information regarding the spread of values of X is given by the distribution function $F_X(x)$, or simply $F(x)$ when X is understood, where

$$F(x) = \Pr(X \leqslant x)$$

Basic properties are:

F(i) $0 \leqslant F(x) \leqslant 1$ for all x

F(ii) $F(x_1) \leqslant F(x_2)$ if $x_1 < x_2$ (i.e. F is increasing)

F(iii) $F(-\infty) = \lim_{x \to -\infty} F(x) = 0, \qquad F(+\infty) = \lim_{x \to +\infty} F(x) = 1$

F(iv) $\Pr(a < X \leqslant b) = F(b) - F(a)$ for $a < b$.

We identify the two types of random variable that occur in practice:

1. X is discrete taking *distinct* values x_1, x_2, ..., with probability function (p.f.) f_X, or simply f, given by $f(x_i) = \Pr(X = x_i)$ for $i = 1, 2, \ldots$, so that

$$\sum_i f(x_i) = 1$$

The definition of f may be completed by putting $f(x) = 0$ for $x \neq x_1, x_2, \ldots$. For such a random variable the distribution function is given by

$$F(x) = \sum_{x_i \leqslant x} f(x_i)$$

and its graph is a step function, with 'jumps' at $x_1, x_2, \ldots$.

2. X is continuous when the distribution function may be represented as

$$F(x) = \int_{-\infty}^{x} f(t)\, dt$$

where the probability density function (p.d.f.) $f(x) \geqslant 0$ for all x and

$$\int_{-\infty}^{+\infty} f(x)\, dx = 1$$

In this case F is continuous and, wherever f is continuous, $F'(x) = f(x)$.

The crucial distinction between the discrete and continuous type random variables is that, when X is of continuous type, $\Pr(X = x) = 0$ for every value x, and consequently

$$\Pr(a \leqslant X \leqslant b) = \Pr(a < X \leqslant b) = \Pr(a < X < b) = F(b) - F(a) = \int_{a}^{b} f(x)\, dx$$

Another, irritating, feature of this distinction in types is that many formulae for calculating purposes have to be stated twice, once for each type. Fortunately, the distinction is uniform in style: we need $\sum$ for discrete X and $\int$ for continuous X.

1.1.3   *Expectation and variance*

For a random variable X, define the expectation $\mu = E(X)$ by

$$E(X) = \begin{cases} \displaystyle\sum_i x_i f(x_i) & \text{for } X \text{ discrete} \\[2mm] \displaystyle\int_{-\infty}^{+\infty} x f(x)\, dx & \text{for } X \text{ continuous} \end{cases}$$

with the definition being valid, as in all the examples we consider, when the series/integral is absolutely convergent.

The variance $V(X) = \sigma_X^2$, or simply σ^2, is defined by

$$V(X) = E((X-\mu)^2 = E(X^2) - \mu^2$$

and $\sigma = \sqrt{V(X)}$ is the standard deviation.

For calculation purposes, when $g(x)$ is any function,

$$E(g(X)) = \begin{cases} \sum_i g(x_i)f(x_i) & \text{for } X \text{ discrete} \\ \int_{-\infty}^{\infty} g(x)f(x)\,dx & \text{for } X \text{ continuous} \end{cases}$$

In particular, with $g(x) = x^2$, we have, in the continuous case,

$$E(X^2) = \int_{-\infty}^{\infty} x^2 f(x)\,dx$$

──── EXAMPLE 1.1 ────

(i) Let X be a discrete random variable taking the values $x_i = i$ for $i = 1, 2, ..., N$, with $f(i) = 1/N$. Then, $F(k) = \sum_{i=1}^{k} f(i) = k/N$ for each $k = 1, 2, ..., N$. Also,

$$E(X) = \sum_{i=1}^{N} if(i) = \frac{N+1}{2}$$

and

$$E(X^2) = \sum_{i=1}^{N} i^2 f(i) = \frac{(N+1)(N+2)}{6}$$

Hence

$$V(X) = E(X^2) - (E(X))^2 = \frac{N^2-1}{12}$$

(ii) Let X be a continuous random variable with p.d.f. $f(x) = n/(x^{n+1})$ for $x \geq 1$. Then

$$F(x) = \int_1^x f(t)\,dt = 1 - \frac{1}{x^n} \text{ for } x \geq 1$$

Also

$$E(X) = \int_1^{\infty} x f(x)\,dx = \frac{n}{n-1} \text{ and}$$

$$E(X^2) = \int_1^\infty x^2 f(x)\,dx = \frac{n}{n-2}, \text{ giving}$$

$$V(X) = \frac{n}{(n-1)^2(n-2)}$$

1.1.4 Joint distribution

For a collection of random variables $X_1, X_2, \ldots, X_n$ we write

$$F(x_1, x_2, \ldots, x_n) = \Pr(X_1 \le x_1, X_2 \le x_2, \ldots, X_n \le x_n)$$

as the joint distribution function. Again we distinguish two types:

1. $X_1, X_2, \ldots, X_n$ are each discrete with joint p.d.f.

$$f(x_1, x_2, \ldots, x_n) = \Pr(X_1 = x_1, X_2 = x_2, \ldots, X_n = x_n)$$

where, for each i, x_i runs through the values of X_i.
2. $X_1, X_2, \ldots, X_n$ are jointly continuous with

$$F(x_1, x_2, \ldots, x_n) = \int_{-\infty}^{x_n} \int_{-\infty}^{x_{n-1}} \ldots \int_{-\infty}^{x_1} f(u_1, u_2, \ldots, u_n)\,du_1\,du_2 \ldots du_n$$

where $f \ge 0$ and integrates to 1 (informally $F(\infty, \infty \ldots \infty) = 1$).

In the case $n = 2$, writing $X_1 = X$ taking values $x_1, x_2, \ldots$ and $X_2 = Y$ taking values $y_1, y_2 \ldots$ in the discrete case, we may calculate

$$f_X(x_i) = \sum_j f(x_i, y_j) \quad \text{and} \quad f_Y(y_j) = \sum_i f(x_i, y_j)$$

to obtain the p.f. of X and Y, respectively. Similarly, in the continuous case

$$f_X(x) = \int_{-\infty}^\infty f(x, y)\,dy \quad \text{and} \quad f_Y(y) = \int_{-\infty}^\infty f(x, y)\,dx$$

——— EXAMPLE 1.2 ———

(i) Let X and Y have the joint probability function

$$f(i, j) = \Pr(X = i, Y = j) = \frac{1}{N(N-1)}$$

for $i \ne j$ and $i, j = 1, 2, \ldots, N$. Then, $f_X(i) = \sum_{j \ne i}^N f(i, j) = 1/N$ and similarly $f_Y(j) = \sum_{i \ne j}^N f(i, j) = 1/N$.

(ii) Let X and Y have the joint probability density function $f(x, y) = 2e^{-x-y}$ for $0 \leqslant y \leqslant x < \infty$. Then,

$$f_X(x) = \int_0^x f(x, y)\, dy = 2e^{-x} \int_0^x e^{-y}\, dy = 2(e^{-x} - e^{-2x})$$

and

$$f_Y(y) = \int_y^\infty f(x, y)\, dx = 2e^{-y} \int_y^\infty e^{-x}\, dx = 2e^{-2y}$$

For two random variables X and Y, the covariance $\mathrm{Cov}(X, Y)$ and correlation coefficient ρ are given by

$$\mathrm{Cov}(X, Y) = \mathrm{E}((X - \mu_x)(Y - \mu_y)) = \mathrm{E}(XY) - \mu_x \mu_y$$

and

$$\rho = \frac{\mathrm{Cov}(X, Y)}{\sigma_X \sigma_Y}$$

Note that $-1 \leqslant \rho \leqslant 1$ with $\rho = \pm 1$ if and only if there are constants a, b such that $Y = aX + b$ (with probability 1).

———— **EXAMPLE 1.3** ————

For X and Y as in Example 1.2(i),

$$\mathrm{E}(XY) = \sum_{i=1}^{N} \sum_{j \neq i}^{N} \frac{ij}{N(N-1)}$$

Now,

$$\sum_{i=1}^{N} \sum_{j \neq i}^{N} ij = \sum_{i=1}^{N} i \sum_{j \neq i}^{N} j = \sum_{i=1}^{N} i \left(\frac{N(N+1)}{2} - i \right) = \frac{N(N^2 - 1)(3N + 2)}{12}$$

and hence

$$\mathrm{E}(XY) = \frac{(N+1)(3N+2)}{12}$$

Since, from Example 1.2,

$$\mathrm{E}(X) = \mathrm{E}(Y) = \frac{(N+1)}{2}$$

we find

$$\mathrm{Cov}(X, Y) = \mathrm{E}(XY) - \mathrm{E}(X)\mathrm{E}(Y) = \frac{-(N+1)}{12}$$

Further, from Example 1.1, $V(X) = V(Y) = (N^2 - 1)/12$ so that $\rho(X, Y) = -1/(N-1)$.

1.1.5   *Sums of random variables*

The distribution of the sum $W = X + Y$ of the random variables X and Y may be calculated from the joint distribution.

(i) Let X and Y have probability density function $f(x, y)$. Then,

$$f_W(w) = \int_{-\infty}^{\infty} f(w - y, y) \, dy \tag{1.1}$$

(ii) Let X and Y each take the integer values $0, 1, \ldots$. Then,

$$\Pr(W = n) = \sum_{k=0}^{n} \Pr(X = n - k, Y = k) \tag{1.2}$$

——— **EXAMPLE 1.4** ———————————————————

(i) Let X and Y have p.d.f. $f(x, y) = e^{-x-y}$ for $x > 0$ and $y > 0$. Then, for $w > 0$, equation (1.1) gives

$$f(w) = \int_{-\infty}^{\infty} f(w - y, y) \, dy = \int_{0}^{w} e^{-(w-y)-y} \, dy = we^{-w}$$

(ii) Let $\Pr(X = m, \ Y = n) = p^2 q^{n+m}$ for integers $n \geq 0$ and $m \geq 0$. Then, for $n \geq 0$, equation (1.2) gives

$$\Pr(W = n) = \sum_{k=0}^{n} p^2 q^{(n-k)+k} = (n + 1)p^2 q^n$$

1.1.6   *Independence*

The important case, when dealing with a collection of random variables X_1, $X_2, \ldots, X_n$, is the concept of independence. Our use of independence will always be in terms of the factorization definition in one of the equivalent forms

I(i) $f(x_1, x_2, \ldots, x_n) = f_{X_1}(x_1)f_{X_2}(x_2)\ldots f_{X_n}(x_n)$

I(ii) $F(x_1, x_2, \ldots, x_n) = F_{X_1}(x_1)F_{X_2}(x_2)\ldots F_{X_n}(x_n)$

I(iii) $\Pr(X_1 \leq x_1, \ldots, X_n \leq x_n) = \Pr(X_1 \leq x_1)\Pr(X_2 \leq x_2)\ldots\Pr(X_n \leq x_n)$

If X and Y are independent it follows that

$$E(XY) = E(X)\,E(Y) \text{ and } \mathrm{Cov}(X, Y) = 0$$

but note that $\mathrm{Cov}(X, Y) = 0$ does not imply that X and Y are independent.

1.1.7   Order statistics

Let $X_1, X_2, \ldots, X_n$ be independent random variables with a common distribution function $F(x)$. We define the order statistics $X_{(1)} \leqslant X_{(2)} \leqslant \cdots \leqslant X_{(n)}$ by

$$X_{(1)} = \min(X_1, X_2, \ldots, X_n) \text{ and } X_{(n)} = \max(X_1, X_2, \ldots, X_n)$$

and, generally, $X_{(k)}$ is the kth smallest of $X_1, X_2, \ldots, X_n$.

Our main interest in applications concentrates on the first- and last-order statistics and their distributions are immediate consequences of the independence assumption. Observing that $X_{(n)} \leqslant x$ is equivalent to $X_i \leqslant x$ for all $i = 1, 2, \ldots, n$ and that $X_{(1)} > x$ is equivalent to $X_i > x$ for all $i = 1, 2, \ldots, n$ we see that

$$\Pr(X_{(n)} \leqslant x) = (F(x))^n \text{ and } \Pr(X_{(1)} \leqslant x) = 1 - (1 - F(x))^n$$

1.1.8   Properties of expectation and variance

E(i)   Expectation is linear. If $X_1, X_2, \ldots, X_n$ are random variables and $a_1, a_2, \ldots, a_n$ are constants then

$$E(a_1 X_1 + \cdots + a_n X_n) = a_1 E(X_1) + \cdots + a_n E(X_n)$$

E(ii)   (a)   Expectation is positive, i.e. $X \geqslant 0$ implies $E(X) \geqslant 0$.
          (b)   (Conversely) $X \geqslant 0$ and $E(X) = 0$ implies $\Pr(X = 0) = 1$.

V(i)   If a and b are constants then

$$V(aX + b) = a^2 V(X)$$

V(ii)   For any random variables $X_1, X_2, \ldots, X_n$ we have

$$V(X_1 + X_2 + \cdots + X_n) = \sum_{i=1}^{n} V(X_i) + 2 \sum_{i<j} \mathrm{Cov}(X_i, X_j)$$

V(iii)   If $X_1, X_2, \ldots, X_n$ are independent (so that $\mathrm{Cov}(X_i, X_j) = 0$), V(ii) simplifies as

$$V\left(\sum_{i=1}^{n} X_i\right) = \sum_{i=1}^{n} V(X_i)$$

V(iv)   Chebychev inequality. If $E(X) = \mu$, $V(X) = \sigma^2$ and $a > 0$

$$\Pr(|X - \mu| \geqslant a) \leqslant \sigma^2 / a^2$$

──────── **EXAMPLE 1.5** ──────────────────────────────────

The discrete random variables $X_1, X_2, \ldots, X_n$ take the values $1, \ldots, N$ with each pair X_i, X_j having the joint probability function

$$\Pr(X_i = k, X_j = \ell) = \frac{1}{N(N-1)} \qquad \text{for } 1 \leqslant k \neq \ell \leqslant N$$

Then, from Examples 1.2 and 1.3,

$$V\left(\sum_{i=1}^{n} X_i\right) = \frac{n(N^2 - 1)}{12} - \frac{n(n-1)(N+1)}{12} = \frac{n(N+1)(N-n)}{12}$$

───

1.1.9   *Large sample properties*

Suppose $X, X_2, \ldots, X_n$ are independent with the same distribution, with finite mean $\mu = E(X_i)$ and variance $\sigma^2 = V(X_i)$. Defining the sample mean $\bar{X}$ and the sample variance S^2 by

$$\bar{X} = \frac{1}{n} \sum_{i=1}^{n} X_i \qquad \text{and} \qquad S^2 = \frac{1}{n-1} \sum_{i=1}^{n} (X_i - \bar{X})^2$$

we note the relations and results:

L(i) $E(\bar{X}) = \mu$ and $V(\bar{X}) = \sigma^2/n$
L(ii) $E(S^2) = \sigma^2$
L(iii) $\bar{X} \to \mu$ as $n \to \infty$ (with probability 1)
L(iv) $S^2 \to \sigma^2$ as $n \to \infty$ (with probability 1)

L(v) $\Pr\left[\dfrac{(\bar{X} - \mu)}{\sigma} \sqrt{n} \leqslant x\right] \to \Phi(x)$ as $n \to \infty$,

for all values of x, where $\Phi(x)$ is the normal distribution function.

L(iii) and L(iv) are laws of large numbers, and L(v) is the central limit theorem.

   In numerical work L(v) is applied to provide the approximations, for large values of n,

(i) $\Pr(\bar{X} \leqslant x) \approx \Phi\left(\dfrac{(x - \mu)\sqrt{n}}{\sigma}\right)$

and

(ii) $\Pr(a < \bar{X} \leqslant b) \approx \Phi\left(\dfrac{(b - \mu)\sqrt{n}}{\sigma}\right) - \Phi\left(\dfrac{(a - \mu)\sqrt{n}}{\sigma}\right)$

1.1.10 Some named distributions

The distributions listed here, specified by the formula for the non-zero values of the p.f./p.d.f., make fleeting appearances in the text – usually as examples or problems.

1. Uniform $U(a, b)$, has p.d.f. $f(x \mid a, b) = 1/(b - a)$ for $a \leqslant x \leqslant b$.

$$E(X) = (a + b)/2, \qquad V(X) = (b - a)^2/12$$

2. Exponential $\text{Exp}(\lambda)$, where $\lambda > 0$, has p.d.f. $f(x \mid \lambda) = \lambda e^{-\lambda x}$ for $x > 0$.

$$E(X) = 1/\lambda, \qquad V(X) = 1/\lambda^2$$

3. Gamma $\Gamma(\alpha, \beta)$ where $\alpha, \beta > 0$, has p.d.f.

$$f(x \mid \alpha, \beta) = \frac{1}{\Gamma(\alpha)\beta^\alpha} x^{\alpha - 1} e^{-x/\beta}$$

$$E(X) = \alpha\beta, \qquad V(X) = \alpha\beta^2$$

4. Beta $B(a, b)$, where $a, b > 0$, has p.d.f.

$$f(x \mid a, b) = \frac{\Gamma(a + b)}{\Gamma(a)\Gamma(b)} x^{a - 1}(1 - x)^{b - 1} \text{ for } 0 < x < 1$$

$$E(X) = a/(a + b), \qquad V(X) = ab/(a + b)^2(a + b + 1)$$

Note: $Y = 1 - X$ has the $B(b, a)$ distribution.

5. Geometric $G(p)$, where $0 < p < 1$, has p.f.

$$\Pr(X = k) = p(1 - p)^k \text{ for } k = 0, 1, 2, \ldots$$

$$E(X) = (1 - p)/p, \qquad V(X) = (1 - p)/p^2.$$

6. Negative binomial $\text{NB}(r, p)$, where $0 < p < 1$, and r is a positive integer, has p.f.

$$\Pr(X = k) = \binom{k + r - 1}{r - 1} p^r (1 - p)^k \text{ for } k = 0, 1, 2, \ldots$$

$$E(X) = r(1 - p)/p \qquad V(X) = r(1 - p)/p^2$$

Note: X may be written as the sum of r independent variables each having the distribution $G(p)$.

7. Poisson $P(\lambda)$, where $\lambda > 0$, has p.f.

$$\Pr(X = k) = \frac{\lambda^k e^{-\lambda}}{k!} \text{ for } k = 0, 1, 2 \ldots$$

$$E(X) = \lambda, \qquad V(X) = \lambda$$

1.1.11   *The gamma function* $\Gamma(\alpha)$

This is defined, for $\alpha > 0$, by

$$\Gamma(\alpha) = \int_0^\infty x^{\alpha-1} e^{-x} \, dx$$

and has the properties:

1. $\Gamma(1) = 1$
2. $\Gamma(\frac{1}{2}) = \sqrt{\pi}$
3. $\Gamma(\alpha) = (\alpha-1)\Gamma(\alpha-1)$, provided $\alpha - 1 > 0$
4. $\Gamma(n) = (n-1)!$, for n a positive integer.

1.1.12   *Problems*

1.1 Let the random variable X have the Poisson distribution $P(\lambda)$.

  (i)   Show that $E(X) = \lambda$.
  (ii) Calculate $E(X(X-1))$ and deduce that $V(X) = \lambda$.

1.2 The continuous random variable X has p.d.f. $f(x) = nx^{n-1}$ for $0 < x < 1$. Calculate $E(X)$, $V(X)$ and find the distribution function of X.

1.3 The discrete random variables X and Y take integer values $0, 1, \ldots$ with joint probability function, given by, for $0 \leqslant m \leqslant n$,

$$f(m, n) = \frac{\lambda^n e^{-2\lambda}}{m!(n-m)!}$$

Find the probability functions of X and Y and deduce that X and Y are not independent.

1.4 The continuous random variables X and Y have joint p.d.f. $f(x, y) = n(n+1)y^{n-1}$ for $0 \leqslant y \leqslant x \leqslant 1$. Calculate $\text{Cov}(X, Y)$.

1.5 Let $X_1, X_2, \ldots, X_n$ be independent random variables with a common mean μ and variance σ^2. If $\bar{X} = (1/n)\sum_{i=1}^n X_i$, show that $E(\sum_{i=1}^n (X_i - \bar{X})^2) = (n-1)\sigma^2$.

1.6 Let $X_1, X_2, \ldots, X_n$ be independent random variables each having the uniform distribution over the interval $(0, \theta)$, where $\theta > 0$.

  (i)   Find the p.d.f. of the first- and last-order statistics.
  (ii) Show, when $\theta = 1$, that $Y = -\log(1 - X_1)$ has the exponential distribution.

1.2   PDF, CDF, INVCDF and RANDOM in MINITAB

MINITAB can supply the values of the probability density function (probability function), the cumulative distribution function and its inverse for each of 15 standard distributions. The corresponding commands PDF, CDF, INVCDF appear in the main command in the form:

    PDF/CDF/INVCDF E1 [E2]

where E1 is a number, a stored constant or a column and indicates the values at which the function is to be evaluated. E2 is an optional storage location, useful when further calculations involving the values are required. Storage entails a PRINT command for viewing. The main command must be used with a sub-command which determines the distribution and its parameters (listed in a specified order).

──── EXAMPLE 1.6 ────────────────────────────────

PDF 3;	displays	K	$P(X = k)$
BINOMIAL 10 0.5.		3.00	0.1172

that is, $f(3 \mid 10, 0.5) = 0.1172$, for the $b(10, 0.5)$ distribution.

──── EXAMPLE 1.7 ────────────────────────────────

If C1 contains 4.0, 7.0 then

CDF C1 C2;	displays	C2	
NORMAL 2 3.		0.747508	0.952210
PRINT C2.			

that is, $\Pr(X \le 4) = 0.747508$, $\Pr(X \le 7) = 0.952210$ when X has the distribution $N(2, 9)$.

───

    In principle, INVCDF finds x such that $P(X \le x) = p$ for given p such that $0 < p < 1$. While there is always such an inverse for a continuous distribution this is not necessarily the case for a discrete distribution. In the discrete case, MINITAB finds a pair of successive values x_1, x_2 such that $\Pr(X \le x_1) = p_1 < p < p_2 = \Pr(X \le x_2)$.

──── EXAMPLE 1.8 ────────────────────────────────

INVCDF 0.975;	supplies	0.9750	7.8799
NORMAL 2 3.			

that is, MINITAB calculates 7.8799 as the value of x such that $\Pr(X \le x) = 0.975$ when X has the distribution $N(2, 9)$.

──── EXAMPLE 1.9 ────────────────────────────────

INVCDF 0.5;	displays	K	$P(X$ LESS OR $= K)$	K	$P(X$ LESS OR $= K)$
POISSON 1.		0	0.3679	1	0.7358

Using INVCDF C1 C2 for a discrete distribution, MINITAB will put the larger value in C2.

For the purpose of simulation, we may need a random sample from a specified distribution. Thus to obtain a random sample of size K, stored in each of seven different columns, the main command is RANDOM K C – C; with a sub-command specifying the distribution.

─────── **EXAMPLE 1.10** ───────────────────────────────

RANDOM 100 C1 – C3;
NORMAL μ σ.

stores three samples of size 100, from the distribution N(μ, σ^2), in each of C1, C2 and C3.

───

1.2.1   Defaults and exceptions

MINITAB arithmetic rounds 0.00005 to 0 and 0.99995 to 1. When using PDF or CDF for a discrete distribution and when there is no call for storage, MINITAB will only print values different from 0 or 1.

Similarly, if the parameter values are not nominated then MINITAB may resort to *default* values (which are listed in the MINITAB manual). Incidentally, a parameter should not be expressed as a result of supplementary arithmetical operations – such as SQRT(K). In such a case MINITAB resorts to a default value.

1.3   Some standard distributions

1.3.1   The normal distribution

If the continuous random variable X has probability density function

$$f(x\,|\,\mu, \sigma) = \frac{1}{\sqrt{2\pi}\sigma}\, e^{-(1/2)(x-\mu)^2/\sigma^2}, \quad -\infty < x < +\infty \tag{1.3}$$

it is said to have the *normal* distribution, with parameters $\mu, \sigma > 0$. In that case we also say that X has the N(μ, σ^2) distribution.

In the special case when the parameters are $\mu = 0$, $\sigma = 1$, the distribution is known as the *standard normal* distribution. The corresponding variable is frequently denoted by Z and its probability density function by

$$\phi(x) = \frac{1}{\sqrt{2\pi}}\, e^{-(1/2)x^2} \quad -\infty < x < +\infty \tag{1.4}$$

The cumulative distribution function of Z,

$$\Phi(x) = \int_{-\infty}^{x} \frac{1}{\sqrt{2\pi}} \, e^{-(1/2)t^2} \, dt \qquad (1.5)$$

has been extensively tabulated. Since $\phi(x)$ is symmetric about zero,

$$\Pr(Z \leqslant -z) = \Pr(Z \geqslant z) = 1 - \Pr(Z \leqslant z)$$

and we need only tabulate $\Phi(x)$ for positive x.

Now $E(Z) = 0$, by immediate integration and $E(Z^2) = 1$, after integration by parts. Hence $V(Z) = E(Z^2) - [E(Z)]^2 = 1$.

If Z is distributed $N(0, 1)$, we show that $X = \sigma Z + \mu$ is distributed $N(\mu, \sigma^2)$. For, since $\sigma > 0$,

$$\begin{aligned}
\Pr(X \leqslant x) &= \Pr(\sigma Z + \mu \leqslant x) \\
&= \Pr(Z \leqslant (x - \mu)/\sigma) \\
&= \Phi((x - \mu)/\sigma)
\end{aligned}$$

After differentiating with respect to x, the p.d.f. of X is

$$\frac{1}{\sigma} \, \phi((x - \mu)/\sigma)$$

which we recognize from (equation 1.3) as $f(x \mid \mu, \sigma)$. Hence X is distributed $N(\mu, \sigma^2)$. Similarly, if X is distributed $N(\mu, \sigma^2)$, then $Z = (X - \mu)/\sigma$ is distributed $N(0, 1)$.

Therefore

$$\Pr(X \leqslant x) = \Pr\left(\frac{X - \mu}{\sigma} \leqslant \frac{x - \mu}{\sigma} \right) = \Pr\left(Z \leqslant \frac{x - \mu}{\sigma} \right)$$

may be evaluated from a table of the c.d.f. of Z.

───── **EXAMPLE 1.11** ─────────────────────────────

X has the distribution $N(2, 9)$. To calculate $\Pr(-1 \leqslant X \leqslant 2.6)$, note that $(X - 2)/3$ has the distribution $N(0, 1)$, and then

$$\begin{aligned}
\Pr(-1 \leqslant X \leqslant 2.6) &= \Pr\left(\frac{-1 - 2}{3} \leqslant \frac{X - 2}{3} \leqslant \frac{2.6 - 2}{3} \right) \\
&= \Pr(-1 \leqslant Z \leqslant 0.2) = \Pr(Z \leqslant 0.2) - \Pr(Z \leqslant -1) \\
&= \Pr(Z \leqslant 0.2) - [1 - \Pr(Z \leqslant 1)] \\
&= 0.5793 + 0.8413 - 1 \\
&= 0.4206
\end{aligned}$$

If X is distributed $N(\mu, \sigma^2)$, we may write $X = \sigma Z + \mu$ as above, where Z is distributed $N(0, 1)$, and hence

$$E(X) = \mu, \qquad V(X) = \sigma^2$$

Hence the mean and standard deviations of X are the parameters of its probability density function. Furthermore, $aX + b$ can be written $a(\sigma Z + \mu) + b = a\sigma Z + (a\mu + b)$, and it follows that $aX + b$ is distributed $N(a\mu + b, a^2\sigma^2)$.

1.3.2   *Sums of independent random variables*

If X_1 has the distribution $N(\mu_1, \sigma_1{}^2)$ and X_2, independent of X_1, has the distribution $N(\mu_2, \sigma_2{}^2)$, then the distribution of $c_1 X_1 + c_2 X_2$ is $N(c_1\mu_1 + c_1\mu_2, c_1{}^2\sigma_1{}^2 + c_2{}^2\sigma_2{}^2)$. For if Z_1, Z_2 are independently distributed $N(0, 1)$, then by equation (1.3) of Section 1.1, the p.d.f. of $W = a_1 Z_1 + a_2 Z_2$ is

$$f_W(w) = \int_{-\infty}^{+\infty} \frac{1}{\sqrt{2\pi a_1}} e^{-(1/2)(w - z_2)^2/a_1^2} \frac{1}{\sqrt{2\pi a_2}} e^{-(1/2)z_2^2/a_2^2} dz_2 \qquad (1.6)$$

The exponent of the integral is

$$-\frac{1}{2}\left[z_2^2 \left(\frac{1}{a_1^2} + \frac{1}{a_2^2} \right) - \frac{2wz_2}{a_1^2} + \frac{w^2}{a_1^2} \right]$$

and, after some tedious but elementary algebra, this is equal to

$$-\frac{1}{2}\left[\frac{a_1^2 + a_2^2}{a_1^2 a_2^2} \left[\left(z_2 - \frac{wa_2^2}{a_1^2 + a_2^2} \right)^2 \right] + \frac{w^2}{a_1^2 + a_2^2} \right]$$

Hence the integral in equation (1.6) becomes

$$\frac{1}{\sqrt{2\pi}\sqrt{a_1^2 + a_2^2}} e^{-(1/2)w^2/(a_1^2 + a_2^2)}$$

which is the p.d.f. of the distribution $N(0, a_1^2 + a_2^2)$.

Now writing $X_1 = \sigma_1 Z_1 + \mu_1$, $X_2 = \sigma_2 Z_2 + \mu_2$ we have $c_1 X_1 + c_2 X_2 = (c_1\sigma_1)Z_1 + (c_2\sigma_2)Z_2 + c_1\mu_1 + c_2\mu_2$ and the result follows.

───────── EXAMPLE 1.12 ─────────────────────────────────

If X_1 is distributed $N(2, 9)$, X_2 is distributed $N(1, 16)$ and X_1, X_2 are independent, find $Pr(X_1 > X_2)$. Now $Pr(X_1 > X_2) = Pr(X_1 - X_2 > 0)$. But $X_1 - X_2$ is distributed $N(1, 25)$, hence $Pr(X_1 > X_2) = Pr[(X_1 - X_2 - 1)/5 > -1/5] = Pr(Z < 1/5) = 0.5793$.

───

1.3.3 *Problems*

1.7   If X is distributed N(4, 25), Y is distributed N(16, 144) and X, Y are independent, calculate (a) $\Pr(X > Y)$, (b) $\Pr(2X + Y > 40)$.

1.8   If 10% of a normal distribution exceeds 80 and 20% is below 60, find the parameters of the distribution.

1.3.4 *Use of MINITAB*

———— EXAMPLE 1.13 ————————————————————————

To evaluate the p.d.f. of the N(1, 4) distribution at the value 2.

> PDF 2;
> NORMAL 1 2.

displays 2.000 0.1760, that is $f(2) = 0.1760$.

The main command will only accept one argument: collections have to be stored, as in Example 1.14.

———— EXAMPLE 1.14 ————————————————————————

To plot $f(x)$ for the distribution N(1, 4). More than 99% of a normal distribution lies between three standard deviations above and below its mean. Hence we may start at -5 and end at $+7$. Suppose we wish to evaluate f( ) at intervals of 0.25. We set C1 with $-5:7/0.25$ and nominate C2 as the location for storing their p.d.f. values. The sequence

> PDF C1 C2;
> NORMAL 1 2.
> PLOT C2 VS C1

will provide the required graph. The reader should study the effect of varying σ and confirm that it is a measure of spread.

———— EXAMPLE 1.15 ————————————————————————

We calculate $\Pr(-1 < X < 2.6)$ when X is N(2, 9). After setting C1 with -1 2.6, the sequence

> CDF C1 C2;
> NORMAL 2 3.
> LET K1 = C2(2)—C2(1)
> PRINT K1

displays 0.420604, which is $\Pr(X \leqslant 2.6) - \Pr(X \leqslant -1)$.

─────── **EXAMPLE 1.16** ───────────────────────────────

We find x such that $\Pr(X \leqslant x) = 0.7$ when X is $N(11, 64)$

    INVCDF 0.7;
    NORMAL 11 8.

displays 0.700   15.1952, that is, $\Pr(X \leqslant 15.1952) = 0.7$.

───

1.3.5   *The binomial distribution*

The discrete random variable X is said to have the *binomial* distribution $b(n, p)$ with index n and parameter p, where $0 < p < 1$, if it has probability function

$$f(x \mid n, p) = \binom{n}{x} p^x (1 - p)^{n-x} \quad x = 0, 1, ..., n$$

$$= 0, \text{ otherwise} \tag{1.7}$$

It is usual to write $q = 1 - p$. This distribution applies to the number of successes x in any *fixed* number of *independent* trials in which the outcome of each trial has *constant* probability of resulting in an event denoted as a 'success'. In this model, let n be the number of trials, p be the probability of a success and $q = 1 - p$ be the probability of a failure.

1.3.6   *Tabulation*

Tables of the cumulative distribution function $F(r \mid n, p)$, where

$$\Pr(X \leqslant r) = F(r \mid n, p) = \sum_{k=0}^{r} \binom{n}{k} p^k q^{n-k}$$

are widely available.

─────── **EXAMPLE 1.17** ───────────────────────────────

If X is $b(4, 0.4)$, then, from tables,

    $\Pr(X \geqslant 2 \mid 4, 0.4) = 1 - F(1 \mid 4, 0.4) = 1 - 0.4752 = 0.5248$, and
    $\Pr(X = 2 \mid 4, 0.4) = F(2 \mid 4, 0.4) - F(1 \mid 4, 0.4) = 0.8208 - 0.4752 = 0.3456$.

───

The tabulation may also be confined to the range of p such that $0 < p \leqslant 0.5$, since if X has the distribution $b(n, p)$ then

$$f(x \mid n, p) = \binom{n}{x} p^x q^{n-x} = \binom{n}{n-x} q^{n-x} p^x = f(n - x \mid n, q)$$

That is, $n - X$ has the distribution $b(n, q)$. This corresponds to interchanging success and failure on the n trials which compose X. Thus

$$F(r \mid n, p) = \Pr(X \leqslant r) = \Pr(n - X \geqslant n - r) = 1 - \Pr(n - X \leqslant n - r - 1)$$

so that

$$F(r \mid n, p) = 1 - F(n - r - 1 \mid n, q)$$

───────── EXAMPLE 1.18 ─────────────────────

To find $F(1 \mid 4, 0.6)$. Hence $n = 4$, $r = 1$, $n - r - 1 = 2$.

$$F(1 \mid 4, 0.6) = 1 - F(2 \mid 4, 0.4) = 1 - 0.8208 = 0.1792.$$

1.3.7 *Properties of the binomial distribution*

If X has the distribution $b(n, p)$ then it can be regarded as the sum of the number of 'successes' in n independent trials. That is,

$$X = \sum_{i=1}^{n} X_i \tag{1.8}$$

where $X_i = 1$ if the ith trial is a success and $X_i = 0$ otherwise. Thus the X_i have independent $b(1, p)$ distributions and it follows that

$$E(X) = np, \quad V(X) = npq \tag{1.9}$$

These results follow, since $E(X_i) = E(X_i^2) = p$ and $V(X_i) = pq$, then using the formulae for the expectation and variance of a sum.

If Y_1 is $b(n_1, p)$, Y_2 is $b(n_2, p)$ and Y_1, Y_2 are independent, then by application of equation (1.8) $Y_1 + Y_2$ is the sum of $n_1 + n_2$ independent $b(1, p)$ variables and, hence, has the distribution $b(n_1 + n_2, p)$.

1.3.8 *Normal approximation*

The central limit theorem asserts that if $X_1, X_2, \ldots, X_n$ are independent random variables such that $E(X_i) = \mu$ and $V(X_i) = \sigma^2$ exist, then

$$\sum_{i=1}^{n} \left(\frac{X_i - \mu}{\sqrt{n\sigma^2}} \right)$$

has a distribution which tends to $N(0, 1)$ as $n \to \infty$. Now if X is $b(n, p)$, we have noted that X can be expressed as the sum of n independent random variables each with expectation p and variance pq. Hence, if n is sufficiently large,

$$\frac{X - np}{\sqrt{npq}}$$

has approximately the distribution $N(0, 1)$.

─────── EXAMPLE 1.19 ───────────────────────────────

If X has the distribution $b(300, 1/4)$, we calculate a normal approximation to $\Pr(X \geq 85)$. Since $np = 75$ and $\sqrt{(npq)} = 15/2$, $\Pr(X \geq 85) = \Pr[(X - 75)/(15/2) \geq (85 - 75)/(15/2)] = \Pr(Z \geq 1.33)$, where Z has approximately the $N(0, 1)$ distribution. From tables this is about 0.092.

───

The accuracy of the approximation depends both on n and p. For instance, if p is close to $1/2$, then two decimal accuracy is obtained for $n \geq 10$. (A crude rule is to require $n \geq 50$ if $0.1 \leq p \leq 0.9$.) Since the binomial distribution is discrete, in applying a normal approximation an improvement can sometimes be obtained by using a continuity correction as shown in Example 1.20.

─────── EXAMPLE 1.20 ───────────────────────────────

If X has the distribution $b(10, 1/2)$, then, from tables, $\Pr(3 \leq X \leq 6) = 0.7734$. If a normal approximation is used, any value in the interval $(2.5, 6.5)$ would correspond to one of the integers 3, 4, 5, 6. Since $np = 5$, $\sqrt{(npq)} \approx 1.58$,

$$\Pr\left(\frac{2.5 - 5}{1.58} \leq Z \leq \frac{6.5 - 5}{1.58}\right) \approx \Pr(-1.58 \leq Z \leq 0.95) \approx 0.7718$$

Without a continuity correction, we obtain 0.632.

───

1.3.9   Use of MINITAB

MINITAB will display all the values of the p.d.f. through the command PDF.

─────── EXAMPLE 1.21 ───────────────────────────────

```
PDF;
BINOMIAL 8 0.5.
```

prints $\Pr(X = x)$ for $x = 0, 1, \ldots, 8$. The value for an individual x, say 3, may be inserted in the command as PDF 3; and $\Pr(X = 3)$ is displayed.

───

The command CDF provides values of the cumulative distribution function.

───── **EXAMPLE 1.22** ──────────────────────────────────

CDF;
BINOMIAL 8 0.5.

will print all the cumulative probabilities for $x = 0, 1, ..., 8$, while CDF 7; only returns $\Pr(X \le 7)$.

───

The command INVCDF attempts to find an integer for which the cumulative probability attains a specified value. Since the binomial distribution is discrete, execution of the command returns two integers which bracket, in terms of cumulative probability, the prescribed value.

───── **EXAMPLE 1.23** ──────────────────────────────────

X is $b(8, 0.5)$, to find x such that $\Pr(X \le x) = 0.4$.

INVCDF 0.4;
BINOMIAL 8 0.5.

prints	K	$P(X \le K)$	K	$P(X \le K)$
	3	0.3633	4	0.6367

If storage is requested, only the greater value is entered.

───

1.3.10   *Fitting a parameter to a tail probability*

The relation between the binomial and beta distributions given in Problem 1.12 below may be used to find the value p such that $\Pr(X \le k \mid n, p) = \alpha$ when X is $b(n, p)$ and α is given. For $\alpha = \Pr(T \le q \mid n, k)$ where T has the beta $(n - k, \ k + 1)$ distribution and we invert the corresponding cumulative distribution function with

INVCDF α;
BETA   N − K   K + 1.

For example, if $n = 2$, $k = 1$, $\alpha = 0.36$, then $q = 0.2$ is returned so that $p = 0.8$. This kind of calculation will be of service in Chapter 3.

1.3.11   *Problems*

1.9   Four hundred and fifty independent trials are divided into 90 blocks of five. The probability of a success on any trial is 0.7. Using a table of the binomial distribution, $b(5, 0.7)$, calculate the expected number of blocks containing $0, 1, 2, 3, 4$ or 5 successes, respectively.

1.10 If X has the distribution $b(n, p)$, show that the probability function has a unique maximum if $(n + 1)p$ is not an integer. What happens when $(n + 1)p$ *is an integer?*

1.11 Let X be a random variable with mean μ and variance σ^2. Suppose the random variable $g(X)$ can be expanded in the form

$$g(X) = g(\mu) + (X - \mu)g'(\mu)$$

assuming that terms of higher degree in $(X - \mu)$ are negligible. Show that

$$E[g(X)] = g(\mu), \qquad V[g(X)] = [g'(\mu)]^2\sigma^2$$

Hence evaluate $E[g(X)]$, $V[g(X)]$ when X is distributed $b(n, p)$ and $g(X) = \sin^{-1}(\sqrt{X/n})$ measured in radians.

   If X is distributed $b(10, \frac{1}{2})$, give a normal approximation for the distribution of $g(X)$ to calculate $\Pr(3 \leqslant X \leqslant 6)$:

(a) using the MINITAB command ASIN
(b) converting the angles to degrees and using

$$\sin^{-1}\sqrt{0.65} = 54°, \ \sin^{-1}\sqrt{0.5} = 45°, \ \sin^{-1}\sqrt{0.25} = 30°.$$

(Hint: rewrite $3 \leqslant X \leqslant 6$ in terms of $g(X)$.)

1.12 If X has the distribution $b(n, p)$, by integration by parts or otherwise, show that

$$\Pr(X \leqslant k \mid n, p) = \frac{n!}{(n - k - 1)!k!} \int_0^q t^{n-k-1}(1 - t)^k \, dt$$

By comparing with the c.d.f. for a beta distribution, deduce that the right-hand side is $\Pr(T \leqslant q)$, where T has a beta $(n - k, \ k + 1)$ distribution. (See Section 1.1.)

1.13 Write a MINITAB program to compare the cumulative probabilities for $b(16, 0.2)$ with those provided by the normal distribution with mean 3.2 and standard deviation 1.6.

1.3.12   *The hypergeometric distribution*

If the discrete random variable X has probability function

$$f(x \mid m, w, n) = \binom{w}{x}\binom{m - w}{n - x} \bigg/ \binom{m}{n}, \quad m \geqslant n, \geqslant w \qquad (1.10)$$

then X is said to have the *hypergeometric* distribution with parameters m, w, n. This probability function is sometimes written $h(m, w, n)$.

   A basic model which leads to a hypergeometric distribution is sampling without replacement. Suppose a bag contains m balls, of which w are white and $m - w$ are black. A random sample of n is drawn *without replacement*. The number of selections which contain just r white (and hence $n - r$ black) balls is

$$\binom{w}{r}\binom{m - w}{n - r}$$

Since there are $\binom{m}{n}$ different, equally likely, samples in all, the probability that a random sample contains exactly r white balls is

$$\binom{w}{r}\binom{m - w}{n - r}\bigg/\binom{m}{n} \tag{1.11}$$

On considering this process, we note

1. r cannot exceed the number of whites available or the number of balls drawn.
2. $n - r$ cannot exceed the number of blacks available or the number of balls drawn.

Violations of these restrictions, by allowing $0 \leqslant r \leqslant n$, lead to zero probabilities in equation (1.11), since, by definition, $\binom{w}{r} = 0$ if $r > w$.

1.3.13   *Tabulation*

The probabilities (1.11) can easily be calculated for small values of the parameters. For example,

$$\mathrm{f}(3 \mid 8, 4, 4) = \binom{4}{3}\binom{4}{1}\bigg/\binom{8}{4} = \frac{4!}{3!1!} \times \frac{4!}{1!3!} \times \frac{4!4!}{8!} \approx 0.2286$$

Most of the applications of the hypergeometric distribution are analogs of the sampling without replacement model. It is helpful to set out the results in a 2×2 table.

	Sample	Non-sample	Total
White	r	$w - r$	w
Black	$n - r$	$m - w - n + r$	$m - w$
Total	n	$m - n$	m

The probability function is usually tabulated for $w \leqslant m/2$, $n \leqslant m/2$ and $n \leqslant w$. We present some simple results which find values for cases outside these restrictions.

1. Since the probability of r whites is also the probability of $n - r$ blacks, we need entries only for $w \leqslant m - w$, for if this is not the case, exchange black with white. In terms of the 2×2 table, this means exchanging the rows. Hence

$$\mathrm{f}(r \mid m, w, n) = \mathrm{f}(n - r \mid m, m - w, n)$$

For example,

$$\mathrm{f}(2 \mid 8, 5, 3) = \mathrm{f}(3 - 2 \mid 8, 8 - 5, 3) = \mathrm{f}(1 \mid 8, 3, 3) = 0.5357$$

2. A similar contraction is available for the sample size – we need entries only for $n \le m - n$. For if $n > m - n$, then $m - n < m - (m - n)$ and the required condition is satisfied by the members of the non-sample (those left behind). To exchange sample with non-sample, we interchange the columns in the 2×2 table. Hence

$$f(r \mid m, w, n) = f(w - r \mid m, w, m - n)$$

For example,

$$f(1 \mid 8, 4, 5) = f(4 - 1 \mid 8, 4, 8 - 5) = f(3 \mid 8, 4, 3) = 0.0714$$

3. More surprisingly, we need entries only for $n \le w$. For if not, since $f(r \mid m, w, n) = f(r \mid m, n, w)$ we exchange n and w. A justification of this operation is provided in Problem 6.1 of Section 6.1. For example, $f(1 \mid 8, 3, 4) = f(1 \mid 8, 4, 3) = 0.4286$.

1.3.14   *Problems*

1.14 If X has the distribution $b(w, p)$, Y has the distribution $b(m - w, p)$ and X, Y are independent, show that the conditional distribution of X, given $X + Y = n$, is hypergeometric. Using this result, with $p = 1/2$, calculate $f(0 \mid 16, 8, 2)$.

1.15 A biologist catches 10 fish from a lake, tags them and returns them to the lake. What is the total number of fish in the lake which maximizes the probability that another sample of 10, drawn without replacement, contains three tagged fish?

1.16 If X has the distribution $h(m, w, n)$, show directly from the probability function that

$$E[X(X - 1)] = w(w - 1)n(n - 1)/[m(m - 1)]$$

and, hence, calculate $V(X)$, assuming that $E(X) = nw/m$.

1.17 A subject is given $2m$ photographs and told that they consist of m notorious criminals and m distinguished academics. The subject puts them at random into separate piles of m and claims that all the criminals are in one pile and all the academics in the other. Calculate the probability that at least $2m - 2$ photos are correctly classified. Find also the expected number classified correctly.

1.3.15   *The trinomial distribution*

The pair of discrete random variables X_1, X_2 is said to have the *trinomial* distribution with index n and parameters p_1 and p_2 if they have joint probability function

$$f(x_1, x_2 \mid n, p_1, p_2) = \binom{n}{x_1}\binom{n - x_1}{x_2} p_1^{x_1} p_2^{x_2} (1 - p_1 - p_2)^{n - x_1 - x_2} \tag{1.12}$$

where x_1, x_2 are non-negative integers such that $0 \le x_1 + x_2 \le n$, n is a positive integer, and p_1, p_2 are positive and satisfy $0 < p_1 + p_2 < 1$.

The trinomial distribution will apply when, on each of any fixed number n of independent trials, there is a probability p_1 of an event of type I, a probability p_2 of an event of type II, and a residual probability $1 - p_1 - p_2$ of an event of type III. Suppose there are x_1, x_2, $n - x_1 - x_2$ events of types I, II, III, respectively. Then

$$p_1^{x_1} p_2^{x_2} (1 - p_1 - p_2)^{n - x_1 - x_2}$$

is the probability of such an outcome of events in some fixed order. But there are $\binom{n}{x_1}$ ways of choosing the trials in which an event of type I may happen and then $\binom{n - x_1}{x_2}$ of choosing from the remaining $n - x_1$ trials in which an event of type II may happen. The remaining $n - x_1 - x_2$ must perforce be of type III.

——— **EXAMPLE 1.24** ———

Suppose a bag contains N balls, w of which are white, b are black and $N - w - b$ red. A random sample of n is drawn with replacement. The probability of a white on any draw is w/N, of a black is b/N, and of a red is $(N - w - b)/N$. Hence the conditions for a trinomial distribution are satisfied and the number of white balls X_1 and the number of blacks X_2, found in the sample, have such a distribution with index N and parameters w/N, b/N.

——— **EXAMPLE 1.25** ———

In particular, if in Example 1.24, $N = 15$, $w/N = 1/3$, $b/N = 1/3$, then, from equation (1.12)

$$\Pr(X_1 = 5 \text{ and } X_2 = 5) = \binom{15}{5}\binom{10}{5}\left(\frac{1}{3}\right)^{15} = 0.0527$$

$$\Pr(X_1 = 6 \text{ and } X_2 = 5) = \binom{15}{6}\binom{9}{5}\left(\frac{1}{3}\right)^{15} = 0.0439$$

We may also enquire as to the probability of six balls of one colour, five balls of another and four of the remaining colour. Since there are $3! = 6$ different orders of the values $6, 5, 4$ this probability is $6 \times 0.0439 \approx 0.2634$.

1.3.16   *Properties of the trinomial distribution*

We consider further the model proposed for this distribution in terms of n independent trials, on each of which an outcome of types I, II or III may result. If we amalgamate types II, III then we may describe the result as 'not type I'. It follows immediately that X_1, the number of outcomes of type I, must have a binomial distribution with index n and parameter p_1. Similarly, the distribution of X_2 is binomial with index n and parameter p_2.

X_1, X_2 are not independent. We obtain their covariance by considering

$$V(X_1 + X_2) = V(X_1) + V(X_2) + 2 \, \text{Cov}(X_1, X_2)$$

Now $V(X_1) = np_1(1 - p_1)$, $V(X_2) = np_2(1 - p_2)$ while the distribution of $X_1 + X_2$ is $b(n, p_1 + p_2)$, hence $V(X_1 + X_2) = n(p_1 + p_2)(1 - p_1 - p_2)$. Substituting these results in the equality we obtain

$$\text{Cov}(X_1, X_2) = -np_1p_2$$

1.3.17   *Conditional distributions*

Suppose in such a sequence of n independent trials, it is *known* that exactly x_2 of the trials resulted in an event of type II. Each of the remaining $N - x_2$ trials may have resulted in either type I or type III. But the probabilities are now conditional on an event of type II *not* taking place and are $p_1/(1 - p_2)$, and $1 - [p_1/(1 - p_2)]$, respectively. That is, the conditional distribution of X_1 given $X_2 = x_2$ is $b(n - x_2, p_1(1 - p_2))$. Similarly, the conditional distribution of X_2 given $X_1 = x_1$ is $b(n - x_1, p_2/(1 - p_1))$.

These results may be used to obtain probabilities for a trinomial distribution by expressing them as the products of two binomial probabilities. Thus if X_1, X_2 have a trinomial distribution with index n and parameters p_1, p_2,

$$\Pr(X_1 = x_1 \text{ and } X_2 = x_2) = \Pr(X_1 = x_1 \mid X_2 = x_2)\Pr(X_2 = x_2)$$

But the distribution of X_1 given $X_2 = x_2$ is $b(n - x_2, p_1/(1 - p_2))$ and the distribution of X_2 is $b(n, p_2)$.

───── EXAMPLE 1.26 ─────

If $n = 15$, $p_1 = \frac{1}{3}$, $p_2 = \frac{1}{3}$, then $p_1/(1 - p_2) = \frac{1}{2}$, and

$$\begin{aligned}
\Pr(X_1 = 5 \text{ and } X_2 = 5) &= \Pr(X_1 = 5 \mid X_2 = 5) \, \Pr(X_2 = 5) \\
&= b(5 \mid 10, 0.5) \, b(5 \mid 15, 0.33) \\
&= 0.2460 \times 0.2143 = 0.0527
\end{aligned}$$

1.3.18   *Problem*

1.18 Let X_1, X_2 have the trinomial distribution with index 18 and parameters $p_1 = \frac{1}{3}$, $p_2 = \frac{1}{3}$. Use the method of Example 1.26 to calculate $\Pr(X_1 = 6 \text{ and } X_2 = 6)$.

1.4   The sampling distributions

1.4.1   *The chi-squared distribution*

If $Z_1, Z_2, \ldots, Z_n$ is a random sample from the distribution $N(0, 1)$, then for $n \geq 1$,

$$Y = \sum_{i=1}^{n} Z_i^2 \tag{1.13}$$

is said to have the *chi-squared* distribution with index n, denoted $\chi^2(n)$.

1.4.2   *Properties*

1. The probability density function of Y is

$$f(y) = \frac{1/2(y/2)^{n/2-1} \exp(-y/2)}{\Gamma(n/2)} \quad 0 < y < \infty \tag{1.14}$$

   which has a single maximum at $y = n - 2$, when $n \geq 2$. (See Appendix 1.)
2. Since Y is the sum of independent random variables, each with mean 1 and variance 2, $E(Y) = n$, $V(Y) = 2n$. Moreover, by the central limit theorem, the random variable

$$(Y - n)/\sqrt{2n} \tag{1.15}$$

   has approximately the $N(0, 1)$ distribution for large n.
3. If Y_1, Y_2 have independent $\chi^2(n_1)$, $\chi^2(n_2)$ distributions respectively, then from equation (1.13) $Y_1 + Y_2$ may be represented as a sum of $n_1 + n_2$ independent random variables and so has the $\chi^2(n_1 + n_2)$ distribution. Similarly, if the sum of two independent random variables has a $\chi^2(n_1 + n_2)$ distribution and one of its components has a $\chi^2(n_1)$ distribution, then the other has a $\chi^2(n_2)$ distribution.

———— EXAMPLE 1.27 ————

If Y has the $\chi^2(25)$ distribution, we can use a normal approximation to calculate $Pr(Y \leq 32)$. Since $E(Y) = 25$, $V(Y) = 50$, using equation (1.15),

$$Pr(Y \leq 32) = Pr\left(\frac{Y - 25}{\sqrt{50}} \leq \frac{32 - 25}{\sqrt{50}}\right) \approx Pr(Z \leq 0.9899)$$

where Z has the distribution $N(0, 1)$. The approximate probability is about 0.839 (compared with a tabulated value of 0.8420).

1.4.3   *Sampling the normal distribution*

We shall show that if $Z_1, Z_2, ..., Z_n$ is a random sample from the distribution $N(0, 1)$ then $\sum_{i=1}^{n} (Z_i - \bar{Z})^2$ has the $\chi^2(n-1)$ distribution. At first it looks as though we may be able to apply the results in (3) above. For

$$\sum_{i=1}^{n} (Z_i - \bar{Z})^2 = \sum_{i=1}^{n} Z_i^2 - n\bar{Z}^2$$

or

$$\sum_{i=1}^{n} Z_i^2 = \sum_{i=1}^{n} (Z_i - \bar{Z})^2 + n\bar{Z}^2 \tag{1.16}$$

Now we know that $\sum_{i=1}^{n} Z_i^2$ has the $\chi^2(n)$ distribution, and since $\bar{Z}$ has the distribution $N(0, 1/n)$ then $\sqrt{n}\bar{Z}$ has the $N(0, 1)$ distribution. Thus $n\bar{Z}^2$ has the $\chi^2(1)$ distribution. However, to apply (3) it remains to be shown that $n\bar{Z}^2$ and $\sum_{i=1}^{n} (Z_i - \bar{Z})^2$ *are* independent. This is proved in reference 1.

The result can be immediately extended to the case where $X_1, X_2, ..., X_n$ is a random sample of n from the distribution $N(\mu, \sigma^2)$. We shall deduce that $\sum_{i=1}^{n} (X_i - \bar{X})^2/\sigma^2$ has the distribution $\chi^2(n-1)$ *whatever* the values of μ, σ.

Since $X_i = \sigma Z_i + \mu$ where Z_i is distributed $N(0, 1)$, then $\bar{X} = \sigma \bar{Z} + \mu$, and $\sum_{i=1}^{n} (X_i - \bar{X})^2/\sigma^2 = \sum_{i=1}^{n} (Z_i - \bar{Z})^2$. By the previous result, $\sum_{i=1}^{n} (X_i - \bar{X})^2/\sigma^2$ has the $\chi^2(n-1)$ distribution.

1.4.4   *Use of MINITAB*

The commands PDF, CDF, INVCDF are available for the χ^2 distribution.

─────── **EXAMPLE 1.28** ───────

CDF 67.5;
CHISQUARE 50.   returns   67.5   0.95

1.4.5   *Problems*

1.19 We have stated that, for large n, $\chi^2(n)$ is approximately distributed $N(n, 2n)$.

   (a) Use this result to find CDF 67.5 for $\chi^2(50)$.
   (b) An improvement is obtained by taking $\sqrt{(2X)}$ as distributed approximately $N(\sqrt{(2n-1)}, 1)$. Use this result to find CDF 67.5 for $n = 50$.

1.20 Find, in each case, an interval which contains 60% of the $\chi^2(10)$ distribution and

(a) Excludes 20% in each tail
(b) Excludes 10% in the lower tail and 30% in the upper tail.

1.21 MINITAB can supply directly a simulated random sample from the χ^2 distribution. In equation (1.13) we have defined this distribution in terms of a function of the sample values from a N(0, 1) distribution. Write a MINITAB program which supplies 1000 random values from the $\chi^2(4)$ distribution, using equation (1.13).

1.22 Write a MINITAB program to plot a graph of the p.d.f. of $\chi^2(n)$ against the integers $1, 2, \ldots, m$. The reader should fix a value of m and examine the effect of varying n.

1.4.6 *The t-distribution*

If Z has the distribution N(0, 1), and if V has the $\chi^2(m)$ distribution and is independent of Z, then

$$T_m = \frac{Z}{\sqrt{V/m}} \qquad (1.17)$$

is said to have *Student's t-distribution* with parameter m, denoted $t(m)$. We show in Appendix 1 that T_m has the probability density function

$$f_m(t) = \frac{1}{\sqrt{m\pi}} \frac{\Gamma[(m+1)/2]}{\Gamma(m/2)} \frac{1}{[1 + (t^2/m)]^{(m+1)/2}} \qquad -\infty < t < \infty \qquad (1.18)$$

The graph of $f_m(t)$ is symmetrical about $t = 0$, where it has a maximum. Note also (see Problem 1.24 below) that as $m \to \infty$, $f_m(t)$ tends to the density of the distribution N(0, 1), For the mean and variance: if $m > 1$ then $E(T_m) = 0$, by symmetry, and if $m > 2$, then $V(T_m) = m/(m-2)$.

——— EXAMPLE 1.29 ———

From tables, for $m = 4$, $Pr(T_4 \leqslant 2.776) = 0.975$. That is, $Pr(T_4 > 2.776) = 0.025$. From symmetry, $Pr(T_4 < -2.776) = 0.025$. Hence $Pr(|T_4| \geqslant 2.776) = 0.05$

1.4.7 *Application of the t-distribution*

(a) If $X_1, X_2, \ldots, X_n$ is a random sample from the distribution $N(\mu, \sigma^2)$, then $\bar{X}$ is distributed $N(\mu, \sigma^2/n)$. That is, $\sqrt{n}(\bar{X} - \mu)/\sigma$ has the distribution N(0, 1).

Furthermore, if

$$S^2 = \sum_{i=1}^{n} (X_i - \bar{X})^2/(n-1)$$

then

$$(n-1)S^2/\sigma^2 = \sum_{i=1}^{n} (X_i - \bar{X})^2/\sigma^2$$

has the $\chi^2(n-1)$ distribution as shown in Section 1.4, and S^2, $\sqrt{n}(\bar{X}-\mu)/\sigma$ are independent.

Hence, by definition,

$$\frac{\sqrt{n}(\bar{X}-\mu)/\sigma}{\sqrt{(n-1)S^2/(n-1)\sigma^2}} = \frac{\bar{X}-\mu}{\sqrt{S^2/n}} = \frac{(\bar{X}-\mu)\sqrt{n}}{S}$$

has the $t(n-1)$ distribution.

(b) If $X_1, X_2, \ldots, X_n$ is a random sample from the distribution $N(\mu_1, \sigma^2)$, while $Y_1, Y_2, \ldots, Y_m$ is an independent random sample from the distribution $N(\mu_2, \sigma^2)$, then $\bar{X}$ is distributed $N(\mu_1, \sigma^2/n)$, $\bar{Y}$ is distributed $N(\mu_2, \sigma^2/m)$ and $\bar{X} - \bar{Y}$ is distributed $N(\mu_1 - \mu_2, \sigma^2/n + \sigma^2/m)$, so that

$$\frac{\bar{X} - \bar{Y} - (\mu_1 - \mu_2)}{\sigma\sqrt{\dfrac{1}{n} + \dfrac{1}{m}}} \qquad\qquad (1.19)$$

has the distribution $N(0, 1)$. Furthermore,

$$\sum_{i=1}^{n} \frac{(X_i - \bar{X})^2}{\sigma^2}, \sum_{i=1}^{m} \frac{(Y_i - \bar{Y})^2}{\sigma^2}$$

have $\chi^2(n-1)$, $\chi^2(m-1)$ distributions, respectively, so that

$$\left[\sum_{i=1}^{n} (X_i - \bar{X})^2 + \sum_{i=1}^{m} (Y_i - \bar{Y})^2 \right]\Big/\sigma^2$$

has the $\chi^2(n+m-2)$ distribution. Hence if

$$S^2 = \left[\sum_{i=1}^{n} (X_i - \bar{X})^2 + \sum_{i=1}^{m} (Y_i - \bar{Y})^2 \right]\Big/(n+m-2)$$

then, as in (a),

$$\frac{\bar{X} - \bar{Y} - (\mu_1 - \mu_2)}{S\sqrt{\left(\dfrac{1}{n} + \dfrac{1}{m}\right)}}$$

has the $t(n + m - 2)$ distribution. It is essential for this result that the two distributions sampled have a common variance.

1.4.8   Use of MINITAB

The commands PDF, CDF, INVCDF may be employed for the t-distribution. MINITAB allows the direct generation of a random sample of values through the command RANDOM, but since this distribution has been derived via the ratio of two random variables, it is interesting to use that route to simulate a random sample (see Problem 1.23 below).

1.4.9   Problems

1.23 Use MINITAB to obtain a random sample of 1000 from the $t(4)$ distribution using equation (1.17).

1.24 Assuming that

$$\frac{1}{\sqrt{m}} \frac{\Gamma(m + 1)/2]}{\Gamma(m/2)} \rightarrow \frac{1}{\sqrt{2}}$$

as m increases, show that the limiting distribution of T_m is $N(0, 1)$. Using tables, compare the upper 2.5% points for $t(m)$ and $N(0, 1)$ as m increases.

1.4.10   The F-distribution

If U has the $\chi^2(n)$ distribution, V has the $\chi^2(m)$ distribution and U, V are independent, then

$$W = \frac{U/n}{V/m} \tag{1.20}$$

is said to have the *F-distribution*, denoted $F(n, m)$, with parameters n, m.

It is shown in Appendix 1 that the definition implies that the probability density function of W is

$$f_{n,m}(w) = \frac{\Gamma((m + n)/2)}{\Gamma(m/2)\Gamma(n/2)} \left(\frac{n}{m}\right)^{n/2} \frac{w^{(n/2) - 1}}{[1 + nw/m]^{(m + n)/2}} \qquad w > 0 \tag{1.21}$$

Provided $n > 2$, the graph of $f_{n,m}(w)$ has a unique mode at $m(n - 2)/n(m + 2)$. It is evident from the definition that if W has the $F(n, m)$ distribution, then $1/W$ has the $F(m, n)$ distribution.

Tables usually give the upper 5%, 2.5%, 1% points of the $F(n, m)$ distribution. Thus the upper 5% point of the $F(3, 4)$ distribution is 6.591, while the upper 5% point of the $F(4, 3)$ distribution is 9.117. The *lower* $100p\%$ point of the $F(n, m)$ distribution is the inverse of the upper $100p\%$ point of the $F(m, n)$ distribution. Hence the lower 5% point of $F(3, 4)$ is $1/9.117 = 0.1097$.

1.4.11   Application

If $X_1, X_2, \ldots, X_n$ is a random sample from the distribution $N(\mu_1, \sigma_1^2)$, then $\sum_{i=1}^{n} (X_i - \bar{X})^2 / \sigma_1^2$ has the $\chi^2(n-1)$ distribution. If $Y_1, Y_2, \ldots, Y_m$ is an independent random sample from the distribution $N(\mu_2, \sigma_2^2)$, then $\sum_{i=1}^{n} (Y_i - \bar{Y})^2 / \sigma_2^2$ has the $\chi^2(m-1)$ distribution. Hence

$$\frac{\displaystyle\sum_{i=1}^{n} (X_i - \bar{X})^2 \Big/ [(n-1)\sigma_1^2]}{\displaystyle\sum_{i=1}^{m} (Y_i - \bar{Y})^2 \Big/ [(m-1)\sigma_2^2]} \tag{1.22}$$

has the $F(n-1, m-1)$ distribution.

1.4.12   Problems

1.25 Show that, if T_m has the $t(m)$ distribution, then $(T_m)^2$ has the $F(1, m)$ distribution.

1.26 $X_1, X_2, \ldots, X_n$ is a random sample of n from the distribution $N(\mu_1, \sigma^2)$, and $Y_1, Y_2, \ldots, Y_m$ is an independent random sample of m from the distribution $N(\mu_2, \lambda\sigma^2)$. If

$$S_1^2 / S_2^2 = \left[\sum_{i=1}^{n} (X_i - \bar{X})^2 \Big/ (n-1) \right] \Big/ \left[\sum_{i=1}^{m} (Y_i - \bar{Y})^2 \Big/ (m-1) \right]$$

and c is the upper $100\alpha\%$ point of the $F(n-1, m-1)$ distribution, show that:

$$\Pr(S_1^2 / S_2^2 > c) > \alpha \text{ if } \lambda < 1$$
$$< \alpha \text{ if } \lambda > 1$$

1.27 If X has the $b(n, p)$ distribution show that

$$\Pr(X \leq k \mid n, p) = \frac{n!}{(n-k-1)!k!} \int_0^q t^{n-k-1}(1-t)^k \, dt$$

By substituting $t = y/(1+y)$ and then $u = y(k+1)/(n-k)$, deduce that $\Pr(X \leq k \mid n, p) = \Pr(U \leq q(k+1)/p(n-k))$ where U has the $F(2n-2k, 2k+2)$

distribution. Use a table of the F-distribution or MINITAB INVCDF to find p such that $\Pr(X \leqslant 1 \mid 2, p) = 0.95$.

1.28 Use MINITAB to find the upper and lower 5% points of the $F(2, 4)$ distribution. Find also the reciprocals of the upper and lower 5% points of the $F(4, 2)$ distribution. Comment on the results.

1.5   Solutions to problems

1.1 (i) $E(X) = \sum_{n=0}^{\infty} n\lambda^n e^{-\lambda}/n! = \lambda e^{-\lambda} \sum_{n=1}^{\infty} \lambda^{n-1}/(n-1)! = \lambda e^{-\lambda} e^{\lambda} = \lambda$

(ii) $E(X(X-1)) = \sum_{n=0}^{\infty} n(n-1)\lambda^n e^{-\lambda}/n! = \lambda^2 e^{-\lambda} \sum_{n=2}^{\infty} \lambda^{n-2}/(n-2)! = \lambda^2$

1.2 $E(X) = \int_0^1 nx^n \, dx = n/(n+1)$ and $E(X^2) = \int_0^1 nx^{n+1} \, dx = n/(n+2)$. Hence

$V(X) = n/(n+2) - (n/(n+1))^2 = n/(n+1)^2(n+2)$

Also, for $0 < x < 1$, $F(x) = \int_0^x nt^{n-1} \, dt = x^n$

1.3 $f_X(m) = \sum_{n=m}^{\infty} \lambda^n e^{-2\lambda}/m!(n-m)! = \lambda^m e^{-2\lambda}/m! \sum_{n=m}^{\infty} \lambda^{n-m}/(n-m)! = \lambda^m e^{-\lambda}/m!$

$f_Y(n) = \sum_{m=0}^{n} \lambda^n e^{-2\lambda}/m!(n-m)! = \lambda^n e^{-2\lambda}/n! \sum_{m=0}^{n} n!/m!(n-m)! = (2\lambda)^n e^{-2\lambda}/n!$

Thus, X and Y have the Poisson distribution with parameters λ and 2λ respectively, but the factorization fails – that is, $f(m, n) \neq f_X(m) f_Y(n)$.

1.4 $f_X(x) = \int_0^x n(n+1)y^{n-1} \, dy = (n+1)x^n$, and $E(X) = \int_0^1 (n+1)x^{n+1} \, dx = n + 1/(n+2)$

$f_Y(y) = \int_y^1 n(n+1)y^{n-1} \, dx = n(n+1)(y^{n-1} - y^n)$, and

$E(Y) = \int_0^1 n(n+1)(y^{n-1} - y^n) \, dy = n/(n+2)$

$E(XY) = \int_0^1 (\int_0^x n(n+1)y \, dy) x \, dx = n/(n+3)$

Hence,

$\text{Cov}(X, Y) = n/(n+3) - ((n+1)/(n+2))(n/(n+2)) = n(n+4)/(n+2)^2(n+3)$

1.5 $E(\bar{X}) = 1/n \sum_{i=1}^{n} E(X_i) = \mu$, and $V(\bar{X}) = 1/n^2 \sum_{i=1}^{n} V(X_i) = \sigma^2/n$

Then $E(\sum_{i=1}^{n} (X_i - \bar{X})^2) = \sum_{i=1}^{n} E(X_i^2) - nE(\bar{X}^2)$

$= \sum_{i=1}^{n} (\sigma^2 + \mu^2) - n(\sigma^2/n + \mu^2) = (n-1)\sigma^2$

1.6 (i) For $0 < x < \theta$, $F(x) = \int_0^x 1/\theta \, dt = x/\theta$. Hence, $F_{X_{(1)}}(x) = 1 - (1 - x/\theta)^n$ and $F_{X_{(n)}}(x) = x^n/\theta^n$ and the p.d.f. may be obtained by differentiating the distribution functions.

(ii) $\Pr(Y \leqslant x) = \Pr(-\log(1 - X_1) \leqslant x) = \Pr(X_1 \leqslant 1 - e^{-x}) = 1 - e^{-x}$

1.7 $\Pr(X > Y) = \Pr(X - Y > 0)$. But $X - Y$ is $N(4 - 16, 25 + 144)$, that is $N(-12, 169)$.

$\Pr(X - Y > 0) = \Pr[(X - Y + 12)/13 > 12/13] \approx \Pr(Z > 0.92) = 0.1780$.

$2X + Y$ is distributed $N(8 + 16, 4 \times 25 + 144)$, that is, $N(24, 244)$.

$\Pr(2X + Y > 40) = \Pr[(2X + Y - 24)/\sqrt{244} > (40 - 24)/\sqrt{244}] \approx \Pr(Z > 1.02) = 0.1539$.

1.8 For $N(0, 1)$, the upper 10% point is 1.2816 and the lower 20% point is -0.8416. Hence $(80 - \mu)/\sigma = 1.2816$, and $(60 - \mu)/\sigma = -0.8416$. Solve for μ, σ (67.93, 9.42, approximately).

1.9   Use tables to find the values of $b(5, 0.7)$,

$Pr(0) = 0.0024$ $Pr(2) = 0.1323$ $Pr(4) = 0.3601$

$Pr(1) = 0.0284$ $Pr(3) = 0.3087$ $Pr(5) = 0.1681$

The expected number of sets are found by multiplying the probabilities by 90.

1.10   If $f(x)$ is the probability function, show that

$$f(x + 1) - f(x) = \left[\frac{(n + 1)p - (x + 1)}{(x + 1)q} \right] f(x), \quad q = 1 - p$$

Hence $f(x + 1) - f(x) > 0$ if the right-hand side is positive, so $f(x)$ is increasing when $(n + 1)p - (x + 1) > 0$. There is a maximum at the largest integer less than $(n + 1)p$ unless $(n + 1)p$ is an integer, when there are two equal maxima at $(n + 1)p$ and $(n + 1)p - 1$.

1.11   For the binomial distribution, $E(X) = np$, $V(X) = npq$.
Hence $E[g(X)] = g(\mu) = \sin^{-1}(\sqrt{p})$. But $g'(\mu) = 1/[2n\sqrt{p(1 - p)}]$,
hence $V[g(X)] = np(1 - p)/[4n^2 p(1 - p)] = 1/4n$.
If angles are measured in degrees, $V[g(X)] = (180/\pi)^2 \, 1/4n \approx 820.7/n$.
Using a continuity correction,
$Pr[3 \leqslant X \leqslant 6] \approx Pr[(30 - 45)/\sqrt{82.07} \leqslant Z \leqslant (54 - 45)/\sqrt{82.07}] = 0.7909$

1.12 $\displaystyle\int_0^q t^{n-k-1}(1 - t)^k \, dt = \frac{p^k q^{n-k}}{n - k} + \frac{k}{n - k} \int_0^q t^{n-k}(1 - t)^{k-1} \, dt$

Hence the right-hand side can be written as

$$\binom{n}{k} p^k q^{n-k} + \frac{n!}{(n - k)!(k - 1)!} \int_0^q t^{n-k}(1 - t)^{k-1} \, dt$$

The first term is $Pr(X = k)$, the second has the same form as the original with k replaced by $k - 1$. Hence by induction.

1.13
```
SET C1
0:16
END
CDF C1 C2;
BINOMIAL 16 0.2.
CDF C1 C3;
NORMAL 3.2 1.6.
LET C4 = C3 - C2
PLOT C4 C1
PRINT C4
```

1.14 $Pr(X = x \mid X + Y = n) = Pr(X = x \text{ and } Y = n - x)/Pr(X + Y = n)$
$\qquad\qquad\qquad\qquad\quad = Pr(X = x) \, Pr(Y = n - x)/Pr(X + Y = n),$

since X, Y are independent. Noting that $X + Y$ has the distribution $b(m, p)$, substitute the binomial probabilities. After cancelling, we obtain

$$\binom{w}{x}\binom{m-w}{n-x}\Big/\binom{m}{n}$$

which is the p.f. of the distribution $h(m, w, n)$. Since X is $b(8, 1/2)$, $\Pr(X = 0) = 0.0039$, Y is $b(8, 1/2)$, $\Pr(Y = 2) = 0.1093$. $X + Y$ is $b(16, 1/2)$, hence $\Pr(X + Y = 2) = 0.0018$. Finally $0.0039 \times 0.1093/0.0018 = 0.237$. The correct value, 0.2333, would require the use of more decimal places in the binomial probabilities.

1.15 Compare the probabilities $f(3 \mid m, 10, 10)$, $f(3 \mid m + 1, 10, 10)$. Show that their ratio is less than 1 provided that $m \leqslant 32$, so that the maximum occurs at $m = 32$.

1.16 $E[X(X - 1)] = \displaystyle\sum_{x=2}^{w} x(x - 1)\binom{w}{x}\binom{m-w}{n-x}\Big/\binom{m}{n}$

$$= w(w - 1)\sum_{x=2}^{w}\binom{w-2}{x-2}\binom{m-w}{n-x}\Big/\binom{m}{n}$$

$$= w(w - 1)\binom{m-2}{n-2}\Big/\binom{m}{n}$$

$$= w(w - 1)n(n - 1)/m(m - 1)$$

Substitute in the identity

$$E[X(X - 1)] = E(X^2) - E(X) = V(X) + [E(X)]^2 - E(X)$$

1.17 Each academic misclassified as a criminal is matched by some criminal misclassified as an academic.

$$\Pr(\text{at least } 2m - 2 \text{ classified correctly}) = \left[\binom{m}{0}\binom{m}{0} + \binom{m}{1}\binom{m}{1}\right]\Big/\binom{2m}{m}$$

$$= (m^2 + 1)\Big/\binom{2m}{m}$$

The number of academics correctly classified has the distribution $h(2m, m, m)$ with expectation $m/2$. Similarly for criminals. Hence m is the expected total number correctly classified.

1.18 $\Pr(X_1 = 6 \text{ and } X_2 = 6) = \Pr(X_1 = 6 \mid X_2 = 6)\Pr(X_2 = 6)$
$= b(6 \mid 12, 0.5)b(6 \mid 18, 0.33) = 0.2256 \times 0.1962 = 0.044$

1.19 (a) CDF 67.5;
   Normal 50 10.   Returns   67.5   0.9599
   (b) CDF 11.6189;
   Normal 9.9498 1.   Returns   11.6189   0.9524

1.20 (a) SET C1
       0.2   0.8
       END
       INVCDF C1 C 2;
       CHISQUARE 10.
       PRINT C2   Returns 6.791   13.442
   (b) SET C1
       0.1   0.7
       END
       INVCDF C1 C2;
       CHISQUARE 10.
       PRINT C2   Returns 4.8652   11.7807

1.21 RANDOM 1000 C1-C4;
     NORMAL 0 1.
     RSSQ C1-C4 C5

1.22 SETC1
     1:m
     END
     PDF C1 C2;
     CHISQUARE n.
     PLOT C2 C1

1.23 RANDOM 1000 C1-C4;
     NORMAL 0 1.
     RMEAN C1-C4 C5
     RSTDEV C1-C4 C6
     LET C7 = (2 * C5)/C6
     HISTOGRAM C7

1.24 Use $(1 + t^2/m)^{m/2} \to \exp(t^2/2)$ as m increases to obtain the result.

1.25 T_m^2 has the form $Z^2/(U/m)$ where Z^2 has the $\chi^2(1)$ and U the $\chi^2(m)$ distribution.

1.26 $[S_1^2/\sigma^2]/[S_2^2/\lambda\sigma^2]$ has the $F(n-1, m-1)$ distribution for any positive λ. But $\lambda^2 S_1^2/S_2^2 \geqslant S_1^2/S_2^2$ only for $\lambda \geqslant 1$, and the result follows.

1.27 Integrating by parts, the integral is

$$\binom{n}{k}p^k q^{n-k} + \frac{n!}{(n-k)!(k-1)!} \int_0^q t^{n-k}(1-t)^{k-1}\, dt$$

and hence by repeated application. Make the suggested substitution and compare with the density in equation (1.20). The upper 5% point of $F(2, 4)$ is 6.944 and after equating to $q(1+1)/[p(2-1)]$, p is ≈ 0.224.

1.28   SET C1   returns   0.050   0.0520
        0.05   0.95   0.950   6.9443
        INVCDF C1;
        F 2 4.

Also

    INVCDF C1 C2;   returns   6.94427   0.051996
    F 4 2.
    LET C3 = 1/C2
    PRINT C3

Reference

1. Hoel, P., Port, S. and Stone, C., *Introduction to Statistical Theory*, Houghton Mifflin, Boston, 1971.

Chapter Two

Data

2.1 The language of data

The word 'data' commonly conjures up a picture of an array of numbers, each number providing some information in a defined context. The aim of this section is to sharpen the picture by reviewing the terms commonly used to describe data.

To fix our ideas and provide examples we start with some familiar illustrations.

1. The full 10-year UK census aims to provide a complete dossier on the resident population identifying the characteristics of age, sex, race, marital status, income, etc.
2. Reports of sporting activity are commonly accompanied by numerical data on best performances, career records (usually in the summary form of averages) and recent form, either for general interest or as a guide for betting.
3. Opinion polls, promoted actively in the media, endeavour to predict the outcome of elections. The members of the electorate are characterized by the candidate they support, and a sample of this information is the base for the prediction.
4. To comply with trading laws a packer of frozen peas must ensure that the packets contain at least the advertised amount of peas (dried weight).
5. An investigation into the relation between left-handedness and artistic ability classifies the population by two characteristics: handedness (in two parts – left and right) and artistic ability (measured on some scale – say, good, average, poor). These characteristics cross-classify the population into six categories, and the investigation is summarized in a two-way table giving the counts in each combined category of the sample of the population examined.
6. A clinician, evaluating a new drug treatment for a particular disease, prescribes the drug to a sample of patients. By examining the patients' reactions to the drug the clinician will hope to decide on the general effectiveness of the drug.
7. An engineer, interested in the relation between the yield in a chemical process and the quantities of two ingredients, runs the process several times with various predetermined amounts of the ingredients. By examining the corresponding yields, the engineer hopes to obtain an optimal combination of ingredients to give maximum yield.

Reviewing the examples and identifying the common structure we arrive at the model for statistical investigations. The first step, not shown in the descriptions given above, is to identify the subject of the investigation and the information required to illuminate the problem. In the general model, the information is provided by the 'measurements' of relevant characteristics of a set of suitable elements – that is, the model consists of:

- A parent or target population of elements
- A specified set of characteristics
- A sample of the parent population, each element of the sample being examined and its characteristics noted.

Conceptually the parent population consists of all elements which possess the identified characteristics. However, this should not be taken to mean that we can list all the elements. For example, the clinician knows only his or her own patients, though he expects the result of his study will apply to the unknown population of all, present and future, people affected with the disease.

The purpose of obtaining data from a sample is to draw conclusions about the characteristics for the whole parent population – in statistical language, to make inferences. Implicit in the inference process is the intuitive feeling that the sample elements are, somehow, typical of the parent population. How to obtain such a sample in practice is a problem we will not address (but see Section 2.3 for a simplistic discussion). Reasons of cost, time and availability may well determine the nature of the sample of elements investigated and the investigator must be aware of the danger involved in extrapolating from the sample. As a trite example, the opinion pollster predicting a national election result will fail badly if all his or her sample information comes from one small area. More formally, the distinction being made is between the parent population and the sampled population (that is, those elements known to the investigator and from which the sample is selected), with the possibility that the sampled population is not representative of the parent population.

As we will soon see, the language used to describe data involves use of the word 'variable' – a natural choice since the data values vary from element to element of the parent population. For the most part the word variable is used as a synonym for characteristic or, almost equivalently, for the values of the characteristic. However, an important distinction will appear.

The most common type of variable (characteristic) is the *quantitative* variable where the characteristic is measured on some appropriate numerical scale. In the examples, age, height and yield are so measured, on a scale which has a natural zero, equal differences between successive integer values and for which the ratio of two measurements is meaningful. Such a scale is termed a *ratio scale*. The alternative type of numerical scale is the *interval scale*, which differs from the ratio scale only in that the zero measurement is arbitrarily defined. A good example of an interval scale is the measurement of temperature, in which 'twice as hot' is not a meaningful statement, but a unit difference anywhere on the scale has the same physical definition.

A further description applied to quantitative variables is to distinguish between continuous and discrete variables. The adjective *continuous* denotes a variable taking values in some interval of real numbers, whereas *discrete* implies that the values are in some (finite or infinite) sequence of numbers – in fact, exactly the same meaning as that used to distinguish between continuous and discrete random variables.

Numerical data are also produced as *count* data, as in the example relating handedness and artistic ability. The characteristics 'handedness' and 'artistic ability' are *categorical* variables with values left or right and good, average or poor, respectively. For a single categorical variable the data set shows the number of elements associated with value of the variable, and for two or more categorical variables the data set takes the form of a cross-classified table – see (5) above, where we obtain a count of the six combined categories in a 2×3 table.

Categorical variables are further described as *nominal* or *ordinal* according to the type of values taken. When the values have no natural ordering, as in the example of handedness (left or right), the variable is termed nominal; and variables for which the values are naturally ordered, as artistic ability (good, average or poor), are termed ordinal. The description *qualitative* variable is also used to imply a nominal categorical variable, but some authors use qualitative to mean simply non-numerical – that is, not quantitative.

For many purposes, indeed essential for the use of some MINITAB commands, it is convenient to code a categorical variable, whether nominal or ordinal, and put it in apparent numerical form. This amounts to nothing more than replacing the values of the variable by a sequence of numbers. For example, in a height/weight problem the male/female characteristic could be coded male = 0, female = 1 and the recorded data for each individual takes the form of three values – height/ weight/code.

As our examples show, the data set may contain values for more than one variable. Data sets are described as *univariate*, *bivariate* or *multivariate* (variate being synonymous with variable) if there are one, two or more than two variables under discussion. In a bivariate or multivariate situation a further distinction may be appropriate – a variable which depends on the other variables is termed a *dependent* or *response* variable, the other variables being *explanatory*, or *predictor*, variables. In the chemical process example, yield is the response and the ingredients are explanatory.

2.2   Exploring the data

The process of collecting data will generally produce a data set with an unorganized, untidy appearance. To produce order out of the chaos, i.e. to move towards obtaining information from the data, the first step in any statistical investigation should be a presentation, in summary form, using the numerical and graphical procedures described later in this section. The aims of this data exploration are to identify unusual or unexpected data values, or to uncover unusual or interesting features of the data, or detect relationships (if they exist) between different variables.

Some simple examples may clarify the general statements.

1. If most of the data are around 50, a value of 25.23 is unexpected and might be a recording error for 52.23. Careful proofreading of the recording/transcription processes, particularly computer entries, will help to reduce this type of error. The value 25.23 may, of course, be a genuine observation which reveals, on further investigation, an unexpected feature of the problem.

2. A histogram of the data may reveal that the data is dichotomized into a set of small values and a set of relatively large values – for example, recording the age of victims of hypothermia would produce a dichotomized data set, since hypothermia tends to affect the very young or the (relatively) old. Basing a statistical analysis on the assumption that the data came from an unimodally distributed population would not be appropriate.

3. Bivariate data may well show a different structure when separated by a third characteristic – for example, adult height/weight data should be further classified by sex.

4. Plots of bivariate data (x, y) may well indicate a functional relationship $y = f(x)$, perhaps a linear relation $y = a + bx$ or a quadratic $y = a + bx + cx^2$. Should the data show, for instance, a tendency for y to increase with x, then any data pair (x, y) not fitting this tendency should be investigated.

Although the numerical measures, termed descriptive statistics, and the graphical presentation of data may all be obtained by hand-calculation, the exposition is combined with the use of MINITAB commands.

2.2.1   *Numerical measures: the DESCRIBE command*

Suppose we have examined n elements of the parent population, a sample of size n, and obtained the quantitative data values $x_1, x_2, ..., x_n$. Entering these values in the column C1, the command DESCRIBE C1 produces a printout showing standard measures of location (mean, median) and measures (standard deviation, range, quartiles) of the spread and variation.

──────── EXAMPLE 2.1 ────────────────────────────────

The examination marks, in the range 0–250, for 48 students are entered in C1, and the sex of the student is entered in C2 using a 1–2 coding. Naming the columns marks and sex, we use the commands

PRINT C1

to obtain a list of the marks

marks

121	43	50	95	122	148	149	141	209	65	83
17	65	71	87	92	46	60	125	129	47	60
74	61	66	44	42	89	30	72	78	70	89
30	76	27	74	174	90	91	43	55	63	142
118	62	59	79							

and

PRINT C2

to view the corresponding coding

sex

1	1	2	1	2	1	1	2	2	1	1	2	1	2	1
2	1	1	2	2	1	1	2	1	1	2	2	2	2	1
2	1	1	2	1	1	1	2	2	2	2	1	2	2	2
2	1	1												

MINITAB provides the standard descriptive statistics, through the command DESCRIBE C1, in the form

	N	MEAN	MEDIAN	TRMEAN	STDEV	SEMEAN
marks	48	81.73	73.00	79.45	40.38	5.83

	MIN	MAX	Q1	Q3
marks	17.00	209.00	56.00	94.25

We review the definition of all the terms in the printout for DESCRIBE, for the general sample $x_1, x_2, \ldots, x_n$. Note first that $N = n$. The mean $\bar{x}$ is the familiar average

$$\bar{x} = \frac{1}{n} \sum_{i=1}^{n} x_i$$

The median m is loosely defined as the middle value. When the data is arranged in increasing order (see Section 1.1) as $x_{(1)} \leqslant x_{(2)} \leqslant \cdots \leqslant x_{(n)}$, the precise definition is

$$m = \begin{cases} x_{(k)} & \text{when } n = 2k + 1 \\ \frac{1}{2}(x_{(l)} + x_{(l+1)}) & \text{when } n = 2l \end{cases}$$

The maximum $x_{(n)}$ and minimum $x_{(1)}$ are self-evidently defined. The difference $r = x_{(n)} - x_{(1)}$ necessarily non-negative, is termed the range, and is not printed.

The sample standard deviation s is defined as the positive square root of the sample variance s^2, where

$$s^2 = \frac{1}{n-1} \sum_{i=1}^{n} (x_i - \bar{x})^2$$

The standard error of the mean, denoted by SEMEAN, is defined as $s/\sqrt{n}$.

The trimmed mean TRMEAN is the average of the data remaining when the 5% smallest and 5% largest values are erased. For example, with $n = 48$ the smallest two and largest two are erased (rounding the percentages to the nearest integer).

The lower and upper quartiles Q_1 and Q_3 are, loosely, values such that a quarter of the data set is less than Q_1 and a quarter is greater than Q_3. For a precise definition, we have, for convenience, selected the version adopted by MINITAB. from the various nearly equivalent formulations. The method is similar to the half-way interpolation for the median when the sample size is even, but for the quartiles is based on multiples of one-quarter.

Using the order statistics $x_{(1)} \leqslant x_{(2)} \leqslant \cdots \leqslant x_{(n)}$, the quartiles are defined by

$$Q_1 = x_{(k)} + \frac{r}{4}(x_{(k+1)} - x_{(k)})$$

where $n + 1 = 4k + r$, with $r = 0, 1, 2$ or 3; and

$$Q_3 = x_{(l)} + \frac{s}{4}(x_{(l+1)} - x_{(l)})$$

where $3(n + 1) = 4l + s$, with $s = 0, 1, 2$ or 3.

To illustrate the definition with $n = 48$ we have $k = 12$ and $r = 1$, and $l = 36, s = 3$ so that

$$Q_1 = x_{(12)} + \tfrac{1}{4}(x_{(13)} - x_{(12)}) \quad \text{and} \quad Q_3 = x_{(36)} + \tfrac{3}{4}(x_{(37)} - x_{(36)})$$

――――― **EXAMPLE 2.2** ―――――

Using the mark data of Example 2.1, classified by sex, the subcommand BY C2 will provide the descriptive statistics for each sex.

    DESCRIBE C1;
    BY C2.

provides the output.

	sex	N	MEAN	MEDIAN	TRMEAN	STDEV	SEMEAN
marks	1	24	74.88	68.00	73.68	29.72	6.07
	2	24	88.58	83.50	86.36	48.48	9.90

	sex	MIN	MAX	Q1	Q3
marks	1	27.00	149.00	59.25	86.00
	2	17.00	209.00	45.50	124.25

A quick comparison shows that there is a marked difference between the sexes – the sex coded 2 has the greater mean, range and standard deviation.

2.2.2   *Graphical procedures in MINITAB*

We describe some of the MINITAB procedures which are helpful in an initial examination of data.

SORT

The command SORT C1 C11 instructs MINITAB to arrange the data of C1 in increasing order and store the rearrangement in C11. Printing C11 will provide an organized view of the data and will immediately identify the minimum and maximum values. The following example extends the use of the SORT command to deal with bivariate data.

———— **EXAMPLE 2.3** ————

Using the mark data (in C1) and sex code (in C2) of Example 2.1 the command SORT C1 C2 C3–C4 arranges the data of C1 in increasing order and stores the rearrangement in C3. The same rearrangement is applied to C2 and stored in C4. Note that this keeps the corresponding mark and code values in the matching rows of C3 and C4. Printing C3 and C4 separately, having named C3 marksort, supplies

```
marksort
 17   27   30   30   42   43   43   44   46   47   50
 55   59   60   60   61   62   63   65   65   66   70
 71   72   74   74   76   78   79   83   87   89   89
 90   91   92   95  118  121  122  125  129  141  142
148  149  174  209

C4
 2   1   2   2   2   1   2   2   1   1   2   1   1   1   1
 1   2   2   1   1   1   1   2   1   2   1   1   2   1   1
 1   2   1   2   2   2   1   2   1   2   2   2   2   2   1
 1   2   2
```

The reader may check that the mark 17 was coded 2 in C2 and the mark 27 coded 1. Since C3 supplies the order statistics, the formulae for the quartiles may be checked for the mark data. For example, $Q_1 = 55 + \frac{1}{4}(59 - 55) = 56$.

DOTPLOT

The command DOTPLOT C1 produces on the screen a horizontal display of the data, with each sample value represented by a dot above its position on a scale showing the range of the data. For large sample sizes, this ideal is approximated by using each printer position to indicate a small range of values.

──────── EXAMPLE 2.4 ────────────────────────────

Using the mark data of Example 1,

>   DOTPLOT C1

produces

```
               ..        :  .  .        ..
 .     .:    ::.  .:::::::..::.      .:..  ....            .                  .
 -----+---------+---------+---------+---------+---------+-marks
      35        70        105       140       175       210
```

and, for a display of each sex,

>   DOTPLOT C1;
>   BY C2.

produces

```
sex                  :  .
1          .    .: .:  :.:....: .          .          ..
     -----+---------+---------+---------+---------+---------+-marks
sex                         .
2   .    :  :..    :  ...   .:        ....  ..          .          .
     -----+---------+---------+---------+---------+---------+-marks
          35        70        105       140       175       210
```

These displays of the data show both any unusual (large or small) value and any clustering in the distribution of the data.

──

The next two commands, STEM and HISTOGRAM, both display the clustering of the data by grouping the data into intervals, more usually called classes, and providing a count of the data values. Although the displays provide similar information, the display given by STEM shows, in essence, the sample values, while HISTOGRAM represents the data by stars.

STEM

This command provides a display called a stem and leaf plot. The grouping of the values is controlled by the subcommand INCREMENT (INCR) which determines the distance between the lines in the display – the distance being the difference between the smallest possible values on successive lines. Note that these values need not be sample values.

──────── EXAMPLE 2.5 ────────────────────────────

Using the mark data of Example 2.1,

>   STEM C1;
>   INCR 10.

produces on the screen

```
          Stem-and-leaf of marks        N = 48
          Leaf Unit = 1.0

     1        1   7
     2        2   7
     4        3   00
    10        4   233467
    13        5   059
    21        6   00123556
    (8)       7   01244689
    19        8   3799
    15        9   0125
    11       10
    11       11   8
    10       12   1259
     6       13
     6       14   1289
     2       15
     2       16
     2       17   4
     1       18
     1       19
     1       20   9
```

There are three parts to the display which are described with reference to the example. The middle column specifies the stem (the tens digit) and the right-hand display shows the leaf values (the units digit). In the first column, there are two cumulative counts, one from the top down and the other from the bottom up, terminating at the line containing the median. The value in brackets indicates that the median lies in that class, and the number is just a count of the sample values lying in that class.

The reader should compare the stem and leaf plot with the data values as given in increasing order in Example 2.3. For example, on line 6 we see the sample values 60, 60, 61, 62, 63, 65, 65 and 66.

To obtain a stem and leaf display for each sex, the subcommand BY may be used (see Example 2.2).

HISTOGRAM

The command HISTOGRAM groups the data into intervals whose width is controlled by the INCREMENT subcommand. The display shows a count of the sample values in each interval and represents the data values by stars. When a sample value falls at a dividing point between adjacent intervals, MINITAB chooses to count the value in the higher range.

——— EXAMPLE 2.6 ———————————————————————

Using the mark data of Example 2.1,

>HISTOGRAM C1;
>INCR 20.

produces

Histogram of marks   N = 48

Midpoint	Count	
20.0	2	* *
40.0	8	* * * * * * * *
60.0	11	* * * * * * * * * * *
80.0	12	* * * * * * * * * * * *
100.0	4	* * * *
120.0	5	* * * * *
140.0	4	* * * *
160.0	0	
180.0	1	*
200.0	1	*

In the display, the intervals are identified by the midpoint column, with 20 representing $10 \leqslant x < 30, 40$ representing $30 \leqslant x < 50$ and so on. The count column is self-explanatory, and the stars represent the sample values.

_____

Again the BY subcommand is available to provide a separate histogram for each sex.

BOXPLOT

This command provides what is known as a box-and-whisker plot. To facilitate the description of the features of BOXPLOT, we view the output for the mark data, augmented by the general notation.

```
                  ------------
         ---------I    +     I---------------*        *              O
                  ------------
+-----+-   ------+-----+---+------+--+---------+--+------+------+---+marks
           35          70        105      140        175         210
O_ℓ    I_ℓ            H_ℓ        H_u              I_u           O_u
```

Using the notation of the diagram, but delaying the specification of the numerical values, the box-and-whisker plot provides:

1. A box, between the lower and upper hinges (H_ℓ and H_u), showing the central 50% of the sample values
2. Inner fences (I_ℓ and I_u), not shown by MINITAB, with the whiskers drawn from the hinges to the extreme sample values within the inner fence, these values being called the adjacent values
3. Outer fences (O_ℓ, and O_u), also not shown by MINITAB, with sample values between the inner and outer fences marked with a star, and sample values outside the outer fences marked with 0; and
4. The + sign to denote the median.

The sample values between the inner and outer fences are described as possible outliers and those beyond the outer fences as probable outliers. 'Outliers' is the technical term for unusual (large or small) sample values, but the adjectives used here to qualify them should not be taken too literally. No measure of probability is assigned to the descriptions – the adjectives merely describe the increasing distance from the boxed data.

All that remains is the definition of the hinges and fences. As (1) indicates, the hinges are practically the lower and upper quartiles, but the definition used by MINITAB generally gives values for the hinges that differ slightly from the quartiles. As for the quartiles, the hinges are defined in terms of the order statistics $x_{(1)} \le x_{(2)} \le \cdots \le x_{(n)}$. First we write $n + 1 = 2k + i$, where $i = 0$ or 1 according whether n is odd or even, and define the depth $d = (k + 1)/2$. Then, when d is an integer,

$$H_\ell = x_{(d)} \qquad \text{and} \qquad H_u = x_{(n+1-d)}$$

and when $d = d_1 + \frac{1}{2}$

$$H_\ell = \tfrac{1}{2}(x_{(d_1)} + x_{(d_1+1)}) \qquad \text{and} \qquad H_u = \tfrac{1}{2}(x_{(n-d_1)} + x_{(n-d_1+1)})$$

Loosely, the hinges are values at depth d from the minimum and maximum values.

The fences are now defined in terms of the length $L = H_u - H_\ell$, essentially the interquartile range, as follows:

$$I_\ell = H_\ell - 3L/2, \quad I_u = H_u + 3L/2$$

and

$$O_\ell = H_\ell - 3L, \quad O_u = H_u + 3L$$

From these definitions it is clear how the range of the central 50% of the sample values is used to classify the possible/probable outliers.

——— **EXAMPLE 2.7** ———————————————————

Using the mark data of Example 2.1, as before,

BOXPLOT C1

produces

Referring to Example 2.3 for the order statistics values, shown in C3, we calculate the hinges and fences. Here $n = 48$, so that $k = 24$ and $d = 12.5$, and we find

$$H_\ell = \tfrac{1}{2}(x_{(12)} + x_{(13)}) = \tfrac{1}{2}(55 + 59) = 57$$
$$H_u = \tfrac{1}{2}(x_{(36)} + x_{(37)}) = \tfrac{1}{2}(92 + 95) = 93.5$$
$$L = H_u - H_\ell = 36.5$$
$$I_\ell = 2.25, I_u = 148.25, O_\ell = -52.5 \text{ and } O_u = 203$$

Thus the adjacent values, the ends of the whiskers are 17 and 148; the values 149 and 174 are shown (by *) as possible outliers; and 209 is shown (by 0) as a probable outlier.

For a breakdown of the mark data by sex

       BOXPLOT C1;
       BY C2.

produces

```
sex
```

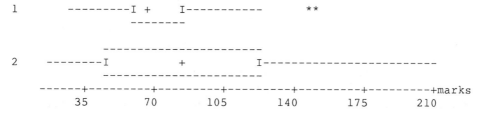

Note that the value 209, identified as a probable outlier for the combined data, is not even identified as a possible outlier for the data coded 2. The explanation is simple – outlier classification is measured in terms of the length of the box, and a longer box (as calculated for sex 2) defines a wider range for the fences.

2.2.3   *Bivariate data*

For a sample $(x_1, y_1), (x_2, y_2), \ldots, (x_n, y_n)$ of bivariate data, we enter $x_1, x_2, \ldots, x_n$ in column C11, say, and $y_1, y_2, \ldots, y_n$ in column C12. Note that it is important that row i of the columns contains x_i and y_i, respectively. In addition to the descriptive statistics obtained by applying the DESCRIBE command separately to each column, the sample correlation coefficient r, given by

$$r = \frac{\sum_{i=1}^{n} (x_i - \bar{x})(y_i - \bar{y})}{\sqrt{\sum_{i=1}^{n} (x_i - \bar{x})^2 \sum_{i=1}^{n} (y_i - \bar{y})^2}}$$

may be extracted from MINITAB by typing the command CORRELATION C11 C12.

From Problem 2.3 we note the properties:

1. $-1 \leqslant r \leqslant 1$
2. $r = \pm 1$ if and only if there is a linear relation
   $y_i = a + bx_i$, for $i = 1, 2, ..., n$, where a and b $(b \neq 0)$ are constant and
3. $b > 0$ for $r = 1$, and $b < 0$ for $r = -1$.

Combining (2) and (3), the exact linear relation which holds when $|r| = 1$, shows that, for $r = 1$, the y-values increase with increasing x-values, and for $r = -1$, the y-values decrease with increasing x-values.

The other special value of the sample correlation coefficient is the value $r = 0$, when the sample values are said to be *uncorrelated*. As the antithesis of an exact linear relation, small (that is, near-zero) values of r are used as an indication of the independence of the characteristics measured by the x- and y-values. However, it is possible to have $r = 0$ when there is an exact (necessarily non-linear) relation between the variables.

PLOT

The PLOT command provides a two-dimensional scatter-plot of the values (x_i, y_i), for $i = 1, 2, ..., n$. This is useful for indicating possible relationships between the variables and for identifying any possible unusual pairs – that is, outliers. The separate data points are shown by stars, but a count is printed for pairs sufficiently close together.

——— EXAMPLE 2.8 ———

An observational study provided 24 bivariate observations (x_i, y_i), for $i = 1, 2, ..., 24$. The data was entered into MINITAB with the y-values in C1 and x-values in C2, and C1 and C2 were named Y and X, respectively. Then

CORRELATION C1 C2

produced

Correlation of Y and $X = -0.984$

indicating a strong decreasing relationship. Further, a visual indication is provided by

> PLOT C1 C2

which produces the scatterplot

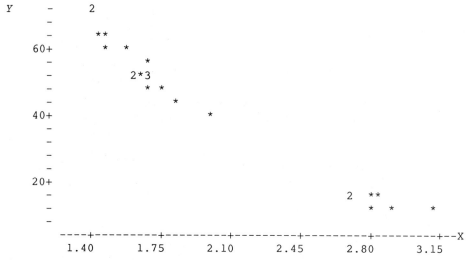

Despite the high correlation ($|r| \approx 1$), indicating a good fit of the data by a linear relation $y_i = a + bx_i$, the hint of curvature suggests that a quadratic relation $y_i = a + bx_i + cx_i^2$ might be preferred. To decide between these model equations is part of the subject matter of regression analysis.

2.2.4   Problems

2.1   Suppose $y_i = ax_i + b$ for $i = 1, 2, \ldots, n$ is a linear change of scale on the sample data $x_1, x_2, \ldots, x_n$. Show that (i) $\bar{y} = a\bar{x} + b$; and (ii) $s_y^2 = a^2 s_x^2$, where, for any set of values $v_1, v_2, \ldots, v_n$, we write

$$s_v^2 = \sum_{i=1}^{n} (v_i - \bar{v})^2 \Big/ (n - 1)$$

2.2   Show, for any set of real numbers $x_1, x_2, \ldots, x_n$ and any real number a, that

$$\sum_{i=1}^{n} (x_i - a)^2 = \sum_{i=1}^{n} (x_i - \bar{x})^2 + n(\bar{x} - a)^2$$

and deduce that

$$\min_{a} \sum_{i=1}^{n} (x_i - a)^2 = \sum_{i=1}^{n} (x_i - \bar{x})^2$$

2.3 For bivariate data $(x_1, y_1), (x_2, y_2), \ldots, (x_n, y_n)$ and any real number λ, show that

$$\sum_{i=1}^{n} (y_i - \bar{y} - \lambda(x_i - \bar{x}))^2 = \sum_{i=1}^{n} (y_i - \bar{y})^2 - 2\lambda \sum_{i=1}^{n} (x_i - \bar{x})(y_i - \bar{y}) + \lambda^2 \sum_{i=1}^{n} (x_i - \bar{x})^2$$

With r denoting the correlation coefficient and

$$a^2 = \sum_{i=1}^{n} (x_i - \bar{x})^2, \quad b^2 = \sum_{i=1}^{n} (y_i - \bar{y})^2$$

show that the right-hand side can be rewritten as $(\lambda a - rb)^2 + (1 - r^2)b^2$. Deduce that $|r| \leq 1$, and, when $|r| = 1$, that $y_i - \bar{y} = r(b/a)(x_i - \bar{x})$ for $i = 1, 2, \ldots, n$.

2.4 Let $Y_1, Y_2, \ldots, Y_n$ be independent random variables with $E(Y_i) = ax_i + b$ and $V(Y_i) = \sigma^2$ for $i = 1, 2, \ldots, n$, where a, b are constants. Write down $E(\bar{Y})$ and $V(\bar{Y})$, and show that

$$E\left(\sum_{i=1}^{n} (Y_i - \bar{Y})^2\right) = (n - 1)\sigma^2 + a^2 \sum_{i=1}^{n} (x_i - \bar{x})^2$$

(Hint: Write $\sum_{i=1}^{n} (Y_i - \bar{Y})^2 = \sum_{i=1}^{n} Y_i^2 - n\bar{Y}^2$, and use the relation $E(X^2) = V(X) + (E(X))^2$.)

2.5 Apply the commands of this section to examine the following data. Enter the data into C1 in MINITAB in the order given by reading the data row by row.

137	143	133	127	43	77	160	187	130	47	47	87	160	200	193	173	203	213
187	227	240	253	237	140	17	70	67	40	47	37	10	10	10	20	43	73
100	123	130	133	140	133	127	127	117	137	153	167	173	167	150	150	93	30
7	77	83	87	87	87	73	67	43	17	7	17	33	43	7	3	40	147
327	380	327	330	253	233	273	283	227	197	180	153	110					

Include a plot of the data against the order – that is, set $1, 2, 3, \ldots, 85$ in C2 and plot C1 against C2.

2.6 (MINITAB) This problem shows how the command HISTOGRAM may suggest the parent distribution.

(i) Use the RANDOM command to generate a (simulated) random sample of size 100 from the binomial distribution $b(20, 0.2)$. Compare the shape of the histogram with that of the corresponding probability function $f(x \mid 20, 0.2)$ – see Section 1.2. Why should you expect the shapes to be similar?

(ii) Use the RANDOM command to obtain in C1 a (simulated) random sample of size 50 from the normal distribution $N(1, 1)$, and in C2 a sample of size 50 from the normal distribution $N(10, 2^2)$. Now stack C1 on C2 via the command >STACK C1 C2 C3 and obtain the histograms and dotplots of the three columns.

2.7 (MINITAB) This problem constructs bivariate data and uses the PLOT command to reveal the possible relation between the variables.

Enter in C11 a random sample of size 10 from the integers $1, 2, \ldots, 100$ by using the RANDOM command with SUBC> INTEGER 1 100. , and let $C2 = C11/10$. Obtain, in C12, a random sample of size 10 from the normal distribution $N(0, 0.1^2)$ and let $C1 = 4 + 2 * C2 + C12$. The data pairs $(C2, C1)$ are the bivariate (x, y) data. A plot of C1 against C2 should suggest the linear relation $y = 4 + 2x$.

Further, let $C3 = 4 + 0.5 * C2 ** 2 + C12$ and plot C3 against C2. What shape do you expect to see?

Examine the effect of increasing the standard deviation of the sample in C12 by considering, successively, $C1 = 4 + 2 * C2 + a * C12$ for $a = 10, 20, 50$.

2.3   The next step

Sample data is collected in order that we may draw conclusions about the nature of the characteristics being examined. These conclusions, termed 'inferences' in statistical language, take the general form of descriptions of properties of the distribution of the values of the characteristics over the whole parent population.

In an ideal situation each member of the parent population is examined and complete information on the chosen measures of the characteristics is obtained. Initial processing of the data may take the summary form of the calculation of descriptive statistics, and, in a multivariate problem, the compilation of cross-classified tables showing aspects of the joint distributions of the variables. Further, the data may be used in future planning (for example, by governments, based on a full census) or for prediction (by gamblers interpreting sporting data) – areas fraught with danger since unforeseen factors (such as those of economic change or human fallibility) may upset otherwise reasonable calculations.

Most frequently, the situation facing the statistician is that of incomplete information – that is, the values of the chosen measures on the characteristics are known only for a small sample of the parent population. Now the initial aim is to extrapolate, from the observed data, the likely nature of the characteristics over the parent population. The process of extrapolation is accomplished by an application of one or more appropriate statistical methods. Evidently, it would be unreasonable to hope that the variations in the sample values can completely reflect the variations in the characteristics for the whole population. However, in order to extrapolate with any degree of confidence, it is essential that the sample must not be atypical of the parent population. A couple of examples may clarify the sort of bias that may occur by ill-chosen selection of the sample elements: the opinion pollster, in a national election, must be aware of regional variation; and a clinician, comparing two treatments, must allot the treatments to patients at random, thereby avoiding any systematic bias which may occur through an unwitting use of clinical judgement. In a slightly different vein, involving experimentation, results obtained under carefully controlled conditions (for example, crop evaluation in a greenhouse) may not be achievable in a wider environment.

The principle that underlies the manner in which the sample elements are selected from the parent population is simply stated: a sample of n elements is said to be a random sample if every sample of size n has an equal probability of being selected.

As a general statement, this principle serves quite well as a guide to intuition, but a little thought reveals a need for greater precision:

1. As a naive mechanism for selecting the sample elements, consider the simple case when the parent population consists of a finite number N of elements. Placing balls, numbered $1, 2, ..., N$ to represent the population elements in a bag, the sample elements are identified by repeated sampling of balls from the bag. The sampling may be with replacement (that is, a ball is selected, its number noted, and replaced in the bag before the next selection) or without replacement (that is, the balls drawn are not replaced). The naivety and impracticality of the mechanism is clear – many populations are not finite, or, even if finite, are too large.
2. For a finite population with N elements, the probability of selecting a given sample of size n is $1/N^n$ with replacement, and $1/N(N-1)$ ... $(N-n+1)$ without replacement. When N is large and n relatively small, probability calculations for the two sampling methods generally differ only slightly, and the conceptually simpler model of sampling with replacement is adopted.
3. When the parent population is infinite the probability suggested in the principle is zero. The practical view of the principle, in this case, is to regard the selection as being made from a large subset of the population – a subset, of course, that is not atypical of the population.

Suppose, now, we are investigating a characteristic using a single quantitative measurement. Although the parent population does not represent a probability space, we may regard the values of the characteristic as having a distribution function F, called the parent distribution function. For each real number x, $F(x)$ is the proportion of the sample space for which the measured value of the characteristic is less than or equal to x. There is no difficulty behind this concept, particularly for a finite population with N elements, when we may write

$$F(x) = (\text{number of elements with value} \leq x)/N$$

but the values $F(x)$ are, of course, unknown.

Following the selection principle, with or without replacement, suppose we have obtained sample values $x_1, x_2, ..., x_n$. The key to statistical analysis is not to consider the obtained values in isolation, but to place them, using distributional arguments, in the context of all possible samples of size n. To this end, we introduce the sample random variables $X_1, X_2, ..., X_n$, defined for all possible samples of size n, by letting X_i be the value for the ith element of the sample. The actual obtained sample values are called the observed values of the sample random variables.

What can be said about the distribution of the random variables $X_1, X_2, ..., X_n$? For the remainder of this section, and in most of the succeeding chapters, we will assume that the population is large enough for the sampling process to be regarded as sampling with replacement. Appendix 2 discusses sampling without replacement from a finite population with applications to the means and variances of some of the statistics we use.

For selections with replacement, the following heuristic argument shows that each random variable has distribution function F, that is, $\Pr(X_i \leqslant x) = F(x)$ for all $i = 1, 2, \ldots, n$ and all real numbers x. To say that $X_i \leqslant x$ is to say that the ith element of the sample was chosen from that part of the parent population where the characteristic has value less than or equal to x – that is, the proportion $F(x)$ of the population. But each element is equally likely to be chosen, so the probability of selecting from this part is the proportionate size $F(x)$ of the part. The second ingredient is the independence of the random variables $X_1, X_2, \ldots, X_n$ – this being nothing more than the assumed independence of the sample elements, as described in the naive mechanism.

To summarize, the model for a statistical analysis involving a single quantitative variable consists of a simple random sample of size n, containing: sample random variables $X_1, X_2, \ldots, X_n$ which are independent with a common distribution, and the obtained sample data $x_1, x_2, \ldots, x_n$.

The model for a statistical analysis involving a single qualitative variable takes a different form. Suppose that we have a population of elements and a single characteristic of the elements which has k values or levels – for example, artistic ability rated at levels 1, 2 or 3 for poor, average and good. A sample of n elements of the population is examined, and the result of the examinations is recorded as counts $n_1, n_2, \ldots, n_k$ of the sample elements exhibiting each of the k levels. The probability structure associated with this situation is, again, to consider all samples of size n and define the sample variables $X_1, X_2, \ldots, X_k$, where $X_1 + X_2 + \cdots + X_k = n$ and X_i denotes the number of sample elements exhibiting level i of the characteristic. When the sampling process is simple random sampling from a large parent population, the joint distribution of $X_1, X_2, \ldots, X_k$ is the multinomial distribution (extending the trinomial distribution of Section 1.3), with

$$\Pr(X_1 = n_1, X_2 = n_2, \ldots, X_k = n_k) = \frac{n!}{n_1! n_2! \ldots n_k!} p_1^{n_1} p_2^{n_2} \ldots p_k^{n_k}$$

where p_i is the unknown proportion of the population showing level i of the characteristic, $0 < p_i < 1$ for $i = 1, 2, \ldots, k$, and $\sum_{i=1}^{k} p_i = 1$.

One further distinction must be noted. Implicit in the above description is the assumption that the investigator has control of the sampling process. For many applications, for example in medicine, psychology and the social sciences, the data presented to the statistician has not been collected by random sampling of population elements, but is simply the set of observed values for the sample at hand. An example where an observational study and subsequent analysis must be employed is a study of the length of time between HIV infection and the onset of AIDS – no volunteers for a controlled experiment are likely to be found. Analysis of observational data is a complex affair and goes beyond the methods we describe in later chapters – methods which usually rely on the assumption of simple random sampling.

In the following chapters we describe a variety of statistical procedures for assessing the information that may be obtained from the sample data. These methods may be classified broadly as follows:

1. Parametric procedures – these are ones in which the distribution function F is partially specified but contains unknown parameters – for example, μ and/or σ^2 in the normal $N(\mu, \sigma^2)$ distribution or p in the binomial $b(1, p)$ distribution. Within this type, the statistician may be interested in estimating a value for the unknown parameter, or assessing the consistency of the observed data with some preassigned value(s) of the parameter(s).
2. Non-parametric procedures – these are methods which are not concerned with the parameters of the distribution – for instance, when investigating the independence of two characteristics.
3. Distribution-free procedures – these apply generally, without requiring any assumption about the parent distribution function, except that some require continuity or symmetry.

Unfortunately, and typically, a broad classification fails to be exact, and borderline examples may be classified differently by various authors. A class of simple examples are procedures involving the mean, variance or median, which are generally treated as parametric population parameters though no functional form for the distribution function F may be assumed. A further complication in the use of language is that some authors do not distinguish between 'non-parametric' and 'distribution-free'.

We conclude this chapter with another broad classification, dividing the class of statistical methods into two parts – estimation and hypothesis testing. Estimation, as indicated above, is essentially a parametric procedure, assigning values determined by the sample data to numerical parameters related to the parent distribution function F. This topic is the subject of Chapter 3. On the other hand, hypothesis testing has the aim of assessing whether the data is reasonably consistent with a conjecture about the nature of the parent distribution. A hypothesis, the statistical synonym for conjecture, may also be parametric (for instance, that the mean μ takes a prescribed value μ_0) or non-parametric (as, for instance, that two measures on a characteristic are independent). As the title suggests, hypothesis testing is the main subject of the book, and occupies Chapters 4–9.

In the parametric set-up, particularly, the methods share the common feature that appropriate functions $g(x_1, x_2, ..., x_n)$ of the sample data, such as, for example, the sample mean or sample variance, are defined, and the analysis proceeds by probability calculations involving the distribution of the corresponding random variable $g(X_1, X_2, ..., X_n)$. The general class of functions $g(x_1, x_2, ..., x_n)$ and corresponding random variables are called statistics. Examples are the sample variables themselves and the measures defined in Section 2.2 as descriptive statistics. It is important to note that although $g(x_1, x_2, \ldots, x_n)$ can be calculated from the sample values and contains no unknown parameters, the distribution of $g(X_1, X_2, \ldots, X_n)$ may contain parameters. For instance, when the distribution function F is normal $N(\mu, \sigma^2)$ the distribution of $\bar{X}$ is normal $N(\mu, \sigma^2/n)$ (see Section 1.3) but $\bar{x} = (x_1 + x_2 + \cdots + x_n)/n$ is obtained from the sample values alone.

2.3.1 Problems

2.8 Choose values for the parameters n, μ and σ and then use the RANDOM command to obtain a (simulated) random sample of size n from the distribution $N(\mu, \sigma^2)$. Examine your sample using the MINITAB commands discussed in Section 2.2. Compare informally the sample mean and sample standard deviation with the chosen values of μ and σ. Repeat for different choices of μ and σ – in particular when σ is (a) small and (b) large, compared to μ (for example, $(\mu, \sigma) = (5, 0.1)$ and $(\mu, \sigma) = (0.5, 2)$).

2.9 Let $X_1, X_2, \ldots, X_n$ be a random sample from the distribution $N(\mu, \sigma^2)$. Show that

$$\Pr(|\bar{X} - \mu| \geq 2\sigma/\sqrt{n}) = 2(1 - \Phi(2))$$

and for the nth-order statistic $X_{(n)}$ (see Section 1.1),

$$\Pr(X_{(n)} \geq \mu + 2\sigma) = 1 - (\Phi(2))^n$$

2.10 Use the RANDOM command to obtain 25 samples of size 100 in C1–C25, from the distribution $N(4, 1^2)$. Treating the rows as 100 samples of size 25, find the row means and row maxima. How many row means satisfy $|\bar{x} - 4| \geq 0.4$, and how many row maxima satisfy $x_{(25)} \geq 6$? How many would you expect in each case?

2.11 Choose a value for the parameter p and then use the RANDOM command to obtain a (simulated) random sample of size 100 from the binomial distribution $b(20, p)$. Examine your sample using the commands discussed in Section 2.2. If the random variable X has the distribution $b(20, p)$ find

$$p_0 = \Pr(X > 20p + 2\sqrt{20p(1 - p)})$$

using tables for your chosen value of p. How many of your sample values exceed $20p + 2\sqrt{20p(1 - p)}$, and how many would you expect to exceed this number?

2.4   Solutions to problems

2.1 $\sum_{i=1}^{n} y_i = \sum_{i=1}^{n} (ax_i + b) = a \sum_{i=1}^{n} x_i + nb$. Now divide by n.

$\sum_{i=1}^{n} (y_i - \bar{y})^2 = \sum_{i=1}^{n} (ax_i + b - a\bar{x} - b)^2 = a^2 \sum_{i=1}^{n} (x_i - \bar{x})^2$, and now divide by $n - 1$.

2.2
$$\sum_{i=1}^{n} (x_i - a)^2 = \sum_{i=1}^{n} (x_i - \bar{x} + \bar{x} - a)^2$$

$$= \sum_{i=1}^{n} (x_i - a)^2 + 2(\bar{x} - a) \sum_{i=1}^{n} (x_i - \bar{x}) + \sum_{i=1}^{n} (\bar{x} - a)^2$$

$$= \sum_{i=1}^{n} (x_i - \bar{x})^2 + n(\bar{x} - a)^2$$

since $\displaystyle\sum_{i=1}^{n} (x_i - \bar{x}) = \sum_{i=1}^{n} x_i - n\bar{x} = 0$

Since all terms are positive, the minimum occurs when $\bar{x} - a = 0$.

2.3 The first relation follows from the expansion

$$(y_i - \bar{y} - \lambda(x_i - \bar{x}))^2 = (y_i - \bar{y})^2 - 2\lambda(x_i - \bar{x})(y_i - \bar{y}) + \lambda^2(x_i - \bar{x})^2$$

The r.h.s. is $b^2 - 2\lambda rab + \lambda^2 a^2 = (\lambda a - rb)^2 + (1 - r^2)b^2 \geqslant 0$ for all values of λ, which implies, by taking $\lambda a = rb$, that $(1 - r^2)b^2 \geqslant 0$, or $|r| \leqslant 1$. Conversely, when $|r| = 1$ and $\lambda a = rb$ then $\sum_{i=1}^{n} (y_i - \bar{y} - \lambda(x_i - \bar{x}))^2 = 0$, so each term must be zero.

2.4 $E(Y_i) = ax_i + b$ implies $E(\bar{Y}) = a\bar{x} + b$, and

$$V(\bar{Y}) = \frac{1}{n^2} \sum_{i=1}^{n} V(Y_i) = \sigma^2/n$$

Also

$$E\left(\sum_{i=1}^{n} (Y_i - \bar{Y})^2 \right) = \sum_{i=1}^{n} E(Y_i^2) - nE(\bar{Y}^2)$$

$$= \sum_{i=1}^{n} (\sigma^2 + (ax_i + b)^2) - n(\sigma^2/n + (a\bar{x} + b)^2)$$

$$= (n - 1)\sigma^2 + \sum_{i=1}^{n} (ax_i + b)^2 - n(a\bar{x} + b)^2$$

$$= (n - 1)\sigma^2 + a^2 \sum_{i=1}^{n} (x_i - \bar{x})^2$$

using Problem 2.2.

2.5 The PLOT command printout reveals that there is some pattern to the variation in the successive data values, suggesting that this is not a simple random sample and that there is some dependence between successive values.

2.6 (i) The expected number of sample values equal to k is $100f(k\,|\,20, 0.2)$ for each $k = 0, 1, 2, \ldots, 20$ and the histogram shows the observed number. For a sample of size n, the normal approximation shows that the difference between the expected and observed values is of order $\sqrt{n}$ with high probability – that is, if X has the binomial distribution $b(n, p)$ then $\Pr(|X - np| < 1.96\sqrt{npq}) = 0.95$. Note that the shape of the p.f. $f(k \mid 20, 0.2)$ may be found from the PDF and PLOT commands.

(ii) The diagrams for C1 and C2 should suggest the corresponding p.d.f., with those for C3 showing the dichotomy in the data values.

2.7 The plot of C3 against C2 should suggest the quadratic $y = 4 + 0.5x^2$. As the value of a increases the variation in the random sample from the normal distribution is likely to obscure the linear relation – the standard deviation is 1, 2, 5 for $a = 10, 20, 50$ respectively.

2.9 $\bar{X}$ has the distribution $N(\mu, \sigma^2/n)$ so that

$$\Pr(|\bar{X} - \mu| \geq 2\sigma/\sqrt{n}) = \Pr(|Z| \geq 2) = 2\Pr(Z \geq 2)$$

The common distribution of $X_1, X_2, \ldots, X_n$ is $F(x) = \Phi((x - \mu)/\sigma)$, and, from Section 1.1,

$$\Pr(X_{(n)} \geq \mu + 2\sigma) = 1 - [F(\mu + 2\sigma)]^n$$

2.10 The required counts may be obtained as follows:

```
RMEANS C1 – C25 C31
LET C32 = SIGNS (ABSO (C31 – 4) – 0.4)
LET C33 = (ABSO (C32) + C32)/2
SUM C33
RMAX C1–C25 C41
LET C42 = SIGNS (C41–6)
LET C43 = (ABSO (C42) + C42)/2
SUM C43
```

Now $2(1 - \Phi(2)) = 0.0456$ and $1 - \Phi(2)^{25} = 0.4382$, so that the expected numbers are 4.56 (for row means) and 43.82 for row maxima.

2.11 The count can be obtained from the HISTOGRAM printouts; the expected number is $100p_0$.

Chapter Three

Estimation

3.1 Introduction

Estimation is the name given to the statistical procedure by which numerical values obtained from a sample (for example, the mean, variance or median) are used as an approximation to the corresponding unknown values for the parent population. Examples of estimation are common in the media (the opinion pollster predicting electoral percentages, the market researcher investigating product preference, the economist studying income or unemployment), and are used frequently in science (zoologists describing the characteristics of species, agriculturalists examining crop-yield for a new cross-breed of seed).

Throughout the chapter we assume that we have a parent population whose characteristic of interest is a quantitative variable, and that we have a random sample of size n with sample random variables $X_1, X_2, ..., X_n$ and observed values $x_1, x_2, ..., x_n$. The random variables are independent, by definition, and their common distribution function is written as $F(x \mid \theta)$ to show the unknown (population) parameter θ. Our goal is to assign a numerical value for θ, and, thereby, obtain information about the characteristic. The symbols μ and σ^2 will be the standard notation for the expectation and variance of the distribution $F(x \mid \theta)$ – that is, $E(X_i) = \mu$, $V(X_1) = \sigma^2$ for $i = 1, 2, ..., n$. Both μ and σ^2 will be assumed to be finite; they will, in general, be functions of the parameter.

As examples of some common situations expressed in this way:

1. The opinion pollster model, in a two-candidate election, can be described by random variables $X_1, X_2, ..., X_n$ taking values 0 or 1 (value 1 for candidate number 1, say), so that the variables have the $b(1, p)$ distribution of Section 1.3. Here $\theta = p$, $\mu = p$, $\sigma^2 = p(1 - p)$ and an estimate for p gives both an estimate for the population mean and, in the form $100p\%$, an estimate of the support for candidate 1.

2. A manufacturer interested in the lifetime to failure of a product may suggest an exponential distribution for the characteristic of failure time, that is, $F(x \mid \theta) = 1 - \exp(-x/\theta)$ for $x > 0$. Here $\mu = \theta$, $\sigma^2 = \theta^2$, with an estimate for θ again being an estimate for the population mean (the average time to failure).

3. In many problems it is assumed that the parent distribution F is normal $N(\mu, \sigma^2)$ – an assumption usually based on data distribution analysis or on the grounds of previous experience, with the parameter $\theta = (\mu, \sigma^2)$, displaying both unknown population parameters, and estimating θ amounts to estimating both μ and σ^2.

A reader who has used the MINITAB RANDOM command to simulate random samples, as in Section 2.2, will be convinced by the numerical evidence that, for the examples above, the sample mean is the intuitive choice as a numerical approximation for the population mean. As in any area of mathematical application, the first question to be asked about an estimate is 'what is the accuracy?' Bearing in mind that, in our context, the estimate is based on a random sample, the question should be re-expressed in terms of what reliance can be placed on the obtained numerical value. A measured response to this question is provided by probability calculations involving the distribution of the sample mean.

For the general model, we define a point estimate $\hat{\theta}$ of the parameter θ to be a numerical function $\hat{\theta} = \hat{\theta}(x_1, x_2, ..., x_n)$ of the sample values, and the statistic $\hat{\Theta} = \hat{\theta}(X_1, X_2, ..., X_n)$ is the corresponding point estimator. The probability calculations which assess the quality of the estimate are made using the distribution of $\hat{\Theta}$.

The preamble to the definition relied on intuition to suggest the sample mean $\bar{X}$ as the choice of estimator for the population mean. However, for a distribution symmetric about the mean, an alternative choice of estimator of the mean is the sample median. Both choices satisfy the definition – indeed, the definition allows any function of the sample variables to be called a point estimator. In order to make further progress, we need to sharpen the definition and provide criteria for choosing an estimator. These are discussed in the next section, where two methods for obtaining estimators also are described. The final section of the chapter addresses the problem of accuracy by replacing the point estimate $\hat{\theta}$ with an interval estimate of 'likely' values for θ.

3.2   Point estimation

In order to select a useful estimator from the range of possibilities allowed by the definition, the first requirement is a set of criteria by which different estimators may be compared. We examine, first, what can be said about the general use of the sample mean $\bar{X}$ as an estimator of the population mean. Since $E(\bar{X}) = \mu$, the average of the values of the sample mean is the true population mean. Unfortunately, however, we have available only the single observed value $\bar{x}$, and we must investigate its place in the distribution of the values of $\bar{X}$. Noting that $V(\bar{X}) = \sigma^2/n$, Chebychev's inequality (see Section 1.1) gives, for any value $a > 0$,

$$\Pr(|\bar{X} - \mu| \geq a) \leq V(\bar{X})/a^2$$

Now, for a large enough sample size n, we have $\sigma^2/na^2 \approx 0$, so that the values of the sample mean are concentrated, with probability close to one, within distance at most *a* from the population mean μ. For a fixed sample size the Chebychev inequality does not provide an accurate estimate of the probability on the left-hand side of the inequality, and more precise results are obtained by making further assumptions about the parent distribution.

───────── EXAMPLE 3.1 ─────────────────────────────

Suppose the random variables $X_1, X_2, \ldots, X_n$ are distributed $N(\mu, \sigma_0^2)$. Tables supply

$$\Pr(|\bar{X} - \mu| \leqslant 1.96\ \sigma_0/\sqrt{n}) \approx 0.95$$

which may be read as saying that, for 95% of samples of size n, the observed value $\bar{x}$ of the sample mean $\bar{X}$ will lie within $1.96\sigma_0/\sqrt{n}$ of the true mean μ, whatever the true value of μ may be.

───

The preceding remarks suggest the criteria for choice of estimator $\hat{\Theta}$ of a numerical parameter θ as:

1. $\hat{\Theta}$ is **unbiased** – that is, $E(\hat{\Theta}) = \theta$
2. Given a choice of unbiased estimators, we should prefer the estimator having smaller variance.

Additionally, though not strictly a criterion, we may impose the condition that we look for estimators among the class of statistics whose distribution may be determined – either exactly or, for large samples, by an approximation. This will facilitate making the choice suggested by (2), and enable precise calculations of probabilities to be made.

It is not our intention to develop the theory of estimation – for this see reference 1. Rather, we wish to outline the general notions of estimation as they appear in the MINITAB printouts of tests of hypotheses. A modest illustration of the comparison of two estimators is given in Example 3.7 of this section. We remark, without proof, that the sample mean is preferred to the sample median as an unbiased estimator for the normal mean, and the sample mean is preferred to the sample variance as an unbiased estimator of the Poisson parameter.

Two general methods by which estimators may be determined are now described. As the examples and problems demonstrate, neither method will automatically produce an unbiased estimator.

3.2.1   *Method 1: The method of moments*

For this method the only requirement on the distribution is that the sample random variables $X_1, X_2, \ldots, X_n$ are assumed to have (common) finite moment $\theta_r = E(X_i^r)$, where r is a positive integer. The method of moments is to estimate θ_r by the corresponding sample moment $\hat{\theta}_r = \sum_{i=1}^{n} x_i^r/n$ with corresponding estimator

$\hat{\Theta}_r = \sum_{i=1}^{n} X_i^r / n$. By definition, the moment estimator is an unbiased estimator for the population moment.

The following examples show how the method of moments may be used, by the device of equating the moments to their expected values, to obtain estimators for other population parameters.

───── EXAMPLE 3.2 ─────────────────────────────────

Suppose $X_1, X_2, \ldots, X_n$ have unknown mean μ and variance σ^2. Since $E(\bar{X}) = \mu$ and

$$E(\hat{\Theta}_2) = E\left(\sum_{i=1}^{n} X_i^2 / n\right) = \sigma^2 + \mu^2$$

the method of moments suggests that we equate

$$\hat{\mu} = \bar{X} \text{ and } \sum_{i=1}^{n} X_i^2 / n = \hat{\sigma}^2 + \hat{\mu}^2$$

obtaining $\hat{\sigma}^2 = \sum_{i=1}^{n} X_i^2 / n - \bar{X}^2 = \sum_{i=1}^{n} (X_i - \bar{X})^2 / n$ as the moment estimator for σ^2. Note, from Section 1.1.9, that $E(\hat{\sigma}^2) = (n-1)\,\sigma^2 / n$ so the moment estimator of σ^2 is not unbiased.

───── EXAMPLE 3.3 ─────────────────────────────────

Suppose $X_1, X_2, \ldots, X_n$ have the uniform distribution $U(\alpha, \beta)$ where α and β are unknown – that is, the probability density function $f(x \mid \alpha, \beta)$ is given by

$$f(x \mid \alpha, \beta) = 1/(\beta - \alpha) \text{ for } \alpha < x < \beta$$

and is zero otherwise. Then, for $i = 1, 2, \ldots, n$, we calculate

$$E(X_i) = \frac{\alpha + \beta}{2}, \; E(X_i^2) = \frac{\alpha^2 + \alpha\beta + \beta^2}{3} \quad \text{and} \quad V(X_i) = \frac{(\alpha - \beta)^2}{12}$$

and, hence, we equate

$$\bar{X} = \frac{\hat{\alpha} + \hat{\beta}}{2} \quad \text{and} \quad \frac{1}{n} \sum_{i=1}^{n} X_i^2 = \frac{(\hat{\alpha} - \hat{\beta})^2}{12} + \frac{(\hat{\alpha} + \hat{\beta})^2}{4}$$

Rewriting the equations in the form

$$\hat{\alpha} + \hat{\beta} = 2\bar{X} \quad \text{and} \quad \hat{\beta} - \hat{\alpha} = \hat{\sigma}\sqrt{12}$$

where

$$n\hat{\sigma}^2 = \sum_{i=1}^{n} (X_i - \bar{X})^2$$

the moment estimators are $\hat{\alpha} = \bar{X} - \hat{\sigma}\sqrt{3}$ and $\hat{\beta} = \bar{X} + \hat{\sigma}\sqrt{3}$.

3.2.2   *Method 2: The maximum likelihood principle*

A frequently used method, available when the distribution $F(x \mid \theta)$ takes a specified form, is to proceed as follows. Writing $f(x \mid \theta)$ for the probability function/ probability density function, form the joint p.f. or p.d.f. of the independent sample variables

$$L(\theta \mid x_1, x_2, ..., x_n) = f(x_1 \mid \theta)f(x_2 \mid \theta)...f(x_n \mid \theta)$$

Regarded as a function of θ for fixed values $x_1, x_2, ..., x_n$, L is called the *likelihood function*.

The principle of maximum likelihood is to choose as the *maximum likelihood estimate* (MLE) the value $\hat{\theta} = \hat{\theta}(x_1, x_2, ..., x_n)$ such that

$$L(\hat{\theta} \mid x_1, x_2, ..., x_n) = \max_{\theta} L(\theta \mid x_1, x_2, ..., x_n)$$

with the corresponding statistic $\hat{\Theta} = \hat{\theta}(X_1, X_2, ..., X_n)$ being the *ML estimator*.

An intuitive understanding of the principle is most readily seen in the discrete case, with an approximation argument extending the idea to the continuous variable case. For discrete sample variables the likelihood function is simply $\Pr(X_1 = x_1, X_2 = x_2, ..., X_n = x_n)$, and with $x_1, x_2, ..., x_n$ fixed, the principle asserts that $\hat{\theta}$ should be taken to maximize the probability of obtaining the sample values $x_1, x_2, ..., x_n$ which actually are obtained.

──────── EXAMPLE 3.4 ────────────────────────────────

Suppose $X_1, X_2, ..., X_n$ have the binomial $b(1, p)$ distribution. Then

$$L(p \mid x_1, x_2, ..., x_n) = p^t(1-p)^{n-t}, \text{ where } t = \sum_{i=1}^{n} x_i$$

Differentiating with respect to p gives, as the condition for L to be a maximum,

$$tp^{t-1}(1-p)^{n-t} - (n-t)p^t(1-p)^{n-t-1} = 0$$

which simplifies to $np = t$. The MLE is therefore $\hat{p} = \bar{x}$.

──────── EXAMPLE 3.5 ────────────────────────────────

Suppose the sample variables are distributed $N(\mu, \sigma^2)$, where μ and σ are unknown. We estimate μ and σ^2 jointly by maximizing the likelihood function

$$L(\mu, \sigma \mid x_1, x_2, ..., x_n) = \left(\frac{1}{\sqrt{2\pi\sigma^2}}\right)^n \exp\left[-\frac{1}{2\sigma^2} \sum_{i=1}^{n} (x_i - \mu)^2\right]$$

as a function of μ and σ. As in many other examples, L may more easily be maximized by differentiating log L with respect to μ and σ.

Now

$$\log L = -n \log \sqrt{2\pi} - n \log \sigma - \frac{1}{2\sigma^2} \sum_{i=1}^{n} (x_i - \mu)^2$$

and equating the partial derivatives to zero, to maximize both L and log L, gives

$$\frac{\partial \log L}{\partial \mu} = n \sum_{i=1}^{n} (x_i - \mu)/\sigma^2 = 0$$

$$\frac{\partial \log L}{\partial \sigma} = -n/\sigma + \sum_{i=1}^{n} (x_i - \mu)^2/\sigma^3 = 0$$

with solutions $\hat{\mu} = \bar{x}$ and $\hat{\sigma}^2 = \sum_{i=1}^{n} (x_i - \bar{x})^2/n$.

Note, as in Example 3.2, that the ML estimator $\hat{\sigma}^2$ for σ^2 is not unbiased.

──────── **EXAMPLE 3.6** ────────────────────────────────

Suppose the sample variables $X_1, X_2, ..., X_n$ have the uniform distribution $U(0, \theta)$ with p.d.f. $f(x \mid \theta) = 1/\theta$ for $0 < x \leqslant \theta$, and $f(x \mid \theta) = 0$ otherwise. In this case

$$L(\theta \mid x_1, x_2, ..., x_n) = 1/\theta^n \qquad 0 < \max(x_1, x_2, ..., x_n) \leqslant \theta$$
$$0 \qquad\qquad \text{otherwise}$$

Observing that L decreases as θ increases, L is maximized by taking $\hat{\Theta} = \max(X_1, X_2, ..., X_n)$. From Section 1.1 Problem 1.6, the p.d.f. of $\hat{\Theta}$ is nx^{n-1}/θ^n for $0 < x \leqslant \theta$ so that

$$E(\hat{\Theta}) = \int_0^\theta x \cdot \frac{nx^{n-1}}{\theta^n} \, dx = n\theta/(n+1)$$

Thus $\hat{\Theta}$ is not unbiased, but $(n+1)\hat{\Theta}/n$ is an unbiased estimator of θ.

───

In all the examples above, the ML procedure has produced a unique estimator – see Problem 3.5 below for an example of non-uniqueness. For the important class of distributions given in Problem 3.7, the ML procedure can be shown to provide a unique MLE and, further, that the MLE becomes (approximately) unbiased with a normal distribution as the sample size increases.

──────── **EXAMPLE 3.7** ────────────────────────────────

We compare the unbiased estimators $\hat{\Theta}_1 = 2\bar{X}$ and $\hat{\Theta}_2 = ((n+1)/n) \max(X_1, X_2, ..., X_n)$ of the parameter θ in the uniform distribution $U(0, \theta)$, by calculating the variance.

Now $E(X_i) = \theta/2$, $V(X_i) = \theta^2/12$, from Section 1.1, so that $V(\hat{\Theta}_1) = 4\ V(\bar{X}) = 4\ \theta^2/12n = \theta^2/3n$. Writing $X_{(n)} = \max(X_1, X_2, \ldots, X_n)$ then

$$E(X_{(n)}^2) = \int_0^\theta x^2 \cdot \frac{nx^{n-1}}{\theta^n}\, dx = n\theta^2/(n+2)$$

so, from Example 3.6,

$$V(X_{(n)}) = n\theta^2/(n+2) - (n\theta/(n+1))^2 = n\theta^2/(n+1)^2(n+2)$$

from which we obtain $V(\hat{\Theta}_2) = ((n+1)/n)^2\, V(X_{(n)}) = \theta^2/n(n+2)$.

Hence, the estimator $\hat{\Theta}_2$ has smaller variance than $\hat{\Theta}_1$ – that is, a multiple of the nth-order statistic is preferred to a multiple of the sample mean.

Our description of the method of maximum likelihood has been in the setting of a simple random sample with independent sample variables. The ML method may be employed in a far more general way: for sample variables $X_1, X_2, \ldots, X_n$, simply take the likelihood function to be the joint p.f./p.d.f. of the variables, and maximize over the unknown parameters. See Problem 3.6 for an application to the trinomial distribution of Section 1.3.

3.2.3   Problems

3.1   Suppose the independent sample random variables $X_1, X_2, \ldots, X_n$ have the common p.d.f.

$$f(x \mid \theta) = \theta/(1+x)^{\theta+1}$$

where $x > 0$, $\theta > 1$. Find the moment estimator of θ.

3.2   Suppose the independent sample random variables $X_1, X_2, \ldots, X_n$ have the common p.d.f.

$$f(x \mid \alpha, \beta) = (1/\alpha) \exp[-(x-\beta)/\alpha]$$

where $\alpha > 0$ and $x \geqslant \beta$. Find the moment estimators for the parameters α and β.

3.3   Suppose the independent sample random variables $X_1, X_2, \ldots, X_n$ have the common p.d.f.

$$f(x \mid \theta) = \exp[-(x-\theta)]$$

where $0 < \theta \leqslant x$. Show that $\hat{\Theta} = \min(X_1, X_2, \ldots, X_n)$ is the MLE of θ and, using Section 1.1, find $E(\hat{\Theta})$.

3.4   Suppose the independent sample random variables $X_1, X_2, \ldots, X_n$ have the common p.d.f.

$$f(x \mid \theta) = 4\theta^4/x^5$$

where $0 < \theta \leqslant x$. Find the moment estimator $\hat{\Theta}_1$ and the MLE $\hat{\Theta}_2$ of the parameter θ. Show that $E(\hat{\Theta}_1) = \theta$ and $E(\hat{\Theta}_2) = 4n\theta/(4n-1)$.

3.5   Suppose the independent sample random variables $X_1, X_2, ..., X_n$ each have the uniform distribution $U(\theta - \frac{1}{2}, \theta + \frac{1}{2})$. Show that any function $\hat{\Theta}$ of the sample variables which satisfies

$$\max(X_1, X_2, ..., X_n) - \tfrac{1}{2} \leqslant \hat{\Theta} \leqslant \min(X_1, X_2, ..., X_n) + \tfrac{1}{2}$$

may be taken as a MLE.

3.6   Let the random variables X_1, X_2 have the trinomial distribution with parameters p_1, p_2 (see Section 1.3). Find the MLE of p_1 and p_2 by maximizing the joint p.f. of X_1 and X_2.

3.7   A p.f./p.d.f. which can be written in the form

$$f(x \mid \theta) = \exp(a(\theta)b(x) + c(\theta) + d(x))$$

where the set of values of x for which $f(x \mid \theta)$ is positive does not depend on the parameter θ is said to belong to the *regular exponential family* of distributions. Show that each of the distributions $N(\theta, 1)$, $N(0, \theta^2)$, $b(1, \theta)$, Poisson $P(\theta)$ and $\Gamma(m, \theta)$ (m known) may be written in this form. Explain why the uniform distribution $U(0, \theta)$ and the translated exponential distribution of Problem 3.2 do not belong to the regular exponential family.

3.3   Interval estimation

The idea of an interval estimate for a parameter θ is to replace the single numerical value given by the point estimator $\hat{\theta}$ with an interval (l, u), say, accompanied by the probability that the interval contains the true value of the parameter. Typically the point estimate $\hat{\theta}$ will lie in the interval (l, u) – for symmetric distributions at the midpoint $(l + u)/2$ – and the length of the interval may be regarded as a bound on the accuracy of the point estimate.

─────── EXAMPLE 3.8 ───────────────────────────

Suppose, as in Example 3.1 of Section 3.2, that the sample random variables have the distribution $N(\mu, \sigma_0^2)$, where σ_0 is known. The probability statement

$$\Pr(\mid \bar{X} - \mu \mid < 1.96\sigma_0/\sqrt{n}) \approx 0.95 \qquad (3.1)$$

or, equivalently,

$$\Pr(\mid \bar{X} - 1.96\sigma_0/\sqrt{n} < \mu < \bar{X} + 1.96\sigma_0/\sqrt{n}) \approx 0.95$$

suggests that for 95% of the observed values $\bar{X}$ the interval $(\bar{X} - 1.96\sigma_0/\sqrt{n}, \bar{X} + 1.96\sigma_0/\sqrt{n})$ will contain the true value of μ. Putting this statement another way, we may say that we have 95% confidence that the point estimate $\bar{X}$ differs from the true value of μ by less than $1.96\sigma_0/\sqrt{n}$.

Writing the probability (3.1) in more general form as

$$\Pr(|\bar{X} - \mu| < a\sigma_0\sqrt{n}) = 1 - \alpha \tag{3.2}$$

where, for any desired α, $0 < \alpha < 1$, the value a can be obtained from a table for the normal distribution, the consideration of Example 3.1 takes the form that the numerical interval $(\bar{x} - a\sigma_0/\sqrt{n}, \bar{x} + a\sigma_0/\sqrt{n})$ gives a $100(1 - \alpha)\%$ confidence interval for the parameter μ.

———— EXAMPLE 3.9 ————

For a numerical illustration take $n = 16$, $\bar{x} = 3.6$ and $\sigma_0 = 1$. Then

$$(3.6 - 1.96/4, \; 3.6 + 1.96/4) = (3.11, 4.09)$$

is a 95% confidence interval, and

$$(3.6 - 1.64/4, \; 3.6 + 1.64/4) = (3.19, 4.01)$$

is a 90% confidence interval for the mean μ.

It is clear from Example 3.2, and can be shown from equation (3.2) to be valid in general, that for the same sample size the length of the confidence interval increases as we require greater confidence.

Returning to the general model, we give the formal definition of what is meant, for any α in the range $0 < \alpha < 1$, by a $100(1 - \alpha)\%$ *confidence interval* for the parameter θ. Suppose $\ell\,(x_1, x_2, \ldots, x_n)$ and $u(x_1, x_2, \ldots, x_n)$ are two functions of the sample values with corresponding statistics $L = \ell(X_1, X_2, \ldots, X_n)$ and $U = u(X_1, X_2, \ldots, X_n)$ satisfying

$$L \leq U$$

and

$$\Pr(L < \theta < U \mid \theta) \geq 1 - \alpha \tag{3.3}$$

Then the numerical interval (ℓ, u) is said to be a $100(1 - \alpha)\%$ confidence interval for θ. We also say that the interval (ℓ, u) – or, more properly, the random interval (L, U) – contains the parameter θ with probability at least $1 - \alpha$.

The following points clarify the definition.

1. When the distribution function $F(x \mid \theta)$ is of discrete type, it may not be possible, for some given values of α, to achieve equality in equation (3.3). The confidence interval should be reported as giving confidence coefficient of at least $100 (1 - \alpha)\%$.

2. The general definition gives no criteria for the choice of the functions ℓ and u, nor even the suggestion that the point estimate $\hat{\theta}$ should lie in the interval. In practice, most types of confidence interval are constructed using a distribution

related to the point estimator – a process which ensures that the point estimate lies in the confidence interval.

3. Equation (3.3) will be satisfied when the statistics L and U satisfy

$$\Pr(L \geqslant \theta \,|\, \theta) \leqslant \alpha_1 \quad \text{and} \quad \Pr(U \leqslant \theta \,|\, \theta) \leqslant \alpha_2$$

where α_1 and α_2 are positive and $0 < \alpha_1 + \alpha_2 = \alpha$.

The choice $\alpha_1 = \alpha_2 = \alpha/2$ provides confidence intervals said to be *symmetric* or *central*, and the confidence intervals we obtain will, in general, be symmetric.

For fixed values of α and n, shorter confidence intervals may be regarded as desirable, since, when they contain the true parameter value, they may be taken to indicate more accurate information about the parameter.. A discussion of this topic would take us too far into the theory of estimation – see Problem 3.1 and Section 9.3, Example 9.14 for two specific examples.

3.3.1   Confidence intervals for the normal distribution parameters

1. μ is unknown, $\sigma = \sigma_0$ known. There is little to add to the earlier discussion in Example 3.8. The central $100\,(1-\alpha)\%$ confidence interval for μ is

$$(\bar{x} - a\sigma_0/\sqrt{n},\ \bar{x} + a\sigma_0/\sqrt{n})$$

where a is obtained, using the normal distribution function Φ, from $\Phi(a) = 1 - \alpha/2$.

2. μ unknown, σ unknown. Again, the natural estimator of μ is $\bar{X}$, but we must now overcome the problem that the distribution of $\bar{X}$ contains σ^2 as an unknown nuisance parameter. Help is at hand from Section 1.4, where it is shown that $(\bar{X} - \mu)\sqrt{n}/S$, where, of course, $S^2 = \sum_{i=1}^{n} (X_i - \bar{X})^2/(n-1)$, has the distribution $t(n-1)$.

For given α, $0 < \alpha < 1$, we may use tables to find t such that

$$\Pr\!\left(-t < \frac{(\bar{X} - \mu)\sqrt{\mu}}{S} < t\right) = 1 - \alpha$$

and obtain the central $100(1-\alpha)\%$ confidence interval for μ

$$(\bar{x} - ts/\sqrt{n},\ \bar{x} + ts/\sqrt{n})$$

Notice that the corresponding random interval $(\bar{X} - tS/\sqrt{n},\ \bar{X} + tS/\sqrt{n})$ has random length $2tS/\sqrt{n}$, in contrast to the fixed length $2a\sigma_0/\sqrt{n}$ when σ_0 is known.

To obtain a confidence interval for σ^2, we use the unbiased estimator S^2. From Section 1.4, the random variable $(n-1)S^2/\sigma^2$ has the $\chi^2(n-1)$ distribution, and from tables we find values a and b, for a central $100(1-\alpha)\%$ confidence interval, by

$$\Pr((n-1)S^2/\sigma^2 \leqslant a) = \Pr((n-1)S^2/\sigma^2 \geqslant b) = \alpha/2$$

Then $((n-1)s^2/b,\ (n-1)s^2/a)$ is the central $100(1-\alpha)\%$ confidence interval for σ^2.

───── **EXAMPLE 3.10** ─────

For a numerical illustration, consider the data

6 3 0 5 10 10 −6 6 10 −2 4 2.

Calculating by hand we obtain $\bar{x} = 4$ and $s = 4.991$.
(i) The central 95% confidence interval for μ, assuming $\sigma = 5$, is

$$\bar{x} \pm 1.96 \times 5/\sqrt{12} \approx -4 \pm 2.829$$

since, for the normal random variable Z, $\Pr(|Z| \leq 1.96) \approx 0.95$.
(ii) The central 95% confidence interval for μ, with σ unknown, is

$$\bar{x} \pm 2.201 \times 4.991/\sqrt{12} \approx 4 \pm 3.171$$

since, for the variable T having the Student's $t(11)$ distribution, tables give $\Pr(|T| \leq 2.201) \approx 0.95$.
(iii) The central 95% confidence interval for σ^2, with μ unknown, is

$$\left(\frac{11 \times 4.991^2}{21.92}, \frac{11 \times 4.991^2}{3.816} \right) \approx (12.5, 71.81)$$

since, for the variable W having the chi-squared distribution $\chi^2(11)$, tables give

$$\Pr(W \leq 3.816) = \Pr(W \geq 21.92) \approx 0.025.$$

3.3.2   Use of MINITAB

The confidence intervals (i) and (ii) may be obtained using the MINITAB commands ZINT and TINT, respectively. Entering the data of Example 3.10 in column C1,

ZINT 95 5 C1

produces

THE ASSUMED SIGMA = 5.00

	N	MEAN	STDEV	SE MEAN	95.0 PERCENT C1
C1	12	4.00	4.99	1.44	(1.17, 6.83)

and

TINT 95 C1

produces

	N	MEAN	STDEV	SE MEAN	95.0 PERCENT C1
C1	12	4.00	4.99	1.44	(0.83, 7.17)

confirming, to two decimal places, the 95% confidence intervals given above. (Note: (1) Replacing 95 by K in the command lines will persuade MINITAB to produce K% confidence intervals. In fact, 95 is the default value of K and may be omitted if a 95% confidence interval is required. (2) The assumed value of the standard deviation must be included in the ZINT command line.)

3.3.3 *Confidence intervals for the binomial parameter p*

We construct $100(1 - \alpha)$% confidence intervals for p in two ways: (1) using the normal approximation for the distribution of $\bar{X}$; and (2) using the binomial distribution and its relation with the beta distribution (see Problem 1.12 of Section 1.3).

Method 1

Using tables of the normal distribution find the value a such that

$$\Pr(-a < (\bar{X} - p)/\sqrt{p(1-p)/n} < a) = 1 - \alpha \tag{3.4}$$

We may rewrite the inequality as

$$n(\bar{X} - p)^2 < a^2 p(1 - p)$$

Using the formula for the roots of a quadratic, and replacing $\bar{X}$ by the observed value $\bar{x}$ we see that p lies between the values

$$(n\bar{x} + \tfrac{1}{2}a^2 \pm a\sqrt{n\bar{x}(1 - \bar{x}) + \tfrac{1}{4}a^2})/(n + a^2) \tag{3.5}$$

giving the $100\%(1 - \alpha)$ confidence interval we seek.

A simplification of this process, providing a more attractive formula, is to rewrite the inequality in equation (3.4), substituting $\bar{x}$ for $\bar{X}$, as

$$\bar{x} - a\sqrt{p(1-p)/n} < p < \bar{x} + a\sqrt{p(1-p)/n}$$

By replacing p with the estimate $\bar{x}$ in the two extremes, the $100(1 - \alpha)$% confidence interval is reported as

$$(\bar{x} - a\sqrt{\bar{x}(1 - \bar{x})/n}, \ \bar{x} + a\sqrt{\bar{x}(1 - \bar{x})/n}) \tag{3.6}$$

——— EXAMPLE 3.11 ———

Taking $n = 1000$, $\bar{x} = 0.65$ and $\alpha = 0.1$, so that $a = 1.64$ approximately, we obtain, substituting in equation (3.5), the 90% confidence interval 0.6496 ± 0.0247, and, substituting in equation (3.6), the almost identical 90% confidence interval 0.65 ± 0.0247.

Method 2

For the second construction we use $S_n = n\bar{X}$, which has the distribution $b(n, p)$ and the observed value k of S_n. The key to the process is the observation that, for each value of x, $\Pr(S_n \leq x \mid p)$ decreases as p increases. This follows from Problem 1.12 of Section 1.3.

Using tables of the binomial distribution, or the beta distribution as described below, we find values p_1 and p_2 such that

$$\Pr(S_n \geq k \mid p_1) = \alpha/2 \quad \text{and} \quad \Pr(S_n \leq k \mid p_2) = \alpha/2$$

It follows that, for $p_1 < p < p_2$,

$$\Pr(S_n \geq k \mid p) > \alpha/2 \quad \text{and} \quad \Pr(S_n \leq k \mid p) > \alpha/2$$

and the interval (p_1, p_2) is taken as a $100(1 - \alpha)\%$ confidence interval for p. Note that p_1 and p_2 are determined by the observed value k of S_n.

We recast these calculations so that they more closely fit the form of the general definition. For each p in the range $0 < p < 1$, and a given value of α, where $0 < \alpha < 1$, we find the largest integer $a(p)$ and smallest integer $b(p)$ such that

$$\Pr(S_n \leq a(p) \mid p) \leq \alpha/2 \quad \text{and} \quad \Pr(S_n \geq b(p) \mid p) \leq \alpha/2$$

Then

$$\Pr(a(p) < S_n < b(p) \mid p) \geq 1 - \alpha$$

and $p_1 < p < p_2$ is equivalent to the inequality $a(p) < k < b(p)$ for the observed value k.

To obtain the values p_1 and p_2 we use the relation between the binomial and beta distributions, and then the MINITAB INVCDF command. For readers who do not have access to MINITAB the values p_1 and p_2 may be obtained approximately by searching in the tables of the binomial or beta distributions.

Problem 1.12 of Section 1.3 provides the relation

$$\Pr(S_n \leq l \mid n, p) = \Pr(T \leq 1 - p \mid n, l) \tag{3.7}$$

where T has the beta $(n - l, l + 1)$ distribution. Noting that $V = 1 - T$ has the beta $(l + 1, n - l)$ distribution, we may rewrite equation (3.7) as

$$\Pr(S_n \leq l \mid n, p) = \Pr(V \geq p \mid n, l)$$
$$= 1 - \Pr(V \leq p \mid n, l)$$

Now

$$\Pr(S_n \geq k \mid n, p_1) = 1 - \Pr(S_n \leq k - 1 \mid n, p_1)$$
$$= \Pr(V_1 \leq p_1 \mid n, k - 1)$$

where V_1 has the beta $(k, n - k + 1)$ distribution, and

$$\Pr(S_n \leq k \mid n, p_2) = 1 - \Pr(V_2 \leq p \mid n, k)$$

where V_2 has the beta $(k + 1, n - k)$ distribution.

In this form we require p_1 and p_2 which satisfy $\Pr(V_1 \leqslant p_1) = \alpha/2$, and $\Pr(V_2 \leqslant p_2) = 1 - \alpha/2$, and these may be obtained from MINITAB.

──────── EXAMPLE 3.12 ────────────────────────

With $n = 1000$, $k = 650$ and $\alpha = 0.1$, we obtain the values p_1 and p_2 as follows:

    INVCDF 0.05;
    BETA 650 351. gives $p_1 = 0.6244$, and
    INVCDF 0.95;
    BETA 651 350. gives $p_2 = 0.6750$.

The central 90% confidence interval $(0.6244, 0.6750)$ may be compared with the corresponding intervals obtained by Method 1 in Example 3.11.

───

The next example shows how the values p_1 and p_2 may be obtained from the tables for the binomial distribution.

──────── EXAMPLE 3.13 ────────────────────────

With $n = 20$, $k = 6$ and $\alpha = 0.05$ we require $\Pr(S_{20} \leqslant 5 \mid p_1) = 0.975$. Tables supply

$$\Pr(S_{20} \leqslant 5 \mid 0.11) = 0.9825 = 0.9750 + 0.0075$$

and

$$\Pr(S_{20} \leqslant 5 \mid 0.12) = 0.9740 = 0.9750 - 0.0010$$

Using linear interpolation in p we approximate p_1 as
$p_1 = (0.11 \times 10 + 0.12 \times 75)/85 \approx 0.1188$.
    In order to obtain $\Pr(S_{20} \leqslant 6 \mid p_2) = 0.025$ from tables we require (see Section 1.3)

$$\Pr(S_{20} \leqslant 13 \mid 1 - p_2) = 0.975$$

In this case tables supply

$$\Pr(S_{20} \leqslant 13 \mid 0.45) = 0.9786 = 0.9750 + 0.0036$$

and

$$\Pr(S_{20} \leqslant 13 \mid 0.46) = 0.9735 = 0.9750 - 0.0015$$

Hence, by linear interpolation, $1 - p_2 = (0.45 \times 15 + 0.46 \times 36)/51$, giving $p_2 \approx 0.5429$. The central 95% confidence interval is therefore $(0.1182, 0.5429)$.

───

The discussion of interval estimation is continued by examining special cases in the problems, and by relating interval estimation and hypothesis testing in Chapters 4 and 9.

3.3.4 Problems

3.8 Suppose a, $0 < a < 1$, is fixed. For each real number a let $b = b(a)$ be determined by $\Phi(a) - \Phi(b) = 1 - \alpha$. Show that $f(a) = a - b$ is a minimum when $b = -a$. Let $X_1, X_2, \ldots, X_n$ be a random sample from a distribution $N(\mu, \sigma_0^2)$, where σ_0 is known. Deduce that the central $100(1 - \alpha)\%$ confidence interval for μ is the shortest $100(1 - \alpha)\%$ confidence interval among intervals of the form $(\bar{x} - a\sigma_0/\sqrt{n}, \bar{x} - b\sigma_0/\sqrt{n})$ where $\Phi(a) - \Phi(b) = 1 - \alpha$.

3.9 The data

1.33	3.23	4.41	3.36	2.87	5.58	8.77
6.11	4.66	7.03	5.48	7.19		

is obtained from a population whose distribution is assumed to be $N(\mu, \sigma^2)$. Find:
(i)   A 95% confidence interval for μ, assuming $\sigma = 2.4$
(ii) A 95% confidence interval for μ, assuming σ is unknown
(iii) A 90% confidence interval for σ^2, assuming μ is unknown.

3.10 (i) Use the RANDOM command to obtain 16 samples of size 100 from the distribution $N(5, 2^2)$, placed in C1 – C16.
(ii) Treating the rows of C1 – C16 as 100 samples of size 16 from a $N(\mu, 2^2)$ distribution, find the 95% confidence intervals for μ, and determine how many contain the value $\mu = 5$.
(iii) Treating the rows of C1 – C16 as 100 samples of size 16 from a $N(\mu, \sigma^2)$ distribution with σ unknown, find the 95% confidence intervals for μ, and determine how many contain the value $\mu = 5$.

3.11 Using equations (3.5) and (3.6) of Method 1, find the 95% confidence intervals for the probability p of obtaining a head in a single toss of a coin if 100 tosses of the coin yielded 35 heads.

3.12 Using (i) the binomial tables and (ii) the beta-distribution, find 90% confidence intervals for the parameter p if seven successes were obtained in 20 trials.

3.4 Solutions to problems

3.1 $E(X_i) = \int_0^\infty \frac{\theta x}{(1 + x)^{\theta + 1}} \, dx = \left[\frac{-x}{(1 + x)^\theta} \right]_0^\infty + \int_0^\infty \frac{1}{(1 + x)^\theta} \, dx = \frac{1}{\theta - 1}$

Equating $\bar{X} = 1/(\hat{\theta} - 1)$ gives $\hat{\theta} = 1 + 1/\bar{X}$.

3.2 $E(X_i) = \int_\beta^\infty \frac{x}{\alpha} \exp[-(x - \beta)/\alpha] dx = [-x \exp[-(x - \beta)/\alpha]]_\beta^\infty + \int_\beta^\infty \exp[-(x - \beta)/\alpha] dx$

$= \beta + \alpha$

Similarly, on integrating by parts,

$$E(X_i^2) = \int_\beta^\infty \frac{x^2}{\alpha} \exp[-(x-\beta)/\alpha]dx = \beta^2 + 2\alpha\beta + 2\alpha^2$$

Equating

$$\bar{X} = \hat{\alpha} + \hat{\beta}, \text{ and } \frac{1}{n}\sum_{i=1}^n X_i^2 = \hat{\beta}^2 + 2\hat{\alpha}\hat{\beta} + 2\hat{\alpha}^2$$

gives

$$\hat{\alpha}^2 = \frac{1}{n}\sum_{i=1}^n X_i^2 - \bar{X}^2$$

and hence $\hat{\beta} = \bar{X} - \hat{\alpha}$.

3.3 The joint p.d.f. is

$$f(x_1, x_2, \ldots, x_n \mid \theta) = \exp\left[-\sum_{i=1}^n x_i + n\theta\right]$$

where all $x_i \ge \theta$. The maximum value is obtained by maximizing θ subject to $\theta \le x_i$ for all i. Hence $\hat{\Theta} = \min(X_1, X_2, \ldots, X_n)$ is the MLE.

From Appendix 3, the p.d.f. of $\hat{\Theta}$ is

$$nf(x \mid \theta)(1 - F(x \mid \theta))^{n-1} = n \exp^{-n(x-\theta)},$$

and $E(\hat{\Theta}) = \theta + 1/n$ using Problem 3.2 (Solution).

3.4
$$E(X_i) = \int_\theta^\infty x \cdot \frac{4\theta^4}{x^5} dx = 4\theta/3$$

Hence $\hat{\Theta}_1 = 3\bar{X}/4$ and $E(\hat{\Theta}_1) = 3E(\bar{X})/4 = \theta$.

Now the joint p.d.f. is

$$f(x_1, x_2, \ldots, x_n \mid \theta) = 4^n \theta^{4n}/(x_1 x_2 \ldots x_n)^5$$

for $0 < \theta \le x_i$ for all i. The maximum occurs when $\theta = \min(x_1, x_2, \ldots, x_n)$, so the MLE is $\hat{\Theta}_2 = \min(X_1, X_2, \ldots, X_n)$.

From Section 1.1, the p.d.f of $\hat{\Theta}_2$ is

$$nf(x \mid \theta)(1 - F(x \mid \theta))^{n-1} = 4n\theta^{4n}/x^{4n+1}$$

and

$$E(\hat{\Theta}_2) = \int_\theta^\infty 4xn \cdot \frac{\theta^{4n}}{x^{4n+1}} dx = \frac{4n}{4n-1}\theta$$

3.5 The joint p.d.f. is $f(x_1, x_2, \ldots, x_n \mid \theta) = 1$ for $\theta - \frac{1}{2} \le x_i \le \theta + \frac{1}{2}$ for $i = 1, 2, \ldots, n$, and zero otherwise. Thus the p.d.f. is a maximum for any value of θ which satisfies $\theta - \frac{1}{2} \le \min(x_1, x_2, \ldots, x_n)$ and $\theta + \frac{1}{2} \ge \max(x_1, x_2, \ldots, x_n)$ – that is, any ML estimator $\hat{\Theta}$ satisfies

$$\max(X_1, X_2, \ldots, X_n) - \tfrac{1}{2} \le \hat{\Theta} \le \min(X_1, X_2, \ldots, X_n) + \tfrac{1}{2}$$

3.6 The joint p.f. of X_1 and X_2 is, from Section 1.3,

$$f(x_1, x_2) = \binom{n}{x_1}\binom{n - x_1}{x_2} p_1^{x_1} p_2^{x_2} (1 - p_1 - p_2)^{n - x_1 - x_2}$$

Equating to zero the partial derivatives of $\log f$ w.r.t. p_1 and p_2 respectively yields

$$\frac{x_1}{p_1} - \frac{n - x_1 - x_2}{1 - p_1 - p_2} = 0 \quad \text{and} \quad \frac{x_2}{p_2} - \frac{n - x_1 - x_2}{1 - p_1 - p_2} = 0$$

giving $\hat{p}_1 = X_1/n$, $\hat{p}_2 = X_2/n$. Note that $E(\hat{p}_i) = p_i$ (from Section 1.3).

3.7 $N(\theta, 1)$: $\exp[-\frac{1}{2}(-2\theta x + (\theta^2 + \log 2\pi) + x^2)]$

$N(0, \theta^2)$: $\exp[-\frac{1}{2}(x^2/\theta^2 + \log 2\pi\theta^2)]$

$b(1, \theta)$: $\exp[x \log[\theta/(1 - \theta)] + \log(1 - \theta)]$

$P(\theta)$: $\exp[x \log \theta - \theta - \log x!]$

$\Gamma(m, \theta)$: $\exp[-x/\theta - (m \log \theta + \log \Gamma(m)) + (m - 1)\log x]$

For the uniform distribution and the translated exponential distribution, the set of x giving non-zero values of the p.d.f. $f(x \mid \theta)$ depends on θ.

3.8 For a minimum

$$0 = \frac{df}{da} = 1 - \frac{db}{da} \quad \text{or} \quad \frac{db}{da} = 1$$

Differentiating $\Phi(a) - \Phi(b) = 1 - \alpha$ with respect to a gives

$$\phi(a) - \phi(b)\frac{db}{da} = 0$$

so that $\phi(a) = \phi(b)$. Hence $b = a$ or $b = -a$, and $b = a$ is excluded since $\Phi(a) - \Phi(b) = 1 - \alpha \ne 0$.

The length of the confidence interval is $(a - b)\sigma_0/\sqrt{n}$, where $\Phi(a) - \Phi(b) = 1 - \alpha$, and this is a minimum when $b = -a$.

3.9 Calculation yields $\bar{x} = 5.002$, $s = 2.115$.

(i) Assuming $\sigma = 2.4$ the 95% confidence interval for μ is $\bar{x} \pm 1.96\sigma/\sqrt{n} = 5.002 \pm 1.96 \times 2.4/\sqrt{12} \approx (3.643, 6.30)$.

(ii) If T has the $t(11)$ distribution $\Pr(|T| \le 2.201) = 0.95$ and the 95% confidence interval for μ is $\bar{x} \pm 2.201 s/\sqrt{n} = 5.002 \pm 2.201 \times 2.115/\sqrt{12} \approx (3.657, 6.346)$.

(iii) If W has the $\chi^2(11)$ distribution then $\Pr(W \leqslant 4.575) = 0.05 = \Pr(W \geqslant 19.68)$ so the 90% confidence interval for σ^2 is $((n-1)s^2/19.68,\ (n-1)s^2/4.575) \approx (2.500, 10.755)$.

3.10 (i) A 95% confidence interval contains $\mu = 5$ if $|\bar{x} - 5| \sqrt{n}/\sigma \leqslant 1.96$, that is, with $n = 16$, $\sigma = 2$, if $|\bar{x} - 5| \leqslant 0.98$. We may obtain the count as follows:

    RMEAN C1 – C16 C21
    LET C22 = SIGNS (0.98 – ABSO (C21 – 5))
    LET C23 = (ABSO (C22) + C22)/2
    SUM C23.

(ii) Using the t-distribution a 95% confidence interval contains $\mu = 5$ if $|\bar{x} - 5| \sqrt{n}/s \leqslant 2.131$. We may obtain the count as follows:

    RMEAN C1 – C16 C31
    RSTDEV C 1 – C16 C32
    LET C33 = SIGNS (2.131 – ABSO ((C31 – 5) * 4/C32))
    LET C34 = (ABSO (C33) + C33)/2
    SUM C34.

3.11 Substituting $\bar{x} = 0.35$, $n = 100$, $a = 1.96$ in equation (3.5) gives the 95% confidence interval $(0.309, 0.402)$, and in equation (3.6) gives $(0.257, 0.443)$.

3.12 (i) For $\Pr(S_{20} \leqslant 6 \mid p_1) = 0.95$ tables supply

$$\Pr(S_{20} \leqslant 6 \mid 0.17) = 0.95 + 0.0091$$
$$\Pr(S_{20} \leqslant 6 \mid 0.18) = 0.95 - 0.0037$$

and linear interpolation gives $p_1 = (0.17 \times 37 + 0.18 \times 91)/128 = 0.177$. For $\Pr(S_{20} \leqslant 7 \mid p_2) = 0.05$, we require $\Pr(S_{20} \leqslant 12 \mid 1 - p_2) = 0.95$, and tables supply

$$\Pr(S_{20} \leqslant 12 \mid 0.44) = 0.95 + 0.0018$$
$$\Pr(S_{20} \leqslant 12 \mid 0.45) = 0.95 - 0.0080$$

Linear interpolation gives $1 - p_2 = (0.44 \times 80 + 0.45 \times 18)/98$ or $p_2 = 0.558$, and the 90% confidence interval is $(0.177, 0.558)$.

(ii) Using the beta-distribution and the INVCDF command

    INVCDF 0.05;   supplies $p_1 = 0.177$
    BETA 7 14.

and

    INVCDF 0.95;   supplies $p_2 = 0.558$
    BETA 8 13.

Reference

1. Beaumont, G. P., *Intermediate Mathematical Statistics*, Chapman and Hall, London, 1980.

Chapter Four

Tests of hypotheses

4.1   Introduction

A statistical hypothesis (see Section 2.3) is a conjecture about the nature of a distribution function. The problem addressed under the heading of tests of hypotheses, or tests of significance, is to decide whether an observed random sample may reasonably be supposed to have been drawn from a population whose distribution is specified by the hypothesis. As the only information available is contained in the observed sample values, the assessment consists of examining the data for evidence to support the hypothesis, or, put in the negative sense, cast doubt on it.

Measuring the doubt, by a probability calculation made under the assumption that the hypothesis is true, is the essence of a statistical test. In this type of investigation, we must recognize that we will never be in a position to claim proof of the truth of the hypothesis – for instance, no amount of tossing a coin will prove the coin is fair. All that can be achieved is a measure of consistency (or inconsistency) between the sample data and the hypothesis.

This section presents the basic concepts of hypothesis testing, using three idealized examples, in an informal way. Section 4.2 largely contains examples of elementary tests concerned with mean values, concentrating on those tests for which MINITAB supplies a command. The difficulties brushed aside in the informal treatment in Sections 4.1 and 4.2 are taken up in Section 4.3. A theoretical base for the test procedures described in this chapter is supplied in Chapter 9.

The hypothesis assessed by a statistical test is called the null hypothesis and is denoted by H_0. With each null hypothesis is associated an alternative hypothesis, denoted by H_1, either naturally defined by the problem under consideration or, simply, the logical opposite 'not H_0'. For instance, Examples 4.1 and 4.3 below both relate to a normal distribution $N(\mu, \sigma^2)$. Example 4.1 discusses the two hypotheses '$\mu = 1$, $\sigma = 1$' and '$\mu = 4$, $\sigma = 1.5$'; and Example 4.3 has the null hypothesis that $\mu = 2$ and the alternative hypothesis is the logical opposite, namely $\mu \neq 2$.

The first example may, more properly, be regarded as a problem in discrimination analysis.

──────── **EXAMPLE 4.1** ────────

A blood test, used to identify a particular disease, produces a reaction X which follows the normal $N(\mu, \sigma^2)$ distribution. Experience has shown that when the disease is present the values $\mu = 1$, $\sigma = 1$ apply, and, otherwise, $\mu = 4$, $\sigma = 1.5$. Based on the observed value x_0 of the blood test reaction, the consultant has to decide whether a course of treatment should be prescribed. There are two hypotheses, which we write as D and W, where D is the hypothesis the disease is present and W (=well) is the hypothesis that the patient does not suffer from the disease. That is, D is the hypothesis that the sample comes from a population with the distribution $N(1, 1)$, and W that it comes from a population with distribution $N(4, 1.5^2)$.

The decision process is clear enough: small values of x_0 suggest D, and large values W. A moment's reflection identifies the problem – what is the cut-off value c, say, such that $x_0 < c$ is taken to indicate D and $x_0 \geq c$ suggests W? Whatever the value chosen for c, there can be no certainty in the diagnosis – all that can be done is obtain a measure of its correctness. In the probability calculations below, Z denotes a variable distributed $N(0, 1)$ and the quoted numerical values may be obtained from tables or the MINITAB CDF command.

Suppose, first, that c is taken to be the halfway value, that is, $c = 2.5$. We calculate the probabilities:

1. That a diseased patient is classified as well and
2. That a healthy individual is diagnosed as being diseased.

For (1),

$$\Pr(X \geq 2.5 \mid \mu = 1, \sigma = 1) = \Pr(Z \geq 1.5) = 0.0668$$

and for (2),

$$\Pr(X < 2.5 \mid \mu = 4, \sigma = 1.5) = \Pr(Z < -1) = 0.1587$$

The error probability (1) may be reduced by increasing the value of c, but only at the expense of increasing the error probability (2). To equalize the two probabilities, the value c must satisfy

$$\Pr(Z \geq c - 1) = \Pr(Z < (c - 4)/1.5)$$

which, from symmetry, requires $c - 1 = -(c - 4)/1.5$, or $c = 11/5$. In this case, both error probabilities are 0.1151, and, whatever the state of health of the individual, there is an 11.51% chance of a misdiagnosis.

The common situation in testing is that one hypothesis is more important than the other, and, rather than equalizing the error probabilities, the concern is to put an upper bound on the error probability for the more important hypothesis. This hypothesis is taken as the null hypothesis. Suppose, then, that the consultant wishes to restrict the probability of wrongly declaring a diseased patient as healthy to be 0.05 – that is, the null hypothesis H_0 is D, and the alternative H_1 is W.

The choice for the cut-off value c must satisfy

$$\Pr(X \geq c \mid \mu = 1, \sigma = 1) = \Pr(Z \geq c - 1) = 0.05$$

with tables supplying $c - 1 = 1.96$, or $c = 2.96$.

Having fixed one error probability and, hence, determined the value of c, the second error probability is calculated as

$$\Pr(X < 2.96 \mid \mu = 4, \sigma = 1.5) = \Pr(Z < -2.08/3) = 0.2451$$

Thus, an observed value $x_0 \geq 2.96$ will lead to the diagnosis that the individual is healthy (that is, the hypothesis H_0 is rejected), and, an observed value $x_0 < 2.96$ will lead to the diagnosis that the disease is present (that is, the hypothesis H_0 is accepted).

4.1.1   Comments and definitions

1. The random variable used in a test is called the *test statistic*, and will be denoted by S (in Example 4.1, $S = X$).
2. The observed value s_0 of S is calculated from the sample data, and the distribution of S, assuming the null hypothesis H_0 is true, must be known – see (5) below.
3. The *critical region*, denoted by R, is the set of values of the statistic such that, if the observed value $s_0 \in R$, the null hypothesis is rejected (in Example 4.1 $R = \{x: x \geq 2.96\}$). Generally R will consist of one or both of the tails of the distribution function of the test statistic S, assuming H_0 is true. The test will be called one-tailed or two-tailed according to the number of tails of the distribution comprising R. The shape of R – which tail or tails make up R – is determined by the relation between the null and alternative hypotheses (in Example 4.1 $R = \{x: x \geq c\}$). Indeed, determining the shape of R is the role of the alternative hypothesis.
4. There are two errors which may be made in expressing the conclusion of a test: the *Type I error* is to reject H_0 when, in fact, H_0 is true; and, the *Type II error* is to accept H_0 when, in fact, H_0 is false.
5. By design, a test aims to put an upper bound on the probability of committing the Type I error. The upper bound, denoted by α (in Example 4.1 $\alpha = 0.05$), is called the *level of significance* of the test. Choosing, in a particular problem, the numerical value for α is at the discretion of the investigator, but later in this section, a standard interpretation scale for values of α is described. The link between the significance level α and the critical region R completing the specification of R (obtaining the value c in Example 4.1), is provided by expressing the probability of Type I error as

$$\Pr(S \in R \mid H_0) \leq \alpha$$

The left-hand side reads as 'the probability that the test statistic S takes values in the critical region R given that S has the distribution implied by the truth of H_0'. The inequality sign appears here in anticipation of Example 4.2. If, as in Example 4.1, the distribution of S is continuous under H_0, the inequality is replaced by equality (in Example 4.1, $\Pr(X \geq c \mid \mu = 1, \sigma = 1) = 0.05$).

6. Having specified the critical region, as in (5), we may now calculate the probability of Type II error as

$$\Pr(S \notin R \mid H_1)$$

which is read as 'the probability that the test statistic S does not take values in the critical region R, given that S has the distribution implied by the truth of H_1' (in Example 4.1, $\Pr(X < c \mid \mu = 4, \sigma = 1.5)$).

7. The conclusion of a test procedure is usually expressed as 'the null hypothesis H_0 is rejected/accepted at the $100\alpha\%$ level of significance', indicating the choice of the upper bound placed on the probability of Type I error. Except when the alternative hypothesis H_1 is the logical opposite of the null hypothesis H_0, the rejection of H_0 should not be taken as acceptance of H_1. The reason for this is that many other alternative hypotheses would result in the same critical region – for instance, the hypothesis '$\mu = 3$, $\sigma = 1$' in Example 4.1 defines the same shape $\{x : x \geq c\}$ for the critical region as does the hypothesis W, and, therefore, the same 5% critical region $\{x : x \geq 2.96\}$.

The second example illustrates the problem that may occur when the distribution of the test statistic is discrete.

———— EXAMPLE 4.2 ————————————————————————

Items from a production line may be classified as good or defective. The production is said to be in control, and the items despatched to customers, if the proportion of defective items in a production batch is judged to be less than or equal to 0.01 – otherwise the process is out of control and the batch scrapped. A possible inspection scheme is to select, at random, 10 items from the batch and classify the selected items. If less than k items in the sample are defective, k to be determined, the process is to be regarded as being in control. Suppose, perhaps for financial reasons, that the manufacturer is prepared to accept a probability of no more than 0.005 that a batch is incorrectly scrapped – that is, that the production will be incorrectly judged to be out of control.

Let p be the (unknown) proportion of defective items in the batch, and assume that the batch size is sufficiently large that the sampling may be treated as sampling with replacement. Under this assumption, the distribution of the number of defective items in the sample is the binomial distribution $b(10, p)$. The description of the problem suggests that the test statistic S be the number of defectives in the sample, the null hypothesis H_0 is that $p \leq 0.01$ and the alternative H_1 is that $p > 0.01$. With the critical region R having the shape $\{x : x \geq k\}$ we require the probability of

Type I error to be bounded by 0.005, that is,

$$\Pr(S \geqslant k \mid p) \leqslant 0.005$$

for all values of p in the range $0 < p \leqslant 0.01$. Since $\Pr(S \geqslant k \mid p)$ increases with p (see Section 1.3 Problem 1.12), the inequality will be satisfied if $\Pr(S \geqslant k \mid p = 0.01) \leqslant 0.005$, and tables supply $k = 2$, with $\Pr(S \geqslant 2 \mid p = 0.01) = 0.0043$, as the smallest suitable integer. The critical region R is $\{x : x \geqslant 2\}$ and the maximum probability of Type I error is 0.0043.

For each value of $p > 0.01$, we may calculate the probability of the Type II error that H_0 is accepted when, in fact, H_0 is false – that is, the production is, mistakenly, accepted as in control and the items produced are dispatched to customers. The required probability is

$$\Pr(S < 2 \mid p) = 1 - \Pr(S \geqslant 2 \mid p)$$

which decreases as p increases. For example, tables supply $\Pr(S < 2 \mid p = 0.05) = 0.9139$ and $\Pr(S < 2 \mid p = 0.1) = 0.7361$. The manufacturer may find these error probabilities disturbingly high and likely to result in damage to the reputation of the company. Using this style of sampling inspection, control over these probabilities can only be exercised by increasing the sample size. See Problem 4.2 for a numerical illustration.

4.1.2   Comments and definitions

1. The hypotheses in Example 4.2 do not completely specify the distribution of the test statistic – the form is known, but the parameter p varies over a range of values. A hypothesis is said to be *simple* if it specifies a single value of the parameter and *composite* if more than one value is allowed. In Example 4.1, the parameter is taken as the pair $\theta = (\mu, \sigma)$ and both hypotheses are simple, with H_0 corresponding to the value $\theta_0 = (1, 1)$ and H_1 the value $\theta_1 = (4, 1.5)$.

2. For a composite null hypothesis, the significance level is used, as in Example 4.2, to bound the probability of Type I error for all possible values of the parameter covered by the null hypothesis. Ideally, the significance level α will be the maximum probability of Type I error, and the test is said to be at exact significance level α. However, as shown by Example 4.2, a critical region with exact significance level α may not exist. The practical solution is to retain α as an upper bound and find a critical region – the largest available – at level $\alpha_1 < \alpha$, and quote α_1 as the significance level. A theoretical alternative, the device of randomization, is discussed briefly in Section 4.3.

The critical region in both the preceding examples was one-tailed. Example 4.3 introduces the problem that occurs when the critical region includes both tails of the distribution of the test statistic.

――――――― EXAMPLE 4.3 ―――――――

Suppose a production process aims to manufacture ball-bearings with a diameter $\mu_0 = 2$. Past experience suggests that the diameter of the ball-bearings follows a normal distribution $N(\mu, 0.5^2)$, whether or not the production is at the target level, $\mu = \mu_0$. To assess the quality of the product, a random sample of size 16 is selected and the sample mean $\bar{x}$ calculated. Deviation, in either direction, from the target value $\mu_0 = 2$ is taken to indicate that the quality is poor. Expressing the problem in the language of a statistical test we define: the test statistic S to be the sample mean $\bar{X}$, having the distribution $N(\mu, 0.5^2/16)$ or $N(\mu, 1/8^2)$; the null hypothesis H_0 to be $\mu = 2$; the alternative hypothesis H_1 to be $\mu \neq 2$; and the critical region R to have the shape $\{x: x \leq a \text{ or } x \geq b\}$. Using the significance level $\alpha = 0.05$ requires the determination of values for a and b that satisfy

$$\Pr(\bar{X} \leq a \text{ or } \bar{X} \geq b \mid \mu = 2) = 1 - \Pr(a < \bar{X} < b \mid \mu = 2) = 0.05$$

that is, since $Z = 8(\bar{X} - \mu)$ has the standard normal distribution $N(0, 1)$,

$$\Pr(8(a - 2) < Z < 8(b + 2)) = 0.95$$

A determination of unique values for a and b satisfying this equation is not possible. In fact, the reader may check that for every value of a, with $8(a - 2) \leq -1.96$, we may determine a (unique) value for b. However, the symmetry of the problem and the symmetry of the distribution of Z suggest that we choose $8(a - 2) = -8(b - 2)$, or $a = 4 - b$. Tables now supply $\Pr(Z < -1.96) = \Pr(Z > 1.96) = 0.025$, and we obtain $8(a - 2) = -1.96$, or $a = 1.755$ and $b = 2.245$. The critical region R is, therefore, $R = \{x: x \leq 1.755 \text{ or } x \geq 2.245\}$ and involves symmetrically the two tails of the distribution of the test statistic, $N(2, 1/8^2)$ under H_0.

For each value of $\mu \neq 2$, the values covered by the alternative hypothesis, we calculate the probability of Type II error – that is, the probability that production at mean level μ is accepted as satisfactory – by

$$\Pr(\bar{X} \notin R \mid \mu) = \Pr(1.755 < \bar{X} < 2.245 \mid \mu)$$
$$= \Pr(8(1.755 - \mu) < Z < 8(2.245 - \mu))$$

For $\mu = 2.01$ we obtain $\Pr(-2.04 < Z < 1.88) = 0.9492$, and for $\mu = 2.3$, $\Pr(-4.36 < Z < -0.44) = 0.3300$.

The resolution of the non-uniqueness of a and b, by putting equal probability $\alpha/2$ in both tails, is not so intuitively appealing when the distribution of the test statistic is not symmetric. The problem is further compounded if the test statistic has a discrete distribution, when equality of probability in the tails may be impossible. A discussion of this problem is given in Section 4.3, with further examples in Section 9.3.

4.1.3   p-values

The descriptions given in the three introductory examples follow the most common approach to hypothesis testing. In this approach a critical region is selected such that the probability of committing a Type I error does not exceed a pre-assigned value, and the investigator reports whether or not the observed data is significant at the chosen level. The examples also illustrate the problems related to the choice of the significance level, namely that the choice may appear to be completely arbitrary (Examples 4.1 and 4.3) and that the level may not be attained (Example 4.2).

An alternative, more informative, interpretation of the probability calculations is to use the *p-value* of the test. The *p*-value may be succinctly defined as the smallest level of significance for which the obtained data leads to the rejection of the null hypothesis. Equivalently, the *p*-value is the probability, calculated assuming the null hypothesis is true, that the test statistic takes values at least as extreme (in the sense determined by the alternative hypothesis) as the observed value of the statistic.

To put these definitions in an operational form, suppose we have a null hypothesis H_0, an alternative hypothesis H_1, a test statistic S and an observed value s_0 of S. Using the alternative hypothesis, we determine the shape of the critical region and, thereby, identify the values of the statistic S which are regarded as being extreme. For a one-tailed test the *p*-value is given by

$$p\text{-value} = \begin{array}{ll} \Pr(S \geq s_0 \,|\, H_0) & \text{for a right tail} \\ \Pr(S \leq s_0 \,|\, H_0) & \text{for a left tail} \end{array}$$

──────── EXAMPLE 4.4 ────────────────────────────────────

For the blood test example (Example 4.1), testing the null hypothesis that the patient has the disease, the test statistic is the measurement X of the reaction, and the critical region has the shape $\{x: x \geq c\}$. Suppose the observed value is 3, then the *p*-value is given by

$$p\text{-value} = \Pr(X \geq 3 \,|\, \mu = 1, \sigma = 1) = \Pr(Z \geq 2) = 0.0228$$

where Z has the standard distribution $N(0, 1)$.

──

For a critical region whose shape consists of two tails of the test statistic, we have to confront the problem, described in Example 4.3, of defining the two tails. For the purposes of Section 4.2, it is sufficient to consider the case when the test statistic has a distribution symmetric about zero. In this case the *p*-value is given by

$$p\text{-value} = \Pr(\,|S| \geq s_0 \,|\, H_0)$$

An extended discussion of the problem of two-sided *p*-values is given in Section 4.3.

―――― EXAMPLE 4.5 ――――

In the setting of Example 4.3, rewrite the test statistic $S = \bar{X} - 2$ so that, assuming the null hypothesis that $\mu = 2$, S is symmetrically distributed about zero. If a sample of size 16 produces an observed sample mean $\bar{x} = 2.13$, then $s_0 = 0.13$, and we calculate the p-value by

$$
\begin{aligned}
p\text{-value} &= \Pr(\,|S| \geq 0.13\,|\,H_0) \\
&= \Pr(\,|Z| \geq 0.13/0.125) = \Pr(\,|Z| \geq 1.04) \\
&= 0.2984
\end{aligned}
$$

since $Z = S/0.125$ has the distribution $N(0, 1)$.

As Examples 4.4 and 4.5 show, the basis of the calculation of the p-value is the same as that used in determining the size of the critical region. Using for illustration a one-tail test, with critical region $\{x : x \geq c\}$ and a test statistic S having a continuous distribution, both calculations are based on the probability relation $\Pr(S \geq c\,|\,H_0) = \alpha$ between the values c and α. In the traditional approach, the value of the significance level α is given and the corresponding value c_α of c is obtained from a table of the distribution of S – that is, $\Pr(S \geq c_\alpha\,|\,H_0) = \alpha$. On the other hand, the p-value calculation puts c equal to the observed value s_0 of the test statistic and obtains the p-value $= \Pr(S \geq s_0\,|\,H_0)$ from the same table. Clearly, the inequality $s_0 \geq c_\alpha$ is equivalent to the inclusion $\{S \geq s_0\} \subseteq \{S \geq c_\alpha\}$, which, in turn, is equivalent to the inequality $p\text{-value} \leq \alpha$. The equivalences in the preceding sentence complete the description of the p-value as the smallest level of significance at which the null hypothesis is rejected. Using p-value reporting we may, more naturally, express the conclusion in the form that the null hypothesis is accepted or rejected at the obtained significance level $100(p\text{-value})\%$.

4.1.4   *Interpretation scale for p-values*

The table given below shows the commonly used meaning attached to the p-value:

p-value	*Interpretation*
≥ 0.1	No evidence to doubt the truth of H_0
≈ 0.05	Some doubt about the truth of H_0
≈ 0.025	Serious doubt about the truth of H_0
≈ 0.01	H_0 almost certainly false

The values of 0.05, 0.025 and 0.01 are the most frequently used choices for preselected significance levels – that is, levels of doubt about H_0. In reports of a test procedure which show the observed value of the test statistic, the reader may find that the value has one, two or three asterisks attached to denote significance at the levels 0.05, 0.025 and 0.01 respectively – showing possibly significant, significant or highly significant by the number of asterisks.

Reporting the p-value – the standard in MINITAB – is clearly more informative than a basic comparison with a preselected significance level. For example, the statement that the null hypothesis is rejected at the 5% level of significance, made after noting that the observed value of the statistic lies in the 5% critical region, may disguise the fact that the value is so extreme as to have a p-value of 0.01. In effect, the p-value allows each individual to form a view of the weight of evidence provided by the data to support (or deny) the null hypothesis, and relate it to a personal choice of the maximum probability of Type I error which may be tolerated.

4.1.5   *Problems*

4.1   Let $X_1, X_2, \ldots, X_n$ be a random sample from a population distribution $N(\mu, \sigma^2)$ where σ is known. Using $\bar{X}$ as the test statistic for the null hypothesis H_0 that $\mu = 3$ against the alternative hypothesis H_1 that $\mu = 1$, explain why the critical region takes the form $\{x: x \leqslant 3 + z\sigma/\sqrt{n}\}$ for a suitable value of z.
     When $\sigma = 2.5$ and $n = 25$ find the value of z which gives a 1% critical region, and calculate the probability of Type II error.

4.2   A sample of size 20 is obtained from a binomial $b(1, p)$ distribution. Using the sample mean to test the null hypothesis H_0 that $p = 0.4$ against the alternative hypothesis H_1 that $p = 0.5$, obtain from tables the largest critical region having the probability of Type I error less than 0.05. What is the probability of Type II error for this region? Find also the p-value if the sample mean is 0.45.

4.3   A sample $X_1, X_2, \ldots, X_{10}$ is drawn from a population distribution $N(\mu, 1.5^2)$. Find the equi-tailed 5% critical region for a test of the null hypothesis H_0 that $\mu = 1.5$ against the alternative hypothesis H_1 that $\mu \neq 1.5$. Find the probability of Type II error when $\mu = 2$ and when $\mu = 3$. What is the p-value if the observed sample mean $\bar{x} = 1.7$?

4.4   (Continuing 4.3) Use the MINITAB CDF and PLOT commands to obtain the 'graph' of the probability of Type II error.

4.2   Tests related to the normal distribution

For reasons which include both the practical and the theoretical, the normal distribution lies at the heart of many statistical methods. The practical importance rests on the fact that many continuous variables that appear in real-life problems have a distribution which is approximately normal – or, by a simple transformation of the measurement scale, can be converted into approximately normal variables. On the theoretical side the attractiveness of the normal distributions relies on the ease with which statistically useful properties may be derived, and the result that, for *any* original distribution which has finite mean and variance, the sample mean $\bar{X}$ has a distribution which tends to normality as the sample size increases.

All except one of the tests described in this section are parametric tests of hypotheses concerned with the values of the means or variances of normally distributed variables. The tests on means may be classified by the corresponding MINITAB commands:

- ZTEST – applied to a univariate characteristic whose distribution has an assumed known standard deviation
- TTEST – applied to a univariate characteristic whose standard deviation is unknown, and also to certain bivariate paired characteristics
- TWOSAMPLE – applied as a test for equality of means for two independent characteristics and
- ONEWAY – applied as a test for equality of means for more than two independent characteristics.

There are no corresponding MINITAB commands for tests on variances of distributions – calculation by hand is required, perhaps aided by the commands DESCRIBE and CDF. The final test in this section has the non-parametric null hypothesis that the random sample data relates to a characteristic whose values are normally distributed.

4.2.1   *Tests on means: univariate data*

For the commands ZTEST and TTEST in basic form, MINITAB supplies the arithmetic for the test of the null hypothesis $\mu = \mu_0$, where μ_0 is a given value, against the two-tailed alternative $\mu \neq \mu_0$. The subcommand ALT may be employed for the one-tailed alternatives in the form ALT = 1 for $\mu > \mu_0$ and ALT = -1 for $\mu < \mu_0$.

ZTEST

Use of the command ZTEST is appropriate when the sample variables $X_1, X_2, \ldots, X_n$ are assumed to have a normal distribution $N(\mu, \sigma^2)$, with a known value for the standard deviation σ. For a test of the null hypothesis H_0 that $\mu = \mu_0$ against the two-tailed alternative H_1 that $\mu \neq \mu_0$, we take (see Section 4.1 Example 4.3) $Z = (\bar{X} - \mu_0)\sqrt{n}/\sigma$ as the test statistic and use a critical region of the shape $\{z : |z| \geq c\}$. Reporting *p*-values, MINITAB calculates the observed value $z_0 = (\bar{x} - \mu_0)\sqrt{n}/\sigma$ and noting that Z has the distribution $N(0, 1)$ when H_0 is true, quotes the *p*-value as $\Pr(|Z| \geq |z_0| \,|\, H_0)$.

───────── **EXAMPLE 4.6** ─────────────────────────────

Suppose that the ball-bearing manufacturer of Example 4.3 has improved the production process by reducing the variance and that the diameter of the ball bearings now follows the distribution $N(\mu, 0.05^2)$. To assess the quality of each production run with regard to the target value $\mu_0 = 2$ for μ, a random sample of size 16 is selected and the measured values of the diameters are found. For a particular run the data values were

$$1.99 \quad 2.01 \quad 2.15 \quad 2.02 \quad 2.01 \quad 1.99 \quad 1.98 \quad 1.89 \quad 2.04 \quad 2.23 \quad 2.11$$
$$2.03 \quad 1.95 \quad 1.99 \quad 2.04 \quad 2.09$$

Setting the data in C1, the command

ZTEST 2 0.05 C1

produces

TEST OF MU = 2.0000 VS MU N.E. 2.0000
THE ASSUMED SIGMA = 0.0500

	N	MEAN	STDEV	SE MEAN	Z	P VALUE
C1	16	2.0325	0.0811	0.0125	2.60	0.0095

Clearly, the values inserted in the command line are the target value of the mean and the known value of the standard deviation. The notation in the printout shows: $N = n =$ sample size; MEAN $= \bar{x} =$ sample mean; STDEV $= s =$ sample standard deviation; and SEMEAN $= \sigma/\sqrt{N}$. The value MINITAB denotes by Z is the observed value $z_0 = (\bar{x} - \mu_0)\sqrt{n}/\sigma$ – in the example $z_0 = (2.0325 - 2)\sqrt{16}/0.05 = 2.60$.

With a p-value of 0.0095, the manufacturer must conclude that the production-run is not of the required quality – that is, the null hypothesis is rejected at an observed significance level of less than 1%.

———— **EXAMPLE 4.7** ————————————————————————

The effect of the assumed value for σ can be seen if we examine the same data using the larger value $\sigma = 0.5$ of Example 4.3. Then

ZTEST 2 0.5 C1

produces

TEST OF MU = 2.000 VS MU N.E. 2.000
THE ASSUMED SIGMA = 0.500

	N	MEAN	STDEV	SE MEAN	Z	P VALUE
C1	16	2.033	0.081	0.125	0.26	0.79

Now, with a p-value of 0.79, the manufacturer will be happy to accept the good quality of the production run. (See also Example 4.9 below, where the data is examined without assuming a value for the standard deviation.)

─────── **EXAMPLE 4.8** ───────

The test information for the one-tailed alternative H_1 that $\mu > 2$ is provided, using $\sigma = 0.5$, by the command

> ZTEST 2 0.5 C1;
> ALT 1.

which produces

> TEST OF MU = 2.000 VS MU G.T. 2.000
> THE ASSUMED SIGMA = 0.500

	N	MEAN	STDEV	SE MEAN	Z	P VALUE
C1	16	2.033	0.081	0.125	0.26	0.40

The only comment required on this printout is to note that the *p*-value $\Pr(Z \geqslant 0.26 \mid H_0)$ is, as expected, half that for the two-tailed test.

─────────────────────────────────────

TTEST

The TTEST command is the replacement for ZTEST when the investigator has no reason to assign a value to the standard deviation. As for ZTEST, large values of $|\bar{X} - \mu_0|$ provide the criteria for rejecting the null hypothesis H_0 that $\mu = \mu_0$ in favour of the two-tailed alternative H_1 that $\mu \neq \mu_0$. From Section 1.4, the statistic

$$T_{n-1} = \frac{(\bar{X} - \mu_0)\sqrt{n}}{S}$$

has the $t(n-1)$ distribution under the null hypothesis, and, since S is unchanged by translations, large values of $|T_{n-1}|$ correspond to large value of $|\bar{X} - \mu_0|$. These observations suggest that T_{n-1} be chosen as the test statistic and that the critical region has the shape $\{t: |t| \geqslant c\}$. Reporting *p*-values, MINITAB calculates the observed value $t_0 = (\bar{x} - \mu_0)\sqrt{n}/s$, and quotes the value $\Pr(|T_{n-1}| \geqslant |t_0| \mid H_0)$ using the $t(n-1)$ distribution function.

─────── **EXAMPLE 4.9** ───────

Using the data and situation of Example 4.6 to test the null hypothesis $\mu = 2$ against the two-tailed alternative $\mu \neq 2$, the command

> TTEST 2 C1

produces

> TEST OF MU = 2.0000 VS MU N.E. 2.0000

	N	MEAN	STDEV	SE MEAN	T	P VALUE
C1	16	2.0325	0.0811	0.0203	1.60	0.13

In the printout the value SEMEAN is now $STDEV/\sqrt{n}$, and T is used to denote the observed value t_0 of the test statistic – that is, $t_0 = (2.0325 - 2)/0.0203$. The reported p-value of 0.13 suggests that the null hypothesis be accepted.

The differing p-values in Examples 4.6, 4.7 and 4.9 show an apparent inconsistency in the weight of evidence that the data supplies in support of the null hypothesis $\mu = 2$. In fact the discrepancies are caused by the three separate assumptions upon which the calculations were made, and which judge the data by three different scales ($\sigma = 0.05$, $\sigma = 0.5$ and, effectively, $\sigma = 0.0811$ in Example 4.9). These observations illustrate the general concept that the effectiveness of any statistical procedure relies heavily on the validity, for the problem at hand, of any assumptions made to justify the use of the procedure.

The subcommand ALT may be employed with TTEST to test the null hypothesis H_0 that $\mu = \mu_0$ against a one-tailed alternative. The reader can check that, as for ZTEST, the p-value for a one-tailed alternative hypothesis is half that for the two-tailed alternative.

4.2.2   *Comparison of means*

Making use of the MINITAB commands ZTEST and TTEST for tests of hypotheses, and (in Chapter 3) the commands ZINT and TINT to construct confidence intervals, we have examined the inferences that can be drawn about the mean of a single normal population. Studies involving two or more characteristics are frequently designed to evaluate the differences between the means rather than the values of the individual means – for example, comparing teaching in schools by the results in standard examinations; comparing the effects of rubber compounds on tyre lifetimes; and comparing the relief obtained from different anti-inflammatory drugs. In any such study the population is said to consist of experimental units and the values of the characteristics are the results of treatments applied to the units. The MINITAB commands TWOSAMPLE, ONEWAY, and a further application of TTEST will be used to test for equality of treatments.

There are two models for the comparison of means – paired comparison and independent populations. We start with the paired comparison model since this analysis uses the TTEST command.

TTEST (for paired comparisons)

Two treatments which may be applied to experimental units are to be compared. The method of paired comparisons is to select matched pairs of similar experimental units, and, by random assignment, apply one treatment to one member of a pair and the second treatment to the other member of the pair. Identical twins would be ideal matched pairs, but pairs of patients of the same age, build and gender are more widely available in clinical trials; similarly litter mates of the same sex could be suitable pairs in animal feeding comparisons. A particular case of pairing is self-pairing when the two treatments are applied to the same unit (in randomized order, and spaced in time to avoid carryover effects). The aim of the pairings is that the differences in responses between the members of the matched pairs will provide a more accurate comparison of the treatments by eliminating any inequalities between the experimental units.

The model for analysis consists of a (bivariate) sample with data (u_1, v_1), $(u_2, v_2), ..., (u_n, v_n)$ and sample random variables (U_1, V_1), $(U_2, V_2), ..., (U_n, V_n)$ where $u_1, u_2, ..., u_n$ are the responses to treatment I, and $v_1, v_2, ..., v_n$ the responses to treatment II. Writing $E(U_i) = \mu_1$ and $E(V_i) = \mu_2$, the differences $X_i = U_i - V_i$ have $E(X_i) = \mu = \mu_1 - \mu_2$. Further, the null hypotheses H_0 that the treatment means are equal (i.e. $\mu_1 = \mu_2$) translates readily to the hypothesis that $\mu = 0$, with the one-tailed alternative hypotheses H_1, that $\mu_1 > \mu_2$ becoming $\mu > 0$.

In order to complete the analysis of these hypotheses, using the criterion that relatively large values of $\bar{X}$ tend to favour the alternative hypothesis H_1, we require a condition on the distribution of the random variable $\bar{X}$. Naturally, in this section, we assume the variables $X_1, X_2, ..., X_n$ have a $N(\mu, \sigma^2)$ distribution (σ unknown). Accordingly, the test statistic $(\bar{X} - \mu)\sqrt{n}/S$ has Student's distribution $t(n-1)$ and the analysis proceeds as for a one-sample TTEST on the data $x_i = u_i - v_i$, $i = 1, 2, ..., n$.

———— EXAMPLE 4.10 ————

Eight matched pairs of male albino rats were the subject of a feeding trial, with the treatments consisting of a standard diet (control) and the same diet with the addition of Ethionine. The following data shows the amount of iron absorbed in the livers at the end of the trial period.

Pair number	1	2	3	4	5	6	7	8
Ethionine Y_1	4.50	3.92	7.33	8.23	2.07	4.90	6.84	6.96
Control Y_2	3.81	2.81	8.42	3.82	2.42	2.85	4.15	5.64

For the null hypothesis that there is no difference in the diets (i.e. $\mu = 0$) against the right tail alternative that Ethionine increases iron deposits (i.e. $\mu > 0$), set the Y_1 data in C2, the Y_2 data in C3 and let C1 = C2 − C3. Then the commands

    TTEST C1;
    ALT 1.

produce

    TEST OF MU = 0.000 VS MU G.T. 0.000

C1	N	MEAN	STDEV	SE MEAN	T	P VALUE
	8	1.354	1.732	0.612	2.21	0.031

From the p-value, we reject the null hypothesis of no difference in diets, and conclude that, at the observed 3.1% significance level, Ethionine increases iron deposits in the liver. Example 5.8 in Section 5.1 views this data in a non-parametric context – that is, omitting the distribution assumption on the variables $X_1, X_2, ..., X_n$.

Applying the TINT command to C1 provides a confidence interval for the mean μ, and, hence, a confidence interval for the difference $\mu_1 - \mu_2$ in the form $\bar{x} \pm ts/\sqrt{n}$ where t is the appropriate $t(n-1)$ percentage point – that is,

$$\Pr(\,|\bar{X} - \mu|\sqrt{n}/S \leq t) = 1 - \alpha, \text{ or } \Pr(\bar{X} - \mu)\sqrt{n}/S \leq t) = 1 - \alpha/2$$

For example,

        TINT 90 C1

produces

C1	N	MEAN	STDEV	SE MEAN	90.0 PERCENT C.I.
	16	2.0325	0.0811	0.0203	(1.9969, 2.0681)

TWOSAMPLE

The TWOSAMPLE command provides the numerical details for a comparison of two treatments when pairing is not appropriate as, for example, when comparing the durability of batteries produced by different manufacturers. As the model for analysis, we have two random samples, one for each treatment, producing a random sample of size n, with sample data $x_1, x_2, ..., x_n$ and sample variables $X_1, X_2, ..., X_n$ having the $N(\mu_1, \sigma_1^2)$ distribution; and a random sample of size m, with sample data $y_1, y_2, ..., y_m$ and sample variables $Y_1, Y_2, ..., Y_m$ having the $N(\mu_2, \sigma_2^2)$ distribution.

In addition to the independence of the sample variables within each population, we assume also that the populations are independent, so that the $n + m$ random variables $X_1, X_2, ..., X_n, Y_1, Y_2, ..., Y_m$ are independent.

Comparing the treatments is effected by testing the null hypothesis H_0 that the means are equal (i.e. $\mu_1 = \mu_2$) against, as appropriate to the problem, one of the alternatives $\mu_1 \neq \mu_2$ (two-tailed), $\mu_1 > \mu_2$ (right-tail) or $\mu_1 < \mu_2$ (left-tail) using a test statistic based on the difference $\bar{X} - \bar{Y}$ which has a $N(\mu, \sigma^2)$ distribution with parameters $\mu = \mu_1 - \mu_2$ and $\sigma^2 = \sigma_1^2/n + \sigma_2^2/m$. A simple solution to the problem of finding a statistic to replace the unknown value of σ^2 is available under the additional assumption that the variances σ_1^2 and σ_2^2 are equal. Assume $\sigma_1^2 = \sigma_2^2$. Writing $(n-1)S_1^2 = \sum_{i=1}^{n} (X_i - \bar{X})^2$ and $(m-1)S_2^2 = \sum_{j=1}^{m} (Y_j - \bar{Y})^2$, we see from Section 1.4 that the statistic

$$T_{n+m-2} = \frac{(\bar{X} - \bar{Y} - (\mu_1 - \mu_2))\sqrt{nm(n+m-2)}}{\sqrt{(n+m)[(n-1)S_1^2 + (m-1)S_2^2]}}$$

has Student's distribution $t(n+m-2)$. Since extreme values of $\bar{X} - \bar{Y}$, in the direction indicated by the alternative hypothesis, correspond to similar extreme values of T_{n+m-2}, the statistic T_{n+m-2} may be employed as the test statistic.

──────── EXAMPLE 4.11 ────────

The resistance to weathering of two types of paint used in highway signs is to be compared. After a period of time, reflectometer readings produced the following sample data:

Paint 1 (x_i) 11.5 10.7 8.9 8.6 9.3 8.6 8.4 10.3 7.7 10.5 9.6 8.7
Paint 2 (y_i) 8.4 10.6 8.7 9.4 5.9 6.3 7.4 6.2 6.0 7.2 11.7

A test of the null hypothesis of equal means against the two-tailed alternative of unequal means is provided by TWOSAMPLE. The assumption of equal variances is entered using the subcommand POOLED. Setting the data in C1 and C2 the command

> TWOSAMPLE C1 C2;
> POOLED.

produces the printout

> TWOSAMPLE T FOR C1 VS C2

	N	MEAN	STDEV	SE MEAN
C1	12	9.40	1.13	0.33
C2	11	7.98	1.96	0.59

> 95 PCT CI FOR MU C2 − MU C2: (0.04, 2.79)
> TTEST MU C1 = MU C2 (VS NE): T = 2.15 P = 0.044 DF = 21
> POOLED STDEV = 1.58

In the printout the values indicated by N, MEAN, STDEV and SE MEAN are the familiar sample values, and DF denotes the degrees of freedom, $n+m-2$. The value T is the observed value of the test statistic calculated assuming H_0 is true (i.e. $\mu_1 = \mu_2$) by

$$T = \frac{(\bar{x}-\bar{y})\sqrt{nm(n+m-2)}}{\sqrt{(n+m)[(n-1)s_1^2+(m-1)s_2^2]}}$$

with calculated value

$$T = \frac{(9.40-7.98)\sqrt{12\times11\times21}}{\sqrt{(12+11)(11\times1.13^2+10\times1.96^2)}} \approx 2.15$$

and the p-value, $\Pr(|T_{n+m-2}| \geq |T| \mid H_0)$ for the two-tailed alternative, is shown by P.

MINITAB also provides an estimate s, denoted by POOLED STDEV, for the common value of σ_1, and σ_2, obtained from

$$S^2 = \frac{1}{n+m-2}[(n-1)S_1^2+(m-1)S_2^2]$$

with a calculated value

$$s^2 = \frac{1}{21}(11 \times 1.13^2 + 10 \times 1.96^2) \approx 1.58^2$$

and a 95% confidence interval for the difference $\mu_1 - \mu_2$ given by

$$\bar{x} - \bar{y} \pm ts\sqrt{(n+m)/nm}$$

where t is obtained from tables of the t-distribution so that $\Pr(T_{n+m-2} \leqslant t) = 0.975$, with a calculated value

$$9.40 - 7.98 \pm 2.08 \times 1.58\sqrt{23/12 \times 11} \approx 1.42 \pm 1.37$$

as shown (except for rounding errors). The p-value 0.044 suggests that the null hypothesis that the means are equal be rejected, and, since $\bar{x} > \bar{y}$, paint 1 is to be preferred.

Using the subcommand ALT, in conjunction with POOLED, and inserting a percentage, 90% say, in the command,

TWOSAMPLE 90 C1 C2
POOLED;
ALT 1.

provides a test of the null hypothesis of equal means against the alternative that $\mu_1 > \mu_2$, and prints a 90% confidence interval for the difference $\mu_1 - \mu_2$.

Comparison of two means without the simplifying assumption of equality of variances is known as the Behrens–Fisher problem. Of the several solutions that have been proposed, the following approximation has the merits of simplicity, the use of the familiar t-distribution tables, and reasonable accuracy.

Under the null hypothesis H_0 that the means are equal, the statistic

$$T_v = \frac{\bar{X} - \bar{Y}}{\sqrt{S_1^2/n + S_2^2/m}}$$

has, approximately, the $t(v)$ distribution where

$$v = \left[\frac{(a+b)^2}{a^2/(n-1) + b^2(m-1)} \right]$$

where $a = s_1^2/n$, $b = s_2^2/m$ and [.] denotes the integer part of the number, for example, $[10.04] = [10.93] = 10$.

Using T_v and t-distribution tables we may test the null hypothesis H_0 against the two-tailed alternative or either one-tailed alternative in the usual way.

——— **EXAMPLE 4.12** ———

Using the data of Example 4.11, the command

TWOSAMPLE C1 C2;
ALT 1.

produces the printout

TWOSAMPLE T FOR C1 VS C2

	N	MEAN	STDEV	SE MEAN
C1	12	9.40	1.13	0.33
C2	11	7.98	1.96	0.59

95 PCT CI FOR MU C1 – MU C2: (−0.02, 2.86)
TTEST MU C1 = MU C2 (VS GT): T = 2.10 P = 0.027 DF = 15

for a test of H_0 against the right-tail alternative hypothesis H_1 that $\mu_1 > \mu_2$.
  In the printout T denotes the observed value of the statistic T_v with calculated value

$$T = (9.40 - 7.98)/\sqrt{1.13^2/12 + 1.96^2/11} \approx 2.104$$

and DF gives the value of v. Substituting $a = 1.13^2/12$, $b = 1.96^2/11$ in the formula above gives $v = [15.699] = 15$.
  MINITAB also provides the 95% confidence interval

$$\bar{x} - \bar{y} \pm t\sqrt{s_1^2/n + s_2^2/m}$$

where $\Pr(T_v \leqslant t) = 0.975$ determines t from tables, for the difference $\mu_1 - \mu_2$ between the means, with calculated value

$$9.40 - 7.98 \pm 2.131\sqrt{1.13^2/12 + 1.96^2/11} \approx 1.42 \pm 1.438$$

The p-value $\Pr(T_v \geqslant 2.10) = 0.027$ suggests that the null hypothesis be rejected at the obtained 2.7% level of significance.

  Although the POOLED procedure is based on an exact distribution, it can cause serious error when the assumption of equal variances is not satisfied. This leads to the general recommendation that the POOLED subcommand should not be applied in most cases.

ONEWAY

The command ONEWAY provides a test for the null hypothesis of equality of the means of several populations against the alternative hypothesis that the means are not all equal. Being part of the area known as analysis of variance, ONEWAY is, strictly, excluded from the scope of this book, but a brief discussion is included to enable the reader to contrast the normal distribution theory with the non-parametric methods and the multi-comparisons described in Chapter 8.
  Suppose, then, that we have m populations and random samples of size n from each population, with sample random variables

$$X_{11}, X_{12}, \ldots, X_{1n}$$
$$X_{21}, X_{22}, \ldots, X_{2n}$$
$$\ldots\ldots\ldots\ldots \quad \ldots$$
$$X_{m1}, X_{m2}, \ldots, X_{mn}$$

where the random variables X_{ij}, for $j = 1, 2, \ldots, n$ are distributed $N(\mu_i, \sigma^2)$ and all the random variables X_{ij}, for $i = 1, 2, \ldots, m$ and $j = 1, 2, \ldots, n$, are independent. Note, not surprisingly, that the variances for the m populations are assumed equal. The null hypothesis H_0 is that $\mu_1 = \mu_2 = \cdots = \mu_m \ (= \mu$, say$)$ and the alternative hypothesis H_1 is that the means are not all equal.

The considerations which lead to the test statistic are based on the values of the quadratic

$$Q(\mu_1, \mu_2, \ldots, \mu_m) = \sum_{i=1}^{m} \sum_{j=1}^{n} (X_{ij} - \mu_i)^2$$

regarded as a function of the means $\mu_1, \mu_2, \ldots, \mu_m$. To be more precise, the test statistic is based on the ratio of two minima for Q – the restricted (equal means) minimum

$$SS_T = \min_{\mu} \ Q(\mu, \mu, \ldots, \mu),$$

evaluated assuming H_0, and the unrestricted (general model) minimum

$$SS_e = \min_{\mu_1, \mu_2, \ldots, \mu_m} \ Q(\mu_1, \mu_2, \ldots, \mu_m)$$

Obtaining the minimum in each case is a simple application of Problem 2.2 of Section 2.2 which shows that, for any set of given values $V_1, V_2, \ldots, V_K$,

$$\min_{a} \sum_{i=1}^{K} (V_i - a)^2 = \sum_{i=1}^{K} (V_i - \bar{V})^2$$

Then

$$SS_T = \min_{\mu} \sum_{i=1}^{m} \sum_{j=1}^{n} (X_{ij} - \mu)^2$$

$$= \sum_{i=1}^{m} \sum_{j=1}^{n} (X_{ij} - \bar{X})^2$$

by taking $K = nm$, and $V_1, \ldots, V_K$ to be the values X_{ij} for $i = 1, 2, \ldots, m$, $j = 1, 2, \ldots, n$, with

$$\bar{X} = \frac{1}{nm} \sum_{i=1}^{m} \sum_{j=1}^{n} X_{ij}$$

the grand mean; and

$$SS_e = \min_{\mu_1, \mu_2, \ldots, \mu_m} \sum_{i=1}^{m} \sum_{j=1}^{n} (X_{ij} - \mu_i)^2$$

$$= \sum_{i=1}^{m} \min_{\mu_i} \sum_{j=1}^{n} (X_{ij} - \mu_i)^2$$

$$= \sum_{i=1}^{m} \sum_{j=1}^{n} (X_{ij} - \bar{X}_i)^2$$

where

$$\bar{X}_i = \frac{1}{n} \sum_{j=1}^n X_{ij}$$

is obtained by taking $K = n$ and $V_j = X_{ij}$ for $j = 1, 2, \ldots, n$.

Now SS_T, the total sum of squares, is the minimum value of Q assuming the null hypothesis H_0, and SS_e, the error sum of squares, is the minimum of Q under the general assumption of the model. Since the minimum defining SS_T is taken over the restricted set of values $\mu_1 = \mu_2 \cdots = \mu_m = \mu$, the inequality $SS_e \leq SS_T$ holds, and the truth of H_0 would suggest that the two minima should be approximately equal. Reversing the last suggestion, the criterion for the rejection of the null hypothesis can be expressed in terms of relatively large values of the ratio SS_T/SS_e. In order to discuss the distribution properties it is convenient to look at relatively large values of $(SS_T - SS_e)/SS_e$.

Observe first that

$$SS_T - SS_e = \sum_{i=1}^m \left[\sum_{j=1}^n (X_{ij} - \bar{X})^2 - \sum_{j=1}^n (X_{ij} - \bar{X}_i)^2 \right]$$

$$= n \sum_{i=1}^m (\bar{X}_i - \bar{X})^2$$

since the relation $\sum_{j=1}^n (V_j - a)^2 = \sum_{j=1}^n (V_j - \bar{V})^2 + n(\bar{V} - a)^2$ may be used with $V_j = X_{ij}$ and $a = \bar{X}$.

Under H_0, the random variables $\bar{X}_i$ have the distribution $N(\mu, \sigma^2/n)$, and $\bar{X}$ the distribution $N(\mu, \sigma^2/nm)$ since $\bar{X} = \sum_{i=1}^m \bar{X}_i/m$. From section 1.4,

$$n \sum_{i=1}^m (\bar{X}_i - \bar{X})^2/\sigma^2 \text{ has the } \chi^2(m-1) \text{ distribution}$$

$$\sum_{j=1}^n (\bar{X}_{ij} - \bar{X}_i)^2/\sigma^2 \text{ has the } \chi^2(n-1) \text{ distribution}$$

and, by the independence of the populations,

$$\sum_{i=1}^m \sum_{j=1}^n (X_{ij} - \bar{X}_i)^2/\sigma^2 \text{ has the } \chi^2(m(n-1)) \text{ distribution}$$

Writing $SS_H = SS_T - SS_e$, and accepting without proof the independence of SS_H and SS_e, Section 1.4 implies that the test statistic

$$F = \frac{m(n-1)}{(m-1)} \frac{SS_H}{SS_e}$$

has the $F(m-1, m(n-1))$ distribution, and the null hypothesis is rejected at the $100\alpha\%$ level of significance if the observed value of F is greater than or equal to f_α, where $\Pr(F \geq f_\alpha) = \alpha$.

Incidentally, the terminology of analysis of variance arises from the decomposition $SS_T = SS_H + SS_e$, since SS_T is nothing more than a multiple (by $nm - 1$) of the sample variance.

──────── **EXAMPLE 4.13** ────────────────────────────────

Suppose we have $m = 4$ populations with $n = 4$ observations on each population, with observed values

x_{1j}	18.95	12.62	11.94	14.42
x_{2j}	10.06	7.19	7.03	14.66
x_{3j}	10.92	13.28	14.52	12.51
x_{4j}	9.30	21.20	16.11	21.41

To apply the ONEWAY command the data is entered in C1 and the population indicator in C2 as shown by the printout:

ROW	C1	C2
1	18.95	1
2	12.62	1
3	11.94	1
4	14.42	1
5	10.06	2
6	7.19	2
7	7.03	2
8	14.66	2
9	10.92	3
10	13.28	3
11	14.52	3
12	12.51	3
13	9.30	4
14	21.20	4
15	16.11	4
16	21.41	4

Then the command

ONEWAY C1 C2

produces the printout

ANALYSIS OF VARIANCE ON C1

SOURCE	DF	SS	MS	F	p
C2	3	111.6	37.2	2.60	0.101
ERROR	12	172.0	14.3		
TOTAL	15	283.6			

INDIVIDUAL 95 PCT CI'S FOR MEAN
BASED ON POOLED STDEV

LEVEL	N	MEAN	STDEV	
1	4	14.483	3.157	(-------- * ---------)
2	4	9.735	3.566	(-------- * ---------)
3	4	12.807	1.506	(-------- * ---------)
4	4	17.005	5.691	(-------- * ---------)

POOLED STDEV = 3.786   10.0   15.0   20.0

In the first part of the printout DF denotes the degrees of freedom of the χ^2-distributions corresponding to the sums of squares (SS) and the MS column denotes the mean squares, where $MS_H = SS_H/(m-1)$ and $MS_e = SS_e/m(n-1)$. Further, F denotes the observed value of the test statistic and p the corresponding p-value $Pr(F \geqslant F)$. Thus: $SS_H = 111.6$, $MS_H = 37.2$, $SS_e = 172.0$ and $MS_e = 14.3$, with observed value

$$F = \frac{MS_H}{MS_e} = 2.60$$

and the p-value 0.101 indicates that the null hypothesis of equality of means be accepted. Additional information given in the second part of the printout consists of:

- The sample mean and standard deviation for each population (denoted by LEVEL)
- The estimate $\hat{\sigma} = \sqrt{MS_e}$ (denoted by POOLED STDEV) for the (assumed common) standard deviation σ and
- A graphical display of individual 95% confidence intervals for the population means μ_i.

The basis of the confidence intervals is the $t(m(n-1))$ distribution. Using the independence of populations and the independence for each population of the sample mean $\bar{X}_i$ and sample variance $\sum_{j=1}^{n} (X_{ij}-\bar{X}_i)^2/(n-1)$, we obtain the independence of all the sample means $\bar{X}_1, \bar{X}_2, ..., \bar{X}_m$ and $MS_e(=\sum_{i=1}^{m} \sum_{j=1}^{n} (X_{ij}-\bar{X}_i)^2/m(n-1))$. Combining the $N(\mu_i, \sigma^2/n)$ distribution of $\bar{X}_i$ with the $\chi^2(m(n-1))$ distribution of SS_e gives the result that $(\bar{X}_i - \mu_i)\sqrt{n}/MS_e$ has the $t(m(n-1))$ distribution, leading to $100(1-\alpha)\%$ confidence intervals of the form

$$\bar{X}_i \pm t\hat{\sigma}/\sqrt{n}$$

where $Pr(T \leqslant t) = 1 - \alpha/2$ is obtained from tables.
Similarly, for any $1 \leqslant k \neq l \leqslant m$, $\bar{X}_k - \bar{X}_l$ has the $N(\mu_k - \mu_l, 2\sigma^2/n)$ distribution, and

$$[\bar{X}_k - \bar{X}_l - (\mu_k - \mu_l)]/\hat{\sigma}\sqrt{2/n}$$

has the $t(m(n-1))$ distribution.

As is seen in the printout for Example 4.13, MINITAB routinely supplies 95% confidence intervals for the means. The next example shows the mechanics of using the subcommand FISHER to obtain confidence intervals for the differences between means.

──────── EXAMPLE 4.14 ────────────────────────────────────

Using for illustration the data of Example 4.13, and choosing $\alpha = 0.05$, the command

    ONEWAY C1 C2;
    FISHER 0.05.

produces, in addition to the printout of Example 4.13, Fisher's pairwise comparisons:

    Family error rate   = 0.184
    Individual error rate = 0.0500

    Critical value   = 2.179

Intervals for (column level mean) – (row level mean)

	1	2	3
2	−1.086		
	10.581		
3	−4.159	−8.906	
	7.509	2.761	
4	−8.356	−13.104	−10.031
	3.311	−1.436	1.636

showing the 95% confidence intervals for $\mu_k - \mu_l$ where $k < l$. For example, the 95% confidence interval for $\mu_1 - \mu_2$ is

$$\bar{x}_1 - \bar{x}_2 \pm t\hat{\sigma}/\sqrt{2} = 14.483 - 9.735 \pm 2.179 \times 3.786/\sqrt{2}$$
$$\approx 4.748 \pm 5.833$$
$$\approx (-1.085, 10.581),$$

where $t = 2.179$, shown as the critical value, is the value, from the $t(12)$-distribution table, corresponding to the chosen individual error rate $\alpha = 0.05$.

──

The *family error rate term* requires a little more explanation. Suppose, in the general case of m populations, that the null hypothesis H_0 of equality of means is true. Using the individual error rate α, each of the family of $N = m(m-1)/2$ confidence intervals will contain the origin (since $\mu_k - \mu_l = 0$) with probability $1 - \alpha$, or, labelling the confidence intervals $1, 2, ..., N$ in some order, the event A_i that the ith confidence interval does not contain the origin has probability $\Pr(A_i \,|\, H_0) = \alpha$.

The family error rate is defined as the probability $\Pr(A_1 \text{ or } A_2 \text{ or } \dots \text{ or } A_N | H_0)$ that at *least one of the confidence intervals does not contain the origin*, calculated assuming the truth of H_0. In Examples 4.13 and 4.14, with $m = 4$ and $\alpha = 0.05$, there is a probability of 0.184 that at least one of the $N = 6$ confidence intervals does not contain the origin – in fact, the confidence interval for the difference between the means of the second and fourth populations does not contain the origin.

Although the preceding paragraph provides a direct definition of the family error rate, it offers no insight into why this probability is of interest. We now correct this deficiency. What is shown by MINITAB as Fisher's pairwise comparisons is known also as the least significant difference (LSD) method for locating those differences between the means which have led (through ONEWAY) to the rejection of the null hypothesis that the means are equal. The LSD principle is to test at level α, separately for each pair (k, l) of integers $1 \leqslant k < l \leqslant m$, the null hypothesis that $\mu_k = \mu_l$ against the two-tailed alternative $\mu_k \neq \mu_l$, and to declare $\mu_k \neq \mu_l$ if the corresponding confidence interval does not include the origin. Equivalently, in more familiar form, the null hypothesis that $\mu_k = \mu_l$ is rejected at the $100\alpha\%$ level of significance if $|\bar{x}_k - \bar{x}_l| \geqslant t\hat{\sigma}\sqrt{2}/n$, where

$$\Pr(|\bar{X}_k - \bar{X}_l| \geqslant t\hat{\sigma}\sqrt{2/n} | H_0) = \alpha$$

The LSD procedure is, therefore, a family of N tests of hypothesis, each considered separately as having probability α for committing the Type I error of incorrectly declaring inequality between the two means. The meaning to be attached to the family error rate is now clear. It is the probability, calculated assuming all the means are equal, that the LSD procedure will incorrectly distinguish (declare unequal) *at least one* pair of means – a Type I error for the complete LSD procedure.

Application of the LSD procedure is a second-stage operation following the rejection, by the F-test of ONEWAY, of the hypothesis that the means are equal. Examples 4.13 and 4.14 show that the F-test may accept the inequality of means while the LSD procedure distinguishes between at least one pair of means. Conversely, it may happen that the F-test rejects equality but the subsequent application of the LSD procedure suggests that equality of means be accepted for every pair of populations!

These inconsistencies and the relatively large family error rate have generated objections to the use of the LSD procedure in multiple comparisons. Since

$$\Pr(A_1 \text{ or } A_2 \text{ or } \dots \text{ or } A_N | H_0) \leqslant \sum_{i=1}^{N} \Pr(A_i | H_0)$$

control of the family error rate, to be less than β, may be exercised by taking $\alpha = \beta/N$, but this is achieved only at the expense of increasing the length of the confidence intervals – for example, $m = 5$, $\beta = 0.05$ gives $N = 10$ and $\alpha = 0.005$, or 99.5% confidence intervals. The literature contains several alternative multiple comparison methods, three of which are covered by subcommands for ONEWAY:

the Tukey method for all pairs of comparisons; Dunnett for comparing $m-1$ populations with a control; and MCB (multiple comparison with the best) for selected pairs of comparisons.

The simplicity of the LSD procedure allows its inclusion here, even though the method for evaluating the family error rate is beyond our scope. Chapter 8 gives further examples of multiple comparison methods.

4.2.3   *Tests on variance*

Under this heading we present two tests relating to the variance parameter in a normal distribution: a test of the null hypothesis that $\sigma = \sigma_0$, where σ_0 is known, in a single population setting; and a test for the equality of variance for two populations.

In the single population, let $X_1, X_2, \ldots, X_n$ be the independent sample random variables each with the distribution $N(\mu, \sigma^2)$, where the mean μ is unknown. A test of the null hypothesis H_0 that $\sigma = \sigma_0$ against the two-tailed alternative that $\sigma \neq \sigma_0$ is based on the statistic $W = (n-1)S^2/\sigma_0^2$, which (see Section 1.5) has the $\chi^2(n-1)$ distribution when H_0 is true. Since the sample variance S^2 is an unbiased estimator for the true variance, relatively large or small values of W will favour the alternative hypothesis, and the null hypothesis is therefore rejected if the observed value of W falls in a critical region of the form $\{w: w \leqslant a \text{ or } w \geqslant b\}$. At the $100\alpha\%$ level of significance we require values a, b which satisfy

$$\Pr(a < W < b \,|\, H_0) = 1 - \alpha$$

and, in this section, we resolve the non-uniqueness of the values a and b with the widely used equi-tailed solution

$$\Pr(W \leqslant a \,|\, H_0) = \Pr(W \geqslant b \,|\, H_0) = \alpha/2$$

(See Section 4.3 for a discussion of this problem.)

──────── **EXAMPLE 4.15** ────────

Suppose a sample of $n = 16$ produces a sample standard deviation 0.081, and we require a test of the null hypothesis $\sigma = 0.05$ at the 1% level of significance. For the $\chi^2(15)$ distribution, tables supply

$$\Pr(W \leqslant 4.601 \,|\, \sigma = 0.05) = \Pr(W \geqslant 32.80 \,|\, \sigma = 0.05) = 0.005$$

giving the critical region $\{w: w \leqslant 4.601 \text{ or } w \geqslant 32.80\}$. Now the observed value of W is $15 \times 0.081^2/0.05^2 = 39.366$, which clearly lies in the critical region, and leads to the rejection of the null hypothesis.

The more frequent application of this test statistic is for the one-tailed alternative $\sigma > \sigma_0$ when quality standard is important and the variability must be controlled. In this case, the critical region takes the form $\{w: w \geqslant b\}$ and the value b is uniquely determined from the relation $\Pr(W \geqslant b \,|\, H_0) = \alpha$.

The probability model for the equality of variances in two independent populations is that specified for testing equality of means, namely, two sets of sample random variables – $X_1, X_2, \ldots, X_n$ a sample from the first population with distribution $N(\mu_1, \sigma_1^2)$, and an independent sample $Y_1, Y_2, \ldots, Y_m$ from the second population with distribution $N(\mu_2, \sigma_2^2)$. With both means μ_1 and μ_2 unknown, we require a test of the null hypothesis that $\sigma_1 = \sigma_2 (= \sigma$, say) against the two-tailed alternative $\sigma_1 \neq \sigma_2$, at the $100\alpha\%$ level of significance. Writing

$$(n-1)S_1^2 = \sum_{i=1}^{n} (X_i - \bar{X})^2 \text{ and } (m-1)S_2^2 = \sum_{j=1}^{m} (Y_j - \bar{Y})^2$$

the statistics $W_1 = (n-)S_1^2/\sigma^2$ and $W_2 = (m-1)S_2^2/\sigma^2$ are independent chi-squared variables, so, from section 1.4, the variance ratio $F = S_1^2/S_2^2$ has the $F(n-1, m-1)$ distribution if H_0 is true. For the two-tailed alternative, a critical region of the form $\{x: x \leqslant a \text{ or } x \geqslant b\}$ is evidently appropriate, and, proceeding directly to the equitailed solution, the values a and b are determined to satisfy

$$\Pr(F \leqslant a \mid H_0) = \Pr(F \geqslant b \mid H_0) = \alpha/2$$

───────── EXAMPLE 4.16 ─────────

Using the data of Example 4.11, we have $n = 12$, $m = 11$ and $s_1 = 1.13$, $s_2 = 1.96$. With $\alpha = 0.05$ we obtain $b = 3.6649$ and $1/a = 3.5257$, i.e. $a = 0.2836$, since $1/F$ has the distribution $F(10, 11)$. Now the critical region is $\{x: x \leqslant 0.2836 \text{ or } x \geqslant 3.6649\}$ which does not contain the observed value $1.13^2/1.96^2 \approx 0.332$, leading to the acceptance of the equality of the variances.

───────────────────────────────

The two tests for variances rely heavily on the assumed normality of the parent distributions, and can be shown to be seriously misleading if the normality assumption fails even quite mildly. On the other hand, it can also be shown that the tests for means lose little in precision if the parent distribution deviates slightly from normality. In consequence, the reader is warned against applying the test for equality of variances as a preliminary to using TWOSAMPLE with the subcommand POOLED.

4.2.4   NSCORES – testing the distribution

All the tests described earlier in this section have been concerned with numerical values of the parameters μ and σ of the underlying $N(\mu, \sigma^2)$ distribution. Using the NSCORES and CORRELATION commands, we now describe a test procedure relating to the nature of the distribution. More precisely, we describe a test of the non-parametric null hypothesis that the sample data $x_1, x_2, \ldots, x_n$ has been drawn from a population which has a $N(\mu, \sigma^2)$ distribution for some values of the parameters μ and σ.

The basis of the test procedure is to compare the sample data $x_1, x_2, \ldots, x_n$ with an ideal sample $e_1, e_2, \ldots, e_n$ from the normal $N(0, 1)$ distribution. In order to remove the variability in the labelling of the data, both sets of data are written in increasing order as $x_{(1)} \leqslant x_{(2)} \leqslant \cdots \leqslant x_{(n)}$ and $e_{(1)} \leqslant e_{(2)} \leqslant \cdots \leqslant e_{(n)}$ using the order statistic notation of Section 1.1, and the pairings in the comparison are $(x_{(1)}, e_{(1)})$, $(x_{(2)}, e_{(2)}), \ldots (x_{(n)}, e_{(n)})$. Having recognized the need to order the sample data, the definition of the ideal sample now follows the standard practice of comparing sample values with their mean. Let $Z_1, Z_2, \ldots, Z_n$ be the sample random variables from the normal distribution $N(0, 1)$ and let $Z_{(1)} \leqslant Z_{(2)} \leqslant \cdots Z_{(n)}$ be the order statistics. The ideal sample values are now defined as the expected values – that is, $e_{(i)} = E(Z_{(i)}) =$ the normal scores. Now that the data has been written in bivariate form, the natural measure of comparison is the sample correlation coefficient

$$r = \frac{\sum_{i=1}^{n} x_{(i)} e_{(i)} - n\bar{x}\bar{e}}{\sqrt{\sum_{i=1}^{n} (x_{(i)} - \bar{x})^2 \sum_{i=1}^{n} (e_{(i)} - \bar{e})^2}}$$

From Problem 2.3 of Section 2.2, the correlation coefficient is unchanged by a linear shift, so that, assuming the null hypothesis is true, the value r is independent of the parameters μ and σ^2 and may be taken as the correlation between observed values of the order statistics and their means for a sample drawn from a normal distribution $N(0, 1)$. In consequence, a high value for r (close to 1) would be anticipated only if the null hypothesis is true, and the test procedure is to reject the null hypothesis if the calculated value of r falls below a tabulated critical value.

The test for normality described in the MINITAB reference manual is a minor modification of the procedure outlined above. In the NSCORES command MINITAB uses a numerical approximation to obtain the values $e_{(i)}$, $i = 1, 2, \ldots, n$, rather than the tabulated normal scores quoted in statistical tables. Using the MINITAB values of $e_{(i)}$ in the calculation of r, 5% critical values are, for selected sample sizes,

n	5	10	15	20	25	30	60
value	0.8804	0.9176	0.9383	0.9511	0.9582	0.9639	0.9799

The reader is referred to the manual for details of the approximation and a larger list of critical values.

——— EXAMPLE 4.17 ———

With the x-data entered in C1 the commands

        NSCORES C1 C2
        PRINT C1 C2

produces

ROW	C1	C2
1	4.50	−0.47097
2	3.92	−0.85042
3	7.33	0.85042
4	8.23	1.43476
5	2.07	−1.43476
6	4.90	−0.15183
7	6.84	0.15183
8	6.96	0.47097

with C2 containing the (MINITAB) normal scores paired row by row with the x-data – for example, row 5 contains the smallest value in each column. The command

CORRELATION C1 C2

prints the observed value of r as 0.974, which, making an appropriate informal interpolation for sample size 8, leads to accepting the null hypothesis at the 5% level.

An alternative, more intuitive, view of the procedure may be obtained by plotting $e_{(i)}$ against $x_{(i)}$. If the sample does come from a normal distribution, the scatterplot should produce points lying roughly on a straight line.

──── EXAMPLE 4.18 ────

The data of Example 4.17 produces, using the command PLOT C2 C1, the scatterplot

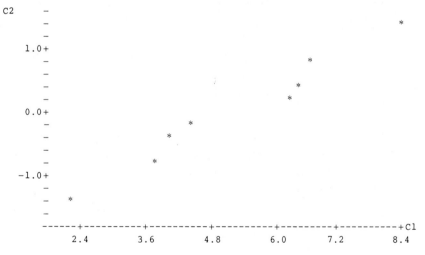

which does have a linear appearance.

Suppose now we have a data set for which the hypothesis of normality has been rejected by NSCORES – what should be done? As noted earlier, tests on means using TTEST or TWOSAMPLE may reasonably be used for small departures from normality but other methods are required in more extreme cases. These methods form the subject matter of the later chapters: the non-parametric methods of Chapters 5–8 apply when almost no distributional assumption is made; or tests may be constructed by the methods of Chapter 9, when an alternative distribution for the parent population is specified.

4.2.6   Problems

In problems 4.5–4.10 assume that each parent population has a normal distribution.

4.5   A manufacturer of binding twine claims that the twine has an average breaking strength of 8 units. Ten randomly selected pieces were found to have breaking strengths

   8.15   7.35   7.55   8.65   7.75   7.10   8.30   7.50   7.90   7.45

Test the manufacturer's claim against the alternative that the mean breaking strength is less than eight:
(i) if the method of production is known, from experience, to have a standard deviation of 0.4 for the breaking strength
(ii) if the standard deviation is unknown.

4.6   Two diets for lambs are compared by examining the weight gain for ten pairs of twins, one lamb of each pair to have diet A and the other diet B. The weight gains were as follows:

Pair	1	2	3	4	5	6	7	8	9	10
Diet A	7.4	6.6	6.6	7.0	7.4	6.4	7.6	9.0	7.4	7.2
Diet B	6.2	5.8	5.6	6.6	6.2	5.8	6.8	8.4	6.0	7.0

Test the hypothesis that the two diets produce equal weight gain.

4.7   An assistant reports that an error was made in the tagging of the lambs in Problem 4.6 and it is not possible to pair the data for the two diets. Re-examine the data as independent sets of observations for the two diets.

4.8   Two diet formulae, diets A and B, and a calorie-counting plan are to be compared. Fifteen volunteers, of approximately equal obesity, were divided at random into groups of five, and the weight losses over a period of one week were as follows:

Diet A	2.36	2.53	1.70	2.24	1.62
Diet B	2.32	1.56	2.25	2.48	1.66
Plan	2.70	3.53	2.92	2.37	2.25.

Compare the three methods at the 5% level of significance. Which method would you recommend?

4.9 Using the data of Problem 4.5, test the null hypothesis that the standard deviation is 0.4 against the two-tailed alternative hypothesis that the standard deviation is not equal to 0.4.

4.10 Using the diet data of Problem 4.6, test equality of the variances of the two populations.

4.11 (MINITAB) Using the data of Problem 4.5, test the normality of the population distribution.

4.3   Matters arising

This section takes up the discussion of the problems of interpretation which have been raised in Sections 4.1 and 4.2, and resolved there in a commonsense way. In particular, we discuss the definition of the tails in a two-tailed test and the associated two-tailed *p*-value problem, and the problem of achieving a predetermined significance level for a test statistic whose distribution under the null hypothesis is of discrete type. Attention is also given to the probability of Type II error, the choice of sample size, and the connection between confidence intervals and tests of hypotheses.

4.3.1   *One tail or two?*

Many textbooks, including this one, introduce tests of hypotheses with examples of one-tailed tests, and then develop the two-tailed structure as a natural combination of the two one-tailed tests – as is required when deviations in either direction of the values of the test statistic cast doubt on the truth of the null hypothesis. In directing attention to the relevant one-tail deviation, the description of the problem is weighted as a declaration of interest only in one type of deviation, with an unstated assumption that deviation in the other direction must be due entirely to random fluctuations. The possibility that doubt may exist about the truth of this assumption has led some statisticians to suggest that the two-tailed test should, in general, be preferred. (Note that this suggestion does not apply to those tests – such as ONEWAY – which measure all deviations from the null hypothesis using a single tail of the test-statistic.)

─────── EXAMPLE 4.19 ───────────────────────────────

Suppose we are to compare a treatment against a control using a paired *t*-test – for example, an application of a fungicide against no application – and that positive values of the test-statistic *S* favour the treatment. In order to contrast the one- and two-tailed alternatives to the null hypothesis of no difference between the treatment and the control, suppose we have a sample of $n = 10$ matched pairs and use a significance level $\alpha = 0.05$. Then the test statistic S has the $t(9)$ distribution, so that tables supply

$$\Pr(S \geq 1.833 \mid H_0) = 0.05$$

and, using the equi-tail solution for the two-tailed test,

$$\Pr(\mid S \mid \geq 2.262 \mid H_0) = 0.05$$

The effect of these calculations is that an investigator confident that the treatment cannot have a negative effect will use the one-tailed version and declare in favour of the treatment if the observed value s_0 of S satisfies $s_0 \geq 1.833$, but the investigator believing a two-tailed test is necessary will require the greater deviation $s_0 \geq 2.262$ to reject H_0 and conclude the treatment is effective. For an observed value s_0 satisfying $1.833 \leq s_0 < 2.262$ the two procedures result in opposite conclusions.

Rephrasing in terms of p-values, suppose that the observed value s_0 is positive. Then the p-value for the one-tailed test is $p_1 = \Pr(S \geq s_0 \mid H_0)$, and, using equi-tails for the two-tailed test, $p_2 = \Pr(\mid S \mid \geq s_0 \mid H_0)$. By the symmetry of the t-distribution, $p_2 = 2p_1$, and for an investigator setting a personal Type I error probability p in the range $p_1 < p < p_2$, the choices of test procedure will determine opposite conclusions.

The view that a two-tailed test is a combination of two one-tailed tests can be seen in its most natural form as follows. Suppose we are comparing two treatments A and B, with a null hypothesis H_0 of no difference between the treatments, using a test statistic S with the two-tailed critical region $\{s: s \leq -c \text{ or } s \geq c\}$, where $\Pr(\mid S \mid \geq c \mid H_0) = \alpha$. Taking positive values of S to favour treatment A, the conclusion of the test, based on an observed value s_0 of S, can be expressed, symmetrically, as: prefer A if $s_0 \geq c$; prefer B if $s_0 \leq -c$; and declare the test inconclusive if $-c < s_0 < c$. In this form the two-tailed test at level α is seen as the combination of two one-tailed tests at level $\alpha/2$, the left tail having the alternative hypothesis that B is preferred, and the right tail that A is preferred.

4.3.2   Defining two tails

Suppose it is determined that a test of the null hypothesis H_0, against the alternative hypothesis H_1, requires a critical region containing both tails of the distribution (under H_0) of the test statistic S. We now describe three principal methods by which a two-tailed critical region achieving the pre-assigned significance level α may be defined, and apply the methods to the definition of a two-tailed p-value based on an observed value s_0 of S.

The three methods (equi-tail, equi-distant and equi-ordinate) are introduced in Examples 4.20 and 4.21 – in Example 4.20 for a test statistic with a continuous type distribution, with Example 4.21 covering the additional problems that arise when the test statistic has a discrete type distribution. As will soon become apparent, the three methods specify the same critical regions (and p-values) when the test statistic has a distribution symmetric around its mean – this explains the choice of a non-symmetric distribution in Example 4.20.

———— EXAMPLE 4.20 ————————————————————

Suppose the distribution under H_0 of the test statistic S has p.d.f. $g(s) = 12s^2(1 - s)$ for $0 < s < 1$, so that the distribution function $G(s) = s^3(4 - 3s)$ for $0 < s < 1$, and the expectation of S under H_0 is

$$E(S|H_0) = \int_0^1 12s^3(1 - s)\, ds = 0.6$$

Suppose, further, that the null hypothesis is such that deviation from the mean 0.6 in either direction cast doubt on the truth of H_0.

The equi-tailed method specifies a critical region $\{s:\ s \leqslant a \text{ or } s \geqslant b\}$ where

$$\Pr(S \leqslant a\,|\,H_0) = G(a) = 0.025$$

and

$$\Pr(S \geqslant b\,|\,H_0) = 1 - G(b) = 0.025$$

The reader may check $G(0.194) \approx 0.02496 \approx 0.025$, and $G(0.932) \approx 0.9747 \approx 0.975$ so the 5% equi-tailed critical region is $\{s:\ s \leqslant 0.194 \text{ or } s \geqslant 0.932\}$.

The equi-distance method is to measure equally deviations on either side of the expectation 0.6 of S under H_0, and take a critical region of the form $\{s:\ |s - 0.6| \geqslant a\}$. Hence we require the value a such that $\Pr(S \leqslant 0.6 - a$ or $S \geqslant 0.6 + a\,|\,H_0) = 0.05$, and trial and error soon yields the values $G(0.234) = 0.0423 = \Pr(S \leqslant 0.6 - a)$, and $G(0.966) = 0.9933 = 1 - \Pr(S \geqslant 0.6 + a) = 1 - 0.0067$, so with $a = 0.366$ we have the equi-distance 5% critical region $\{s:\ s \leqslant 0.234$ or $s \geqslant 0.966\}$.

The equi-ordinate method specifies the critical region to be of the form $\{s:\ s \leqslant a$ or $s \geqslant b\}$ where $g(a) = g(b)$ – that is, equal ordinates for the p.d.f. of the test statistic. A small amount of algebra is useful. To obtain $g(a) = g(b)$ we require

$$a^2(1 - a) = b^2(1 - b)$$

or

$$(a - b)(a + b - a^2 - ab - b^2) = 0$$

so that, given the value a, we calculate b as the root of the quadratic

$$b^2 - (1 - a)b + a^2 - a = 0$$

that is, $b = (1 - a \pm \sqrt{1 + 2a - 3a^2})/2$, using the positive root to obtain $0 < b < 1$.

Thus we find, for $a = 0.229$, that $b \approx 0.956$, $G(a) \approx 0.0398$, and $1 - G(b) \approx 1 - 0.989 \approx 0.011$. Hence the 5% equi-ordinate critical region is $\{s:\ s \leqslant 0.229$ or $s \geqslant 0.955\}$.

Calculating the two-tailed p-value for an observed $s_0 = 0.8$, say, presents no problem for the nimble fingered.

The equi-tailed p-value is $2(1 - G(0.8)) = 2(1 - 0.8192) = 0.3616$ (since $s_0 = 0.8 > 0.6$ is in the right tail); the equi-distant p-value is given by $Pr(S \geqslant 0.8 \mid H_0) + Pr(S \leqslant 0.4 \mid H_0) = 1 - G(0.8) + G(0.4) = 1 - 0.8192 + 0.1792 = 0.36$; and, using $a = s_0 = 0.8$, then $b \approx 0.512$ and the equi-ordinate p-value is $1 - G(0.8) + G(0.512) = 1 - 0.8192 + 0.3307 = 0.51$.

———— EXAMPLE 4.21 ————————————————————

This example illustrates the interpretation of two-tailed critical regions and p-values when the test statistic S has a discrete distribution. Suppose that S has the binomial distribution $b(20, 0.4)$ under the null hypothesis H_0. Using the PDF and CDF commands for the distribution $b(20, 0.4)$, and the relation $Pr(S \geqslant k \mid H_0) = 1 - Pr(S \leqslant k - 1 \mid H_0)$ to determine the right-tailed probabilities, we obtain

k	$Pr(S = k \mid H_0)$	$Pr(S \leqslant k \mid H_0)$	$Pr(S \geqslant k \mid H_0)$
0	0.00004	0.00004	1.00000
1	0.00049	0.00052	0.99996
2	0.00309	0.00361	0.99948
3	0.01235	0.01596	0.99639
4	0.03499	0.05095	0.98404
5	0.07465	0.12560	0.94905
6	0.12441	0.25001	0.87440
7	0.16588	0.41589	0.74999
8	0.17971	0.59560	0.58411
9	0.15974	0.75534	0.40440
10	0.11714	0.87248	0.24466
11	0.07099	0.94347	0.12752
12	0.03550	0.97897	0.05653
13	0.01456	0.99353	0.02103
14	0.00485	0.99839	0.00647
15	0.00129	0.99968	0.00161
16	0.00027	0.99995	0.00032
17	0.00004	0.99999	0.00005
18	0.00000	1.00000	0.00001
19	0.00000	1.00000	0.00000
20	0.00000	1.00000	0.00000

A survey of these columns of probabilities immediately identifies the problems – there are no equi-ordinate values of the probability function, no (non-trivial) equi-tail probabilities, and no pair of tails with combined probability 0.05. In particular, therefore, any attempt to define a 5% two-tailed critical region is doomed to failure. One solution is to adapt the conservative approach and find a critical region R of the required type such that $\alpha = Pr(S \in R \mid H_0)$ is maximal subject to $\alpha \leqslant 0.05$.

Of the three methods, only the equi-tail version presents a serious problem. We observe, by inspecting the last two columns, that the region $R = \{s: s \leqslant 3 \text{ or } s \geqslant 13\}$, with $\text{Pr}(S \leqslant 3 \text{ or } S \geqslant 13 \mid H_0) = 0.01596 + 0.02103 = 0.03699$ is the closest approximation to an equi-tailed 5% critical region.

The equi-distance critical region is most readily interpreted. Since $E(S \mid H_0) = 8$, we require the smallest value a such that

$$\text{Pr}(\mid S - 8 \mid \geqslant a \mid H_0) = \text{Pr}(S \leqslant 8 - a \text{ or } S \geqslant 8 + a \mid H_0) \leqslant 0.05$$

and the tables indicate $a = 5$, with $R = \{s: s \leqslant 3 \text{ or } s \geqslant 13\}$ as the equi-distance critical region.

In order to apply the equi-ordinate method, we first require the concept of equi-ordinate regions for a discrete distribution. These are defined in terms of the probability function to be regions of the form $\{s: \text{Pr}(S = s \mid H_0) \leqslant x\}$, where $0 < x < 1$. In practice, the only values x that need be examined are $x = \text{Pr}(S = k \mid H_0)$ where k ranges over the values of the test-statistic. For example, $k = 3$ defines the region $\{s: \text{Pr}(S = s \mid H_0) \leqslant 0.01235\} = \{s: s \leqslant 3 \text{ or } s \geqslant 14\}$, and $k = 13$ provides $\{s: \text{Pr}(S = s \mid H_0) \leqslant 0.01456\} = \{s: s \leqslant 3 \text{ or } s \geqslant 13\}$. Checking the probabilities of these regions for $k = 0, 1, 2, \ldots, 20$ reveals that the conservative approximation to the equi-ordinate 5% critical region is given by $k = 13$, that is $\{s: s \leqslant 3 \text{ or } s \geqslant 13\}$.

Note: it is fortuitous that the three methods define the same critical region – see problem 4.13 below.

We illustrate the three definitions of two-tailed p-values using an observed value $s_0 = 12$. Again only the equi-tailed method gives rise to any difficulty. A strict application of the equi-tail notion suggests that the two-tailed p-value be taken as $2\text{Pr}(S \geqslant 12 \mid H_0) = 0.11306$, but this number is not achieved as the probability of any region of the form $\{s: s \leqslant a \text{ or } s \geqslant 12\}$. To ensure that the p-value is an achievable probability we define the two-tailed p-value, based on $s_0 = 12$, to be $\text{Pr}(S \leqslant a \mid H_0) + \text{Pr}(S \geqslant 12 \mid H_0)$ where a is the largest value for which $\text{Pr}(S \leqslant a \mid H_0) \leqslant \text{Pr}(S \geqslant 12 \mid H_0)$ – that is, the obtained tail probability plus the largest probability in the other tail not exceeding the obtained tail probability. Thus, the achievable two-tailed p-value, modifying the equi-tail method, is $0.05095 + 0.05653 = 0.10748$ – obtained with $a = 4$.

As with the equi-distance critical region the equi-distance p-value is readily obtained as $\text{Pr}(\mid S - 8 \mid \geqslant \mid 12 - 8 \mid \mid H_0) = \text{Pr}(S \leqslant 4 \text{ or } S \geqslant 12 \mid H_0) = 0.10748$, as above.

Finally, for the equi-ordinate p-value we require the probability attached to the region $\{s: \text{Pr}(S = s \mid H_0) \leqslant \text{Pr}(S = 12 \mid H_0)\} = \{s: s \leqslant 3 \text{ or } s \geqslant 12\}$ – that is, the collection of values each less probable than the observed value. Hence, the equi-ordinate two-tail p-value is $0.01596 + 0.05653 = 0.07249$.

The discussion in Examples 4.20 and 4.21 extends naturally to any test statistic whose p.d.f./p.f. is (essentially) unimodal, and is an increasing function to the left of the mode and decreasing to the right – as is typically the case with the standard distributions met in practice. Under this condition on the distribution the equi-ordinate critical regions are easily seen to be of the expected two-tailed form $\{s: s \leq a \text{ or } s \geq b\}$ for suitable values a and b.

The reader anxious for a recommendation of preference between the three methods is referred to the subsection on Type II error and to Chapter 9. For a test statistic whose p.d.f./p.f. is symmetric about the mean, it is clear that the three methods are equivalent – a quick sketch of a symmetric p.d.f. may help – and no choice is necessary. As a general suggestion, particularly for discrete distributions, the equi-ordinate method is preferred when the distribution is markedly non-symmetric – see, for example, the legend of Table 26 in the Cambridge Tables. It is amusing to note that, whenever a normal distribution approximation is used, the symmetry of the normal p.d.f. produces an equi-tailed solution.

The three methods discussed above for resolving the two-tail problem each have an intuitive appeal, and the reader is likely to find any one of them invoked in an informal discussion. Other methods of defining a two-tailed critical region arise, in special circumstances, from the application of a general principle for specifying a test procedure – see Section 9.3.

4.3.3   *Randomized test*

Example 4.21 shows that, when the test statistic has a discrete distribution, it may not be possible to define, for a given level of significance α_0, a critical region R_0 achieving the value α_0 as the probability of Type I error. The solution offered was to adopt the practical and conservative approach – that is, to obtain a $100\alpha\%$ critical region R of the required type where α is maximal subject to $\alpha \leq \alpha_0$.

An alternative to the conservative method is the randomized test structure which provides a rejection procedure having exact probability α_0 of committing a Type I error. In this method, the null hypothesis is rejected if the observed value s_0 of the test statistic lies in the region R, and a supplementary rejection rule is applied if s_0 lies 'just outside' R. The form of the rejection rule may be visualized as the rejection of the null hypothesis if the toss of a suitably biased coin results in a head. The following example shows the method in action.

——— EXAMPLE 4.22 ———

Suppose it is asserted in the media that 40% of the population are opposed to a proposed new road development. To test this claim the opinion of 20 randomly selected individuals was obtained. For our illustration we test the null hypothesis H_0 that the binomial parameter $p = 0.4$ against the alternative hypothesis H_1 that $p > 0.4$, using the number S in the sample who oppose the development as the test statistic. Then, S has the distribution $b(20, 0.4)$ under H_0, and the tables of Example 4.21 give $\Pr(S \geq 12 \mid H_0) = 0.05653 > 0.05$ and $\Pr(S \geq 13 \mid H_0) = 0.02103 < 0.05$. With

$\alpha_0 = 0.05$, this gives $R = \{s : s \geqslant 13\}$ and $\alpha = 0.02103$. The null hypothesis is rejected, based on the observed value s_0, if

1. $s_0 \geqslant 13$ – that is, $s_0 \in R$, or
2. $s_0 = 12$, and the toss of the coin results in a head.

If γ denotes the probability that a single toss of the coin results in a head, then the probability of Type I error is
$\Pr(S \geqslant 13 \,|\, H_0) + \gamma \Pr(S = 12 \,|\, H_0) = 0.02103 + 0.0355\gamma$, and γ is determined to ensure that this probability is the required value $\alpha_0 = 0.05$ – that is, $\gamma = 0.816$.

Although the randomized procedure generates a unified theory, the emphasis it places on the supplementary rejection rules ensures it is not used in practical decision making. In Example 4.22 approximately 60% of the probability of rejection of the null hypothesis rests on the toss of a coin.

4.3.4   *Type II error*

Suppose we are testing a null hypothesis H_0 against an alternative hypothesis H_1, using a sample of size n, a test statistic S and an exact $100\alpha\%$ critical region R, so that $\Pr(S \in R \,|\, H_0) = \alpha$. The conclusion of the test procedure is to accept or reject the null hypothesis – a decision which embraces two possible types of error:

- Type I error – reject H_0 when H_0 is true or
- Type II error – accept H_0 when H_0 is false.

In the formulation of the test, the alternative hypothesis determines the shape of the critical region, and the complete specification is obtained from the relation $\Pr(\text{Type I error}) = \Pr(S \in R \,|\, H_0) = \alpha$. In the examples of Section 4.1 we calculated the probability $\Pr(S \notin R \,|\, H_1)$ of Type II error, and noted that this probability is not always pleasingly small. The next example relates the probability of Type II error to the sample size, and shows that the probability may be made arbitrarily small by increasing the sample size.

——— **EXAMPLE 4.23** ———

Suppose the population distribution function is $N(\mu, 1)$ and that we test the null hypothesis H_0 that $\mu = 0$, against the alternative hypothesis H_1 that $\mu = 1$, at the 5% level of significance. Using a sample of size n, the test statistic $S = \sqrt{n}\bar{X}$ has the distribution $N(0, 1)$ under H_0, and the critical region R of the form $\{s : s \geqslant a\}$ is specified by $\Pr(S \geqslant a \,|\, \mu = 0) = 0.05$, so that $a = 1.64$.

For the probability of Type II error we calculate

$$\Pr(S < 1.64 \mid \mu = 1) = \Pr((\bar{X} - 1)\sqrt{n} < 1.64 - \sqrt{n} \mid \mu = 1)$$
$$= \Pr(Z < 1.64 - \sqrt{n}) = \Phi(1.64 - \sqrt{n})$$

where Z and Φ are the standard $N(0, 1)$ random variable and distribution function, respectively. Clearly, as the sample size increases, the value $\Phi(1.64 - \sqrt{n})$ decreases to zero. Some numerical values obtained from tables:

n	$\Phi(1.64 - \sqrt{n})$
4	$\Phi(-0.36) \approx 0.3594$
9	$\Phi(-1.36) \approx 0.0869$
16	$\Phi(-2.36) \approx 0.0091.$

Thus, with $n = 9$, for example, 95% of the values of the test statistic lie in the region $\{s: s < 1.64\}$ and 5% in the critical region $\{s: s \geq 1.64\}$ when H_0 is true, but, when the alternative hypothesis H_1 is true, the weight of the distribution of values is reversed – 8.69% lie in the region $\{s: s < 1.64\}$ and 91.31% in the critical region.

Similar considerations, involving only additional algebra, will apply to a $N(\mu, \sigma^2)$ distribution to test the null hypothesis that $\mu = \mu_0$ against the alternative $\mu = \mu_1$, provided the numerical values of μ_0, μ_1 and σ are known – see problem 4.14.

──────── EXAMPLE 4.24 ────────────────────────────────

Suppose we test the binomial parameter p, taking the null hypothesis H_0 to be $p = 0.5$ and the alternative hypothesis to be $p = 0.64$, using a sample of size n and the level of significance $\alpha = 0.01$. The test statistic S has the binomial distribution $b(n, 0.5)$ under H_0, and the critical region has the shape $\{s: s \geq a\}$. Applying a normal approximation, without a continuity correction, we require

$$\Pr(S \geq a \mid p = 0.5) = \Pr\left(Z \geq \frac{a - n/2}{\sqrt{n/4}}\right) = 0.01$$

and tables supply $(2a - n)/\sqrt{n} \approx 2.3263$, giving a relation between the values a and n.

Suppose we now require that the probability of Type II error is also less than 0.01, then

$$\Pr(S < a \mid p = 0.64) = \Pr\left(Z < \frac{a - 0.64n}{\sqrt{n \times 0.64 \times 0.36}}\right) = 0.01$$

requires $(a - 0.64n)/0.48\sqrt{n} = -2.3263$.

Solving these two equations in the values a and n, we find, on eliminating a, that $0.28\sqrt{n} \approx 2.3263(1 + 0.96)$ giving $n \approx 265$.

──

The Type II error probability has an important role to play in the comparison of test procedures. Suppose that we have identified two critical regions, both at the same level of significance, for a test of the null hypothesis H_0 against the alternative hypothesis H_1. How should we discriminate between the regions? The criterion we adopt is suggested by the remarks in Example 4.23 – we prefer the region which provides the smaller Type II error probability. This criterion is the key to the Neyman–Pearson theory of hypothesis testing introduced in Chapter 9.

In the next example we find the Type II error probabilities for the equi-tailed, equi-distance and equi-ordinate critical regions obtained in Example 4.20.

──────── **EXAMPLE 4.25** ────────────────────────────────

The three types of critical region were obtained, in Example 4.20, for a test statistic S whose distribution under the null hypothesis had the p.d.f. $g(x) = 12x^2(1 - x)$, for $0 \le x \le 1$. This p.d.f. is the case $\theta = 2$ of the family of p.d.f.

$$g_\theta(x) = (\theta + 1)(\theta + 2)x^\theta(1 - x)$$

where $0 \le x \le 1$ and $\theta > 0$ – that is, the beta distribution $B(\theta + 1, 2)$.

In order to calculate the probability of Type II error, we require the distribution of the test statistic under the alternative hypothesis H_1. Suppose, first, that S has the beta distribution $B(2, 2)$ under the alternative hypothesis, with p.d.f. $g_1(x) = 6x(1 - x)$. Then, for any critical region $R = \{s: s \le a \text{ or } s \ge b\}$, the probability of Type II error is given by

$$\Pr(a < S < b | H_1) = \int_a^b 6x(1 - x)\, dx = [x^2(3 - 2x)]_a^b$$

Hence, substituting the values a and b, the probability of Type II error for the three critical regions are

- 0.8885 – equi-tail $(a = 0.194, b = 0.932)$
- 0.8580 – equi-distance $(a = 0.234, b = 0.966)$ and
- 0.8608 – equi-ordinate $(a = 0.229, b = 0.955)$.

Similarly, if the p.d.f. under the alternative hypothesis is $g_3(x) = 20x^3(1 - x)$, the probability of Type II error is given by

$$\int_a^b 20x^3(1 - x)\, dx = [x^4(5 - 4x)]_a^b$$

with values 0.9537 (equi-tail), 0.9770 (equi-distance) and 0.9703 (equi-ordinate).

For each alternative hypothesis, there are only small differences in the Type II error probabilities – by a short head, the equi-distance region is preferred for $g_1(x)$, and the equi-tail region for $g_2(x)$.

───

Example 9.14 in Section 9.3 gives a more realistic example – a comparison of the equi-tail region (Example 4.15 of Section 4.2) and the likelihood ratio region for a test on the variance of a normal distribution.

4.3.5   *Confidence intervals and tests of hypothesis*

The reader will have noted that the test command ZTEST and the confidence interval command ZINT are based on a common statistic, as also are the commands TTEST and TINT. We re-examine ZTEST and ZINT in Example 4.26, and then propose a general method of defining a confidence interval from a hypothesis test.

--------- EXAMPLE 4.26 ---------

Suppose the sample random variables $X_1, X_2, \ldots, X_n$ have the distribution $N(\mu, \sigma_0^2)$, where σ_0 is known, and let α be a prescribed level of significance. With the value c defined by $\Pr(|\bar{X} - \mu| \geqslant c\sigma_0/\sqrt{n}) = \alpha$, and obtained from tables of the normal distribution, the language of Section 3.3 asserts that the random interval $(\bar{X} - c\sigma_0/\sqrt{n}, \bar{X} + c\sigma_0/\sqrt{n})$ contains the true value of μ with probability $1 - \alpha$, and that the obtained interval $(\bar{x} - c\sigma_0/\sqrt{n}, \bar{x} + c\sigma_0/\sqrt{n})$ is a $100(1 - \alpha)\%$ confidence interval for μ.

Turning now to a test of the null hypothesis H_0 that $\mu = \mu_0$, where μ_0 is given, against the two-tailed alternative hypothesis that $\mu \neq \mu_0$, we accept H_0 if the observed value $\bar{x}$ satisfies $|\bar{x} - \mu_0| \leqslant c\sigma_0/\sqrt{n}$. Regarding the observed value $\bar{x}$ as fixed, we identify the confidence interval with the set of values μ_0 for which the null hypothesis H_0 that $\mu = \mu_0$ is accepted at the $100\alpha\%$ level of significance.

Suppose, more generally, that the distribution function contains a numerical parameter θ, and that we have constructed a test procedure for the null hypothesis H_0 that $\theta = \theta_0$ against the two-tailed alternative hypothesis that $\theta \neq \theta_0$. Then a $100(1 - \alpha)\%$ confidence interval is defined to be the set (assumed to be an interval!) of all values θ_0 such that the observed data does not lead to the rejection of the null hypothesis that $\theta = \theta_0$ at the $100\alpha\%$ level of significance.

4.3.6   *Problems*

4.12 Find, for Example 4.20, the equi-tailed, equi-distance and equi-ordinate 2.5% critical regions. Find also the three corresponding *p*-values if the observed value of the test statistic is $s_0 = 0.5$.

4.13 Use the method discussed in Example 4.21 to define the equi-tail, equi-distance and equi-ordinate 5% critical regions when the test statistic has the binomial distribution $b(20, 0.15)$ under the null hypothesis. Find also the three corresponding two-sided *p*-values when the observed value of the test statistic is $s_0 = 4$.

4.14 Let $X_1, X_2, \ldots, X_n$ be a random sample from a population distribution $N(\mu, \sigma_0^2)$, where σ_0 is known. Write down the 5% critical region for a test of the null hypothesis H_0 that $\mu = \mu_0$ against the alternative hypothesis H_1 that $\mu = \mu_1$, where $\mu_1 < \mu_0$. Show that the probability of Type II error is

$$1 - \Phi(-1.64 + \sqrt{n}(\mu_0 - \mu_1)/\sigma_0)$$

and obtain its value when $\mu_0 = 3$, $\mu_1 = 1$ and $\sigma_0 = 2$ in the cases $n = 9$, $n = 16$.

4.15 For Example 4.25 find the probability of Type II error for the equi-tail, equi-distance and equi-ordinate critical regions when the test statistic had the p.d.f. $g_4(x) = 30x^4(1-x)$ under the alternative hypothesis.

4.16 Show that the probability of Type II error for the randomized test of Example 4.22 is

$$\Pr(S \leqslant 11 \mid H_1) + (1-\gamma)\Pr(S = 12 \mid H_1)$$

and obtain the numerical value when $p = 0.6$.

4.17 In the context of Example 4.22, find the value γ which determines a randomized test at the 2.5% level of significance.

4.4   Solutions to problems

4.1 Small values of $\bar{X}$ favour the alternative hypothesis, so that the critical region R has the shape $\{x: x \leqslant c\}$. Now $\Pr(\bar{X} \in R \mid H_0) = \alpha$ implies

$$\alpha = \Pr(\bar{X} \leqslant c \mid H_0) = \Pr\!\left(Z \leqslant \frac{(c-3)}{\sigma}\sqrt{n}\right) = \Pr(Z \leqslant z)$$

where $Z = (\bar{X}-3)\sqrt{n}/\sigma$ has the $N(0,1)$ distribution. Hence $c = 3 + z\sigma/\sqrt{n}$.
For $\alpha = 0.01$, $z = -2.3263$, and thus $c \approx 1.837$. The probability of Type II error is given by

$$\Pr(\bar{X} \geqslant c \mid \mu = 1) = \Pr(Z \geqslant (1.837-1) \times 2)$$
$$= \Pr(Z \geqslant 1.674) \approx 0.0471$$

4.2 The critical region has the shape $\{x: x \geqslant c\}$ or, with $S = n\bar{X}$, $\{s: s \geqslant nc\}$. Now S has the distribution $b(20, 0.4)$ under the null hypothesis and tables supply

$$\Pr(S \geqslant 12 \mid H_0) \approx 1 - 0.9435 = 0.0565$$
$$\Pr(S \geqslant 13 \mid H_0) \approx 1 - 0.9790 = 0.0210$$

so the required critical region is $\{s: s \geqslant 13\}$.
The probability of Type II error is given by $\Pr(S \leqslant 12 \mid H_1) = 0.8684$ from tables, and the p-value $= \Pr(S \geqslant 9 \mid H_0) = 1 - 0.5956 = 0.4044$.

4.3 Test statistic $S = \bar{X}$; critical region $R = \{s: |s - 1.5| \geqslant c\}$.

$$0.05 = \Pr(S \in R \mid H_0) = \Pr(|Z| \geqslant c\sqrt{10}/1.5), \text{ and tables supply}$$
$$c = 1.96 \times 1.5/\sqrt{10} \approx 0.9297.$$

Type II error probability is $\Pr(S \notin R \mid H_1)$ – that is,

$$\Pr(-c + 1.5 < S < c + 1.5) = \Pr((0.5703 - \mu)\sqrt{10}/1.5 < Z < (2.4297 - \mu)\sqrt{10}/1.5)$$
$$= \Phi((2.4297 - \mu)\sqrt{10}/1.5) - \Phi((0.5703 - \mu)\sqrt{10}/1.5)$$

when $\mu = 2$: $\Phi(0.9059) - \Phi(-3.041) = 0.8162$
        $\mu = 3$: $\Phi(-1.2023) - \Phi(-5.1223) = 0.1146$

The p-value when $\bar{x} = 1.7$ is $2\Pr(\bar{X} \geqslant 1.7) = 2\Pr(Z > (1.7 - 1.5)\sqrt{10}/1.5)$
$$= 2(1 - \Phi(0.4216)) = 0.6732$$

4.4 Let C1 contain the numbers 0 0.1 0.2 ... 3.0 by entering 0:3/0.1
Let $C2 = (2.4297 - C1)^*(10^{**}0.5)/1.5$ and

$$C3 = (0.5703 - C1)^*(10^{**}0.5)/1.5$$

Then CDF C2 C4;   and   CDF C3 C5;
        NORMAL 0 1.   NORMAL 0 1.

Let $C6 = C4 - C5$ gives the probability of Type II error
PLOT C6 C1 produces the 'graph'

4.5 Calculation reveals that $\bar{x} = 7.77$, $s = 0.479$.
(i) In the notation of ZTEST, calculate $z_0 = (\bar{x} - 8)\sqrt{10}/0.4 \approx -1.82$ and find
the p-value $= \Pr(Z \leqslant -1.82) = 0.035$. Hence reject the claim at the 5% level;
(ii) In the notation of TTEST, calculate $t_0 = (\bar{x} - 8)\sqrt{10}/0.479 \approx -1.52$ and
find the p-value $= \Pr(T_9 \leqslant -1.52) = 0.082$. Hence accept the claim at the 5%
level.

4.6 As a paired TTEST let the x-data be the diet A − diet B values, and calculate
$\bar{x} = 0.82$ and $s = 0.382$. Conducting a two-tail test on the x-data, calculate
$(\bar{x} - 0)\sqrt{10}/s = 6.78$ and find the p-value $= \Pr(|T_9| \geqslant 6.78) = 0.0001$. Equality
of weight gain is rejected and diet A is preferred (as it gives greater weight
gain).

4.7 Analyse as in TWOSAMPLE, with diet A as the x-data and diet B the y-data.
Then, $\bar{x} = 7.26$ and $s_1 = 0.737$, and $\bar{y} = 6.44$ and $s_2 = 0.826$. Testing equality of
the means against the two-tailed alternative:
(i) Assuming equal standard deviations, calculate

$$T = \frac{(\bar{x} - \bar{y})\sqrt{nm(n + m - 2)}}{\sqrt{(n + m)[(n - 1)s_1^2 + (m - 1)s_2^2]}} = \frac{(7.26 - 6.44)\sqrt{10.10.18}}{\sqrt{20 \times 9(0.737^2 + 0.826^2)}} \approx 2.34$$

with p-value $= \Pr(|T_{18}| \geqslant 2.34) = 0.031$. Hence reject at the 5% level.
(ii) Without assuming equal standard deviations, calculate

$$T_v = \frac{(\bar{x} - \bar{y})}{\sqrt{\dfrac{s_1^2}{n} + \dfrac{s_2^2}{m}}} = \frac{7.26 - 6.44}{\sqrt{\dfrac{0.737^2 + 0.826^2}{10}}} \approx 2.34$$

with the degrees of freedom v given by

$$v = \left[\frac{(0.737^2 + 0.826^2)/100}{(0.737^4 + 0.826^4)/100 \times 9}\right] = [17.77] = 17$$

Hence the p-value $= \Pr(|T_{17}| \geqslant 2.34) = 0.032$, and equality is rejected at the
5% level.

4.8   In the notation of ONEWAY, calculate $\bar{x}_1 = 2.09$, $\bar{x}_2 = 2.054$, $\bar{x}_3 = 2.754$ and $\bar{x} = 2.299$. Also, the observed values $SS_H = \sum (\bar{x}_i - \bar{x})^2 = 1.554$ and $SS_e = \sum\sum (x_{ij} - \bar{x}_i)^2 = 2.386$, and the value of the F-statistic $= 3 \times 4 \times 1.554/ 2 \times 2.386 = 3.908$, giving the p-value $\Pr(F(2, 12) > 3.908) = 0.049$. Hence equality of the three means is rejected at the 5% level.

From t-distribution tables, $\Pr(|T_{12}| > 2.179) = 0.05$, and for pairwise comparisons we reject equality of the i and j level means if $|\bar{x}_i - \bar{x}_j| > 2.179 \sqrt{MS_e/2} = 2.179 \sqrt{SS_e/24} = 0.687$. Hence, we declare that calorie counting produces a greater weight loss than diet B $(2.754 - 2.054 = 0.70 > 0.68)$, but detect no difference (*each* at the 5% level) between the other pairs. Note that the *family error rate* is 0.116.

4.9   The test statistic here is $W = (n - 1) S^2/\sigma^2$ which has the chi-squared distribution with $(n - 1)$ degrees of freedom. From tables, with $n = 10$, $\Pr(W < 2.70) = 0.025 = \Pr(W > 19.02)$, and the observed value of $W = 9 \times (0.479)^2/0.4^2 = 12.96$ does not lie in either tail. Hence accept $\sigma = 0.4$ at the 5% level.

4.10   The test statistic here is $F = (m - 1)S_1^2/(n - 1)S_2^2$ which has the $F(n - 1, m - 1)$ distribution. MINITAB'S CDF command gives, with $n = m = 10$, $\Pr(F < 0.2484) = 0.025 = \Pr(F > 4.026)$, and the observed value of $F = 0.737^2/0.826^2 = 0.7961$ does not lie in either tail. Hence, accept equality of the variances at the 5% level.

4.11   With the data in C1 the commands NSCORES C1 C2 and then CORRELATION C1 C2 give the value 0.982. This value is greater than the 5% tabulated value of 0.9176, so that the normality of the distribution is accepted at the 5% level.

4.12   Trial and error yields:

equi-tailed region $= \{s: s \leqslant 0.1522 \text{ or } s \geqslant 0.9528\}$, since $G(0.1522) \approx 0.0125$ and $G(0.9528) \approx 0.9875$;
equi-distance region $= \{s: |s - 0.61| \geqslant 0.4056\}$, since $G(0.1944) \approx 0.025$, and $1 - G(1.0056) \approx 0$.
equi-ordinate region $= \{s: s \leqslant 0.18 \text{ or } s \geqslant 0.9719\}$, since $G(0.18) \approx 0.0202$, and $G(0.9719) \approx 0.9954$.
equi-tailed p-value $= 2G(0.5) \approx 0.625$.
equi-distance p-value $= G(0.5) + 1 - G(0.7) \approx 1.3125 - 0.6517$
                           ≈ 0.6608.
equi-ordinate p-value $= G(0.5) + 1 - G(0.809) \approx 1.3125 - 0.833$
                           ≈ 0.4795.

4.13   The equi-tailed region is $\{s: s = 0 \text{ or } s \geqslant 9\}$, with achieved significance level $\Pr(S = 0 | H_0) + \Pr(S \geqslant 9 | H_0) = 0.0387 + 1 - 0.9987 = 0.04$.

Similarly, the equi-distance region is $\{s: |s - 3| \geqslant 4\}$ with achievable probability $1 - \Pr(S \leqslant 6 | H_0) = 0.0220$. Note that this region is effectively $\{s: s \geqslant 7\}$, since $S \leqslant -1$ is impossible.

The equi-ordinate region with achieved probability $\leqslant 0.05$ is obtained from the p.d.f. and c.d.f tables. It is $\{s: s = 0 \text{ or } s \geqslant 9\}$.

The p-values based on $s_0 = 4$ are given by:

equi-tailed: $\Pr(S \leq 1 \text{ or } S \geq 4 \mid H_0) = 0.1756 + 0.3523 = 0.5279$
equi-distance: $\Pr(S \leq 2 \text{ or } S \geq 4 \mid H_0) = 0.3523 + 0.4049 = 0.7572$
equi-ordinate: $\Pr(S \leq 1 \text{ or } S \geq 4 \mid H_0) = 0.5279$

4.14 Test statistic $S = \sqrt{n}(\bar{X} - \mu_0)/\sigma$ with critical region $R = \{s: s \leq -1.64\}$ for a 5% left tail test.

Probability of Type II error $= \Pr(S \notin R \mid H_1)$

$$= \Pr(S \geq -1.64 \mid H_1) = \Pr(Z \geq -1.64 + \sqrt{n}(\mu_0 - \mu_1)/\sigma_0)$$
$$= 1 - \Phi(-1.64 + \sqrt{n}(\mu_0 - \mu_1)/\sigma_0)$$

where $Z = \sqrt{n}(\bar{X} - \mu_1)/\sigma_0$

Calculation: $n = \ \ 9$ requires $1 - \Phi(1.36) \approx 0.0869$
$n = 16$ requires $1 - \Phi(2.36) \approx 0.0091$

4.15 The distribution function of the test statistic under the alternative hypothesis is

$$G_4(x) = \int_0^x 30t^4(1 - t)\,dt = x^5(6 - 5x) \text{ for } 0 < x < 1$$

and, for a critical region $\{s: s \leq a \text{ or } s \geq b\}$, the probability of Type II error is $\Pr(a < S < b \mid H_1) = G_4(b) - G_4(a)$. Substituting for a and b, the three probabilities are:

equi-tailed: $G_4(0.932) - G_4(0.194) \approx 0.9409$
equi-distance: $G_4(0.966) - G_4(0.234) \approx 0.9808$
equi-ordinate: $G_4(0.955) - G_4(0.229) \approx 0.9700$

4.16 The null hypothesis is accepted if $S \leq 11$ or $S = 12$ and the coin falls tail up. Hence the result. Under H_1, with $p = 0.6$, S has the distribution $b(20, 0.6)$, so that $T = 20 - S$ has the distribution $b(20, 0.4)$.

Then $\Pr(S \leq 11 \mid H_1) + (1 - \gamma)\Pr(S = 12 \mid H_1)$
$$= \Pr(T \geq 9 \mid H_1) + (1 - 0.816)\Pr(T = 8 \mid H_1)$$
$$= 0.4044 + 0.184 \times 0.1747 \approx 0.4375$$

4.17 From Example 4.22 we see that $\{s: s \geq 13\}$ is the critical region with achieved significance $0.02103 < 0.025$. Thus the value of γ for a randomized test is given by $0.02103 + \gamma 0.0355 = 0.025$ – that is, $\gamma \approx 0.1118$.

Chapter Five

Tests related to the binomial distribution

5.1 The sign test

We first discuss a simple distribution-free statistic which may be used to draw inferences concerning the median of a continuous distribution. We recall that, for a continuous random variable X, the median of the distribution is a number m, such that $\Pr(X \leqslant m) = \Pr(X \geqslant m) = 1/2$.

Suppose $X_1, X_2, \ldots, X_n$ is a random sample of size n from a continuous distribution. Then the probability that any member exceeds the median is $1/2$ and hence the number S_+ of the sample members which exceed the median has a binomial distribution with index n and parameter $1/2$. This is the case *whatever the distribution* and thus the statistic S_+ is distribution-free. It can be used to test hypotheses or construct confidence intervals for the median, m. Similarly, the number S_- of the sample members below the median has a binomial distribution with index n and parameter $1/2$.

5.1.1 *One-sided tests of shift*

Suppose the null hypothesis is that $m = m_0$ and the alternative is that $m = m_1 > m_0$ – that is, the median is shifted to the right of the conjectured value m_0 (see Figure 5.1).

From Figure 5.1 it is clear that for any X_i, $\Pr(X_i > m_0 \mid m = m_1) > \Pr(X_i > m_1 \mid m = m_1)$, so it is more likely that an observation exceeds m_0 if the alternative is true. Hence the null hypothesis can be tested by examining if there is a significant number of sample values greater than m_0. If m_0 is subtracted from each sample value, a positive residue indicates that the value exceeds m_0. That is, if $z_i = x_i - m_0$, then S_+ is the number of positive z_i – this accounts for the name '*sign test*'.

─────── EXAMPLE 5.1 ───────────

A random sample of six from a continuous distribution yielded the values 0.6, 3.2, 1.7, 5.4, 2.3, 4.1. We use S_+ to test the hypothesis that the median is 1 against the alternative that the median is 2.

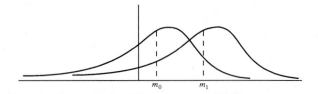

Figure 5.1

If 1 is subtracted from the sample values, the residues are −0.4, 2.2, 0.7, 4.4, 1.3, 3.1. Hence $S_+ = 5$. If the null hypothesis is true, S_+ has a binomial distribution with index 6 and parameter $1/2$:

$$\Pr(S_+ \geq 5 \mid 6, 1/2) = \binom{6}{5}\left(\frac{1}{2}\right)^6 + \binom{6}{6}\left(\frac{1}{2}\right)^6 = \frac{7}{64} \approx 0.11$$

Since the distribution of S_+ is discrete, it is not in general possible to attain prescribed significance levels. In Example 5.1, $\Pr(S_+ \geq 6 \mid 6, 1/2) = 0.0156 < 0.05$. Thus $S_+ = 5$ and $S_+ = 6$ bracket the 5% level.

5.1.2   *Power of the test*

An important characteristic of any test is its behaviour when the alternative hypothesis is true (see Chapter 4). We require the probability that the null hypothesis is rejected when the alternative is true. Suppose $\Pr(X > m_0 \mid m = m_1) = \theta$, then the distribution of S_+, when the alternative is true, is again binomial with index n and parameter θ.

Thus, if we have a random sample of size n from a continuous distribution with median m let the null hypothesis be $m = m_0$, the alternative be that $m = m_1 > m_0$ and the critical region be $S_+ \geq k$, then the significance level of the test is

$$\sum_{i=k}^{n} \binom{n}{i}\left(\frac{1}{2}\right)^n \tag{5.1}$$

and the power of the test is

$$\sum_{i=k}^{n} \binom{n}{i}\theta^i(1-\theta)^{n-i} \tag{5.2}$$

Thus to evaluate equation (5.2) we must know θ. This cannot be calculated without further information, for the alternative can be *any* continuous distribution with median m_1.

─────── EXAMPLE 5.2 ───────────────────────────────────

Suppose in Example 5.1 we take the critical region to be $S_+ \geq 5$. To judge the performance of this critical region when in fact the distribution of X is exponential with parameter λ, we have for $m_1 = 2$ say

$$\int_2^\infty \lambda\, e^{-\lambda x}\, dx = 1/2 \text{ or } \lambda = (1/2)\log_e 2$$

and

$$\theta = \int_1^\infty \lambda\, e^{-\lambda x}\, dx = e^{-\lambda} = 1/\sqrt{2} \approx 0.71$$

Hence $\Pr(S_+ \geq 5 \,|\, 6, 0.71) = \Pr(S_- \leq 1 \,|\, 6, 0.29) \approx 0.442$.

───

There is little to add concerning testing the null hypothesis $m = m_0$ against $m = m_1 < m_0$. In such a case, after subtracting m_0 from the sample values, a paucity of positive residues will suggest departure from m_0 – that is, we use the left tail of the distribution of S_+.

5.1.3   Two-sided tests of shift

Suppose $X_1, X_2, \ldots, X_n$ is a random sample of size n from any continuous distribution. The null hypothesis is that the median m of the distribution is m_0, the alternative that $m \neq m_0$. That is, we are testing whether this measure of location is shifted either to the right or to the left.

After subtracting m_0 from the sample values, either an excessive or an insufficient number of positive residues will cast doubt on the null hypothesis. Just as in the one-sided case, we can either just calculate the significance level of a result at least as extreme as the one observed or adopt a more definite attitude which causes us to reject or accept the null hypothesis at a (nominal) significance level.

─────── EXAMPLE 5.3 ───────────────────────────────────

The following observations are the result of drawing a random sample of 20 from a continuous distribution:

    37.0   31.4   34.4   33.3   34.9   31.6   31.3   34.6   32.6   31.6

    36.2   31.0   33.5   33.7   33.4   33.4   32.1   33.3   32.7   31.5

It is required to test the null hypothesis that the median of the distribution is 32 against the alternative that it is not 32. After subtracting 32 from each observation, we have 14 positive residues, and for the two-sided alternative and a symmetric distribution, we calculate

$$\Pr(S_+ \geq 14 \,|\, 20, 1/2) + \Pr(S_+ \leq 6 \,|\, 20, 1/2)$$

From tables, this is about $2(0.0577) = 0.115$.

For large n we may approximate the distribution $b(n, 1/2)$ with the distribution $N(\mu, \sigma^2)$, where $\mu = n/2$, $\sigma^2 = n/4$ (see Section 1.3).

─────── **EXAMPLE 5.4** ───────────────────────────────

For Example 5.3, $\mu = 10$, $\sigma = \sqrt{5}$ and using a continuity correction of $1/2$

$$\Pr(S_+ \geq 14 \,|\, 20, 1/2) = \Pr(Z \geq 1.565) = 0.0589$$

where Z has the $N(0, 1)$ distribution. This gives a significance level of $2(0.0589) = 0.118$, not very different from the result obtained in Example 5.3.

5.1.4   *Confidence intervals*

The order statistics may be used to construct confidence intervals. Suppose $X_1, X_2, \ldots, X_n$ is a random sample of size n from a continuous distribution with median m. Let $X_{(1)} \leq X_{(2)} < \cdots \leq X_{(n)}$ be the order statistics. If we take $(X_{(1)}, X_{(n)})$ as the confidence interval, then it will contain m unless:

1. $X_{(1)} > m$, that is, all n sample values exceed m – which happens with probability $(1/2)^n$.
2. $X_{(n)} < m$, that is, all n sample values are less than m – which happens with probability $(1/2)^n$.

Hence the confidence coefficient of the interval $(X_{(1)}, X_{(n)})$, being the probability that it *does* contain m, is $1 - 2(1/2)^n$, *regardless of the distribution sampled.*

─────── **EXAMPLE 5.5** ───────────────────────────────

For Example 5.1, we have $n = 6$, hence the confidence coefficient of the interval $(X_{(1)}, X_{(6)})$ is $1 - 2(1/2)^6 = 31/32 \approx 0.97$. When the obtained sample is put in numerical order, we have

$$0.6 \quad 1.7 \quad 2.3 \quad 3.2 \quad 4.1 \quad 5.4$$

Hence a 97% confidence interval from *this* sample is $(0.6, 5.4)$.

More generally, suppose we take $(X_{(i+1)}, X_{(j)})$ as the confidence interval. Then it will contain the median unless:

1. $X_{(i+1)} > m$, when there are, at most, i values below m, that is, at least $n - i$ above m. This happens with probability

$$\sum_{k=n-i}^{n} \binom{n}{k}\left(\frac{1}{2}\right)^n = \Pr(S_+ \geq n - i \,|\, n, 1/2) \tag{5.3}$$

2. $X_{(j)} < m$, when there are at least j values below m, that is, at most, $n - j$ above m. This happens with probability

$$\sum_{k=0}^{n-j} \binom{n}{k}\left(\frac{1}{2}\right)^n = \Pr(S_+ \leq n - j \,|\, n, 1/2) \tag{5.4}$$

Hence the confidence coefficient is

$$1 - \left[\sum_{k=n-i}^{n} \binom{n}{k}\left(\frac{1}{2}\right)^n + \sum_{k=0}^{n-j} \binom{n}{k}\left(\frac{1}{2}\right)^n\right] \tag{5.5}$$

For a central confidence interval, the probabilities (5.3) and (5.4) are made equal and not exceeding $\alpha/2$ if the coefficient is to be at least $1 - \alpha$. The shortest such confidence interval is found by making i as large as possible and j as small as possible.

─────── **EXAMPLE 5.6** ───────

For Example 5.3, construct a central confidence interval with coefficient at least 99%. Thus $\alpha = 0.01$, $\alpha/2 = 0.005$. Here $n = 20$. Hence, from tables, the smallest integer j such that $\Pr(S_+ \leq 20 - j \,|\, 20, 1/2) \leq 0.005$ is 17, from a table of $b(20, 1/2)$. Similarly the largest integer i such that $\Pr(S_+ \geq 20 - i \,|\, 20, 1/2) \leq 0.005$, is 3. Hence $(X_{(4)}, X_{(17)})$ is the required confidence interval. For the particular sample drawn, this interval is $(31.5, 34.6)$.

───────────────────────────────

It is usual to estimate an unknown median of a continuous distribution by the median of the sample values. If this estimate $\hat{m}$ is subtracted from the sample values, then about half the differences $x_i - \hat{m}$ will be positive. The slight ambiguity springs from samples of odd size necessarily yielding a zero. In any case, the number of positive signs is close to the expected number and in that sense best supports $\hat{m}$ as the median of the distribution.

5.1.5   *Sign test in the presence of zeros*

For a continuous random variable X, $\Pr(X = x) = 0$ for each x, so that, for a sample from such a distribution, there is zero probability of obtaining a value equal to the median. In practice this will occasionally happen, due to limitations in recording the data.

If one or more values are reported as being equal to the hypothetical median, then when applying the sign test these values will produce zero residues. There are three possible ways of dealing with such zeros:

1. Discard the zeros and perform the test on the reduced sample.
2. Assign independently for each zero, a positive or negative sign with probability 1/2.
3. Replace the zeros by the signs which favour the null hypothesis.

─────── EXAMPLE 5.7 ───────────────────────────────

We illustrate the three methods when testing the null hypothesis that the median is 1 against the alternative that it exceeds 1 using the sample 0.6, 3.2, 1.7, 5.4, 1, 2.3, 4.1, 1.5.

The number of values which exceed 1 is 6 but one value falls at the median.

1. If the zero is discarded, $\Pr(S_+ \geqslant 6 \,|\, 7, 1/2) = 0.0625$, for the reduced sample of size 7.
2. If the zero is replaced by a plus (with probability 1/2),

$$\Pr(S_+ \geqslant 7 \,|\, 8, 1/2) = 0.035$$

but if replaced by a minus (with probability 1/2),

$$\Pr(S_+ \geqslant 6 \,|\, 8, 1/2) = 0.145$$

3. From (2), the conservative choice is to replace the zero with a minus and report an observed significance level of 0.145.

Notice that, for this relatively small sample, the choice of procedure may lead to drastically different conclusions. The most common practice is to recommend discarding the zeros.

───

5.1.6   *Paired comparisons*

We have seen, in Chapter 4, that one way of limiting uncontrolled variation when comparing the effects of two treatments is to use matched pairs of elements. One member of each pair – chosen at random – receives treatment A, the other receives treatment B. Let X be the response to treatment A and Y be the response to treatment B. Then $D = X - Y$ is a measure of the difference in the treatment effects. Since X, Y are correlated, through the matching, to claim that the treatment effects are equal requires an assumption about their joint distribution.

Suppose that, when the treatments have equal effects,

$$\Pr(X \leqslant x \text{ and } Y \leqslant y) = \Pr(X \leqslant y \text{ and } Y \leqslant x) \tag{5.6}$$

This asserts that the chance of reaching prescribed levels of response does not depend on the choice of treatment. The interchangeability expressed in equation (5.6) is enough to guarantee that the distribution of D is symmetric and has median zero. Let the joint p.d.f. of X, Y be $f(x, y)$, so that from equation (5.6), $f(x, y) = f(y, x)$. Thus for any pair (x, y) satisfying $x - y < k$, the pair (y, x) satisfies $y - x > -k$ and has the same probability density function. Hence $\Pr(X - Y < k) = \Pr(X - Y > -k)$. That is, the distribution of $X - Y$ is symmetric about zero.

If we have n independent matched pairs, with corresponding pairs of responses X_i, Y_i, then the differences $D_i = X_i - Y_i$, $(i = 1, 2, \ldots, n)$ are a random sample of n from a continuous symmetric distribution, which has, under the null hypothesis of equal treatment effects, a median of zero. Here the sign test may be applied to the sample of differences. For this situation, no fresh considerations arise concerning the treatment of zeros.

———— EXAMPLE 5.8 ————

Sixteen male albino rats of approximately equal weight were divided at random into eight pairs. The extent of feeding differed between pairs but was the same for rats in the same pair, except that one member of each pair had Ethionine added to its diet. The data consists of the amount of iron absorbed in the livers of the rats after a fixed time.

	Ethionine	Control	Differences
1st pair	4.50	3.81	0.69
2nd pair	3.92	2.81	1.11
3rd pair	7.33	8.42	-1.09
4th pair	8.23	3.82	4.41
5th pair	2.07	2.42	-0.35
6th pair	4.90	2.85	2.05
7th pair	6.84	4.15	2.69
8th pair	6.96	5.64	1.32

The null hypothesis is that Ethionine has no effect and the alternative is that it tends to increase the amount of iron absorbed. If the sign test is used, we have obtained six positive differences in eight and $\Pr(S_+ \geqslant 6 \mid 8, 1/2) = 0.1445$.

5.1.7   *Paired comparisons, discrete distributions*

In some experiments the method employed to make comparisons may be relatively crude so that it is only possible to observe that the effect of one treatment is superior, indistinguishable or inferior to another. Essentially we are then sampling a discrete distribution of differences and the probability of a zero difference may now be positive. We may in this case feel reluctant to 'discard the zeros', since they might have occurred as an inescapable feature of the way the data has been recorded.

Nevertheless, there are good theoretical grounds for retaining only the non-zero comparison. We have a trinomial distribution for the responses, the probability of a positive, zero, or negative difference being p_+, p_0, p_- respectively such that $p_+ + p_0 + p_- = 1$. Suppose for n pairs the number of positive, zero and negative differences are S_+, S_0, S_-. Then, in Section 1.3, we have shown that the conditional distribution of S_+, *given* $S_0 = s_0$ is $b(n - s_0, p_+/(1 - p_0)) = b(n - s_0, p_+/(p_+ + p_-))$. If $p_+ = \lambda p_-$, where λ is known, then the parameter of this binomial distribution is $\lambda/(\lambda + 1)$ and does not depend on p_+. By an extension of the ideas to be discussed in Chapter 9, it can be shown that the statistic S_+, given S_0, is the basis for best tests of $p_+ = \lambda p_-$. To test $p_+ = p_-$, set $\lambda = 1$ and the conditional distribution becomes $b(n - s_0, 1/2)$.

──────── **EXAMPLE 5.9** ────────────────────

A sample of 30 subjects were invited to express a preference between a new blend of coffee and a well-established blend marketed by the same firm. Cups of each type were offered in a random order to each subject and it was found that 14 subjects preferred the new blend, six the old and 10 were indifferent. It is required to test whether, other things (such as cost) being equal, the new blend will be preferred by consumers.

The null hypothesis is that the probabilities of preference for the new and old are equal. Discarding the 10 indifferent responses the conditional distribution of S_+ (preference for the new blend), given 10 zeros, is $b(20,1/2)$ when the null hypothesis is true. Now $\Pr(S_+ \geq 14 \mid 20, 1/2) = 0.058$, a result which, if not decisive, might encourage further work on the product.

───

5.1.8   *Use of MINITAB*

We have seen that the sign test is based on the binomial distribution. A count of the number of sample values above (or below) a conjectured median is provided by the command STEST. The general form

$$\text{STEST} \quad K \quad C, \ldots\ldots, C$$

carries out a sign test on each of the requested columns of data for the stipulated median K. The test will be two-sided, unless otherwise stipulated by the sub-command ALT.

―――― **EXAMPLE 5.10** ――――

We revisit Example 5.1. With 0.6 3.2 1.7 5.4 2.3 4.1 set in C1, the command

STEST 1 C1;

ALT + 1.

produces

N	BELOW	EQUAL	ABOVE	P-VALUE	MEDIAN
6	1	0	5	0.1094	2.75

As the reader will observe, the printout includes the sample median which is a point estimate of the distribution median. Note that if the median tested lies between 0.6 and 1.7 then the same result will be obtained. If a sample value falls at the conjectured median, then it will be omitted and the test carried out on the reduced sample. This may have a dramatic effect on the p-value for a small sample. Thus STEST 1.7 C1; with the sub-command ALT = 1. for the last example, produces

N	BELOW	EQUAL	ABOVE	P-VALUE	MEDIAN
6	1	1	4	0.1875	2.75

The MINITAB command SINTERVAL constructs confidence intervals for an unknown median. This has the form

SINTERVAL K C,........., C

and presents three central confidence intervals, for each column of data, for a requested coefficient of confidence K. The first of these has achievable confidence coefficient just below K and the third just above K. The second is derived from a non-linear interpolation sub-routine (see MINITAB Manual for reference). For instance, if the data of Example 5.3 is placed in C2, SINTERVAL 99 C2 prints

	N	MEDIAN	ACHIEVED CONFIDENCE	CONFIDENCE INTERVAL	POSITION
C2	20	33.3	0.9882	(31.60 34.40)	5
			0.9900	(31.59 34.41)	NLI
			0.9974	(31.50 34.60)	4

The position entry 5 tells us that if the 20 sample values are put in increasing order, then the end-points of the confidence interval starts at the fifth value from the bottom and ends at the fifth value from the top. The reader should confirm this by sorting the data and printing the sorted values.

5.1.9 *Problems*

5.1 A random sample of size 13 is drawn from a continuous distribution. It is desired to test the null hypothesis that the median of the distribution is 1/4.

Explain why small values of S_- are suitable for testing against the alternative that the median exceeds $1/4$. Calculate $\Pr(S_- \leqslant 3)$, when the null hypothesis is true. If, in fact, the p.d.f. of the distribution is of the type

$$f(x \mid \theta) = \theta x^{\theta-1}, \quad 0 < x < 1$$

check that when $\theta = 1/2$, the median is $1/4$. Show that $\Pr(X < 1/4 \mid \theta = 3/2) = 1/8$. Hence calculate $\Pr(S_- \leqslant 3)$ when $\theta = 3/2$.

5.2 A random sample of size 20 is drawn from a continuous distribution. Use a sign test statistic to construct a central confidence interval for the median with confidence coefficient at least 95%.

5.3 A single value is drawn from the distribution $b(16, p)$ and found to have the value 9. Calculate, using the appropriate distribution function, a central confidence interval for p with confidence coefficient at least 95%.

5.4 For the discrete random variable X, $\Pr(X > 0) = p_+$, $\Pr(X = 0) = p_0$, $\Pr(X < 0) = p_-$. Show that $\Pr(X > 0) \mid X \neq 0) = p_+/(p_+ + p_-)$. A random sample of size n is drawn from the distribution of X and S_+, S_0, S_- are the numbers of positive, zero and negative values in the sample. If $p_+ = \lambda p_-$ where λ is a known positive constant, explain why the distribution of S_+ given $S_0 = s_0$ is $b(n - s_0, \lambda/(1 + \lambda))$. Suppose further that any zero is independently resolved as 'positive' with probability $1/2$ or as 'negative' with probability $1/2$. Show that the probability that a sample value is finally recorded as positive is $\frac{1}{2} + \frac{1}{2}(p_+ - p_-)$. If T_+ is the ultimate number of positives, deduce that the distribution of T_+ is $b(n, \frac{1}{2}(1 + p_+ - p_-))$. State the distributions of S_+, given $S_0 = s_0$, and T_+ when $\lambda = 1$.

5.5 For problem 5.4, assume $n = 7$ and let the null hypothesis be $\lambda = 1$, the alternative being that $\lambda = 3$. Calculate the size of the critical region $T_+ \geqslant 6$ and the conditional power when $s_0 = 2$.

5.6 For Problem 5.4 show that $E(S_+ - S_-) = n(p_+ - p_-)$, $V(S_+ - S_-) = n(p_+ + p_-) + n(p_+ - p_-)^2$. Show further that

$$\frac{n}{n-1}\left[S_+ + S_- - \frac{(S_+ - S_-)^2}{n}\right]$$

is an unbiased estimator of the variance of $S_+ - S_-$.

5.7 Apply the result of Problem 5.6 to Example 5.9, taking $S_+ - S_-$ as the test statistic and assuming that it is approximately normally distributed.

5.2   The Wilcoxon signed-rank distribution

The 'sign test' takes no account of the absolute magnitude of the sample values obtained. In this section we lead up to a distribution-free statistic which takes some account of their relative magnitudes. To do this, we make one further assumption about the distribution sampled – namely that it is symmetric, necessarily about the median m. If the distribution is symmetric, then for any $x_0 > 0$, the probability that any observation will exceed $m + x_0$ is equal to the probability that it will be less than $m - x_0$. In particular, for any observation there is probability $1/2$ that the deviation from the median has positive sign.

Suppose that a random sample of 4 was drawn from a continuous distribution and yielded 2.1, 1.2, −0.6, 0.9. To test that the median is zero against the alternative that it is positive, the sign test would only record that three observations are positive. Since $\Pr(S_+ \geqslant 3 \mid 4, 1/2) = 5/16$, this would not be deemed significant at the 5% level. Yet the only negative observation is relatively small compared to the positive values. It would appear that we should attempt to compare the sum of the positive values with the sum of the negative values This idea forms the basis of a *permutation* test. In such a test, when $m = 0$, all samples with the same numerical values, but different signs are regarded as having the same probability. Each of the four values could be assigned a positive or a negative sign. Thus there are $2^4 = 16$ ways of assigning the signs. For each of these, we calculate the sum of the sample values and the actual sum observed is compared to this array of possible sums. Table 5.1 shows the 16 possible patterns and the resulting sample sums.

When the sums are written in increasing order of magnitude, we find that the obtained value, $2.1 + 1.2 - 0.6 + 0.9 = 3.6$ is only exceeded by 4.8. Since each sum has probability $1/16$, the conditional probability given the observations, that an observed sum is greater than or equal to 3.6 is $2/16 = 0.125$ The evidence against the median being zero is more impressive, but not decisive.

Table 5.1

2.1	1.2	0.6	0.9	Sum
+	+	+	+	+4.8
−	+	+	+	+0.6
+	−	+	+	+2.4
+	+	−	+	+3.6
+	+	+	−	+3.0
−	−	+	+	−1.8
−	+	−	+	−0.6
−	+	+	−	−1.2
+	−	−	+	+1.2
+	−	+	−	+0.6
+	+	−	−	+1.8
−	−	−	−	−4.8
+	−	−	−	−0.6
−	+	−	−	−2.4
−	−	+	−	−3.6
−	−	−	+	−3.0

5.2.1   Using ranks

The procedure we have just illustrated is distribution-free, but only conditionally – and there's the rub. For another random sample of four from the same continuous distribution, we would obtain a different table of sums. This difficulty can be overcome by replacing the data values with their signed-ranks. Assume, for the initial discussion, that there are no ties. That is, the absolute values are distinct. For our sample 2.1, 1.2, −0.6, 0.9, the data is arranged in increasing order of absolute value, with negative sample values in italic:

	0.6	0.9	1.2	2.1
rank	1	2	3	4
signed-ranks	−1	2	3	4

The signs are now replaced and we arrive at the *signed-ranks* $-1, 2, 3, 4$ with sum 8. To evaluate the significance of this result, we must compare it with the other 15 signed-rank sums obtainable by considering all possible assignments of signs to the ranks $1, 2, 3, 4$. The sums range from $+10$, obtained from $1 + 2 + 3 + 4$, down to -10, obtained from $-1 - 2 - 3 - 4$. There are only two greater than or equal to 8. Hence the (one-sided) significance level is again 2/16.

It will be clear that the technique outlined above is unconditionally distribution-free, since it uses the same set of ranks $1, 2, 3, 4$ for *every* sample of four values. In a similar way the distribution of the sum of the signed-ranks can be tabulated for each sample size. In general, to test whether the median (or mean) of a continuous *symmetric* distribution is m_0, first subtract m_0 from each sample value before finding the signed-rank sum.

5.2.2   Distribution of the Wilcoxon signed-rank statistic, W_+

The tabulation required may be reduced by noting that, if W_+ is the sum of the ranks of those observations which have positive signs and W_- is the sum of the ranks of those observations which have negative signs, then

$$W_+ + W_- = n(n + 1)/2$$

since this is the sum of all the ranks $1, 2, 3, \ldots, n$. Hence all questions relating to the magnitude of $W = W_+ - W_-$ can be equally answered by considering either W_+ or W_-. Note that by reversing the signs the values of W_+ and W_- are interchanged but the probabilities are unchanged. Since $W_+ + W_-$ is constant, the distributions of W_+ and W_- are symmetric and equal.

──────── EXAMPLE 5.11 ────────────────────────────────

A random sample of 7 from a continuous symmetric distribution yielded the values

$$0.4 \quad -1.2 \quad -2.3 \quad -0.9 \quad 0.7 \quad -2.1 \quad -1.1$$

The null hypothesis is that the median is zero, the alternative that it is negative. If the alternative is true, a sample value is more likely to be negative than positive and this tends to depress the number and magnitude of the elements which comprise W_+.

Critical values for this one-sided test will consist of small values of W_+. The order of the absolute magnitudes is

	0.4	*0.7*	0.9	1.1	1.2	2.1	2.3
with ranks	*1*	*2*	3	4	5	6	7

The sum of the ranks of the (italic) positive ranks is 3.

There are 2^7 ways in which the $\pm$ signs could be assigned to the 7 ranks. The ways in which the sum of the positive ranks does not exceed 3 are

1. All signs negative, sum zero – one possibility
2. One sign positive, ranks 1, 2, or 3 – three possibilities and
3. Two signs positive, one with rank 1, other with rank 2 – one possibility.

Table 5.2   Table 5.2 displays cases 1–3.

Rank	1	2	3	4	5	6	7	W_+	W_-
1	–	–	–	–	–	–	–	0	28
	+	–	–	–	–	–	–	1	27
2	–	+	–	–	–	–	–	2	26
	–	–	+	–	–	–	–	3	25
3	+	+	–	–	–	–	–	3	25

Hence $\Pr(W_+ \leqslant 3) = 5/2^7 \approx 0.039$, significant at the 5% level.

Notice that to calculate the significance level it was not necessary to find all possible values of W_+. Table A3.1 in Appendix 3 displays the lower tail of the distribution function of W_+ for $4 \leqslant n \leqslant 20$.

――――― **EXAMPLE 5.12** ―――――

For $n = 10$, from Table A3.1, $\Pr(W_+ \leqslant 8) = 0.02441$ and, since the maximum value of W_+ is 55, $\Pr(W_+ \geqslant 55 - 8) = \Pr(W_+ \geqslant 47) = 0.02441$. Clearly, a two-sided test of size 0.0488 of the hypothesis that the sample of ten has been drawn from a distribution with median zero can be carried out by calculating W_+ and comparing with 8 and 47.

If we are concerned with significance only at nominal levels such as 1%, $2\frac{1}{2}$% and 5%, then a table of critical levels for W_+ will suffice. If a table lists the largest integer i such that $\Pr(W_+ \leqslant i) \leqslant \alpha$, then, to find the least integer j such that $\Pr(W_+ \geqslant j) \leqslant \alpha$, set $j = \frac{1}{2}n(n+1) - i$.

5.2.3   *Normal approximation*

When a random sample, $X_1, X_2, \ldots, X_n$ of size n is drawn from a continuous symmetric distribution, with median zero, then the probability that a particular rank is assigned a positive value is $\frac{1}{2}$, independently for each rank. Thus we may write

$$W_+ = \sum_{i=1}^{n} iU_i$$

where $U_i = 1$ if $X_i > 0$, $U_i = 0$ otherwise. Hence U_i has the binomial distribution $b(1, \frac{1}{2})$, so that $E(U_i) = \frac{1}{2}$ and $V(U_i) = \frac{1}{4}$. But

$$E(W_+) = \sum_{i=1}^{n} i E_i(U_i) = \frac{1}{2} \sum_{i=1}^{n} i = \frac{1}{4} n(n+1) \tag{5.7}$$

and

$$V(W_+) = \sum_{i=1}^{n} [i^2 V(U_i)] = \frac{1}{4} \sum_{i=1}^{n} i^2 = \frac{n(n+1)(2n+1)}{24} \tag{5.8}$$

It can be shown that, for sufficiently large n, W_+ is approximately normally distributed with mean $n(n+1)/4$, and variance $n(n+1)(2n+1)/24$.

––––––– EXAMPLE 5.13 ––––––––––––––––––––––––––––––––

When $n = 24$ then $E(W_+) = 150$, $V(W_+) = 1225$. Hence $\sqrt{V(W_+)} = 35$. Since the lower 2.5% point of the $N(0, 1)$ distribution is -1.96, and $-1.96 \sqrt{V(W_+)} = -68.6$, this suggests that the lower $2\frac{1}{2}\%$ point of the distribution of W_+ is $150 - 68.6 = 81.4$. The greatest integer such that $Pr(W_+ \leq i) \leq 0.025$, when $n = 24$, is 81.

5.2.4   *Paired comparisons*

The idea of using matched pairs of elements for comparing two treatments has been introduced in Section 5.1. We recall that it is essential that the element of each pair which is to receive a particular treatment is chosen at random. That is, the probability that the first element of each pair receives treatment A is $\frac{1}{2}$ and that it receives treatment B is also $\frac{1}{2}$, independently for each pair. With n pairs, there are thus 2^n ways of randomizing the assignments of the treatments. If the treatments are equally effective, then each of the n differences, on some continuous response scale, is equally likely to be recorded in favour of either of the treatments. In these circumstances, we have n independent observations from a continuous symmetric distribution. We can then apply the Wilcoxon signed-rank test in the manner already described.

─────── EXAMPLE 5.14 ───────

There are seven matched pairs; one member of each pair was assigned treatment A, at random, the other received treatment B. The results, when listed according to *treatment*, were

A	B	A–B	|A–B|	Signed rank
7.3	6.9	0.4	0.4	+1
6.4	7.6	−1.2	1.2	−5
5.1	7.4	−2.3	2.3	−7
5.9	6.8	−0.9	0.9	−3
4.0	3.3	0.7	0.7	+2
6.2	8.3	−2.1	2.1	−6
4.3	5.4	−1.1	1.1	−4

Note the subtractions must always be performed in the same order. If the null hypothesis is that the treatments are equivalent, the alternative that B is superior, then Example 5.1, which examines the same differences, evaluates the significance level as about 0.039.

─────────────────────────────

5.2.5   *Confidence interval for the median*

Suppose we have a random sample $X_1, X_2, \ldots, X_n$ from a continuous symmetric distribution and we wish to construct a central confidence interval with coefficient at least $1 - \alpha$, for the unknown median, m. The routine solution, based on the method of testing hypotheses about m, is to find the largest c_1 and the smallest c_2 such that $\Pr(W_+ \leqslant c_1) \leqslant \alpha/2$, $\Pr(W_+ \geqslant c_2) \leqslant \alpha/2$, so that c_1, c_2 are two possible values satisfying $\Pr(c_1 < W_+ < c_2) \geqslant 1 - \alpha$, that is,

$$\Pr(c_1 + 1 \leqslant W_+ \leqslant c_2 - 1) \geqslant 1 - \alpha$$

─────── EXAMPLE 5.15 ───────

If a random sample of six values is available, suppose we require $1 - \alpha \geqslant 0.9$, that is, $\alpha/2 \leqslant 0.05$. Then from tables, the largest integer c_1 such that

$$\Pr(W_+ \leqslant c_1) \leqslant 0.05 \quad \text{is } c_1 = 2$$

The maximum value of W_+ is $\frac{1}{2}6(6+1) = 21$. Hence by symmetry, the smallest integer c_2 such that

$$\Pr(W_+ \geqslant c_2) \leqslant 0.05 \quad \text{is } c_2 = 21 - 2 = 19$$

From Table A3.1 in fact $\Pr(W_+ \leqslant 2) \approx 0.047$. Hence $1 - \alpha$ is actually about $1 - 2(0.047) = 0.906$.

We have still to locate the end-points of the interval at which the values of c_1 and c_2 are attained. For a sample of size n, the maximum value of W_+ is $\frac{1}{2}n(n+1)$ and is attained when all the sample values exceed m. Correspondingly W_+ is 0 when m exceeds all the sample values. As m increases, W_+ decreases – not smoothly – with downward jumps of one or more units.

Consider a sample which arranged in increasing order commences $3, 4, \ldots$. When m is slightly less than $3, 3 - m$ is positive and $|3 - m|$ will have rank 1. As m increases to slightly more than $3, 3 - m$ switches to negative, still has rank 1 but no longer contributes to W_+. That is, W_+ suffers a loss of 1. Another change occurs as m crosses the mid-point 3.5. Below the mid-point $|3 - m| < |4 - m|$ and $4 - m$ contributes to W_+. But when m is just above the mid-point, $|3 - m| > |4 - m|$ and, although $(4 - m)$ is still positive, $|3 - m|$, $|4 - m|$ have now changed places in the list of rankings by absolute value, and W_+ again loses 1. Thus the changes to W_+ happen at the samples values and at the averages of pairs of sample values. For a sample of size n, there will be n sample values and $\binom{n}{2} = n(n-1)/2$ averages at which a drop in W_+ will happen. Should a sample value and an average coincide, there will be a decrease of 2 in W_+. It will be necessary to arrange the $n + n(n-1)/2 = n(n+1)/2$ change points, known as the *Walsh* averages, in increasing order of magnitude and then count off from each end an appropriate number of change points.

--------- EXAMPLE 5.16 --

The sample values, arranged in order of magnitude are 1, 2.5, 3, 4, 7, 10.5. We list the sample values followed in each row by their averages with subsequent members of the sample. The figures are spaced to provide the rank order.

1	1.75	2	2.5		4		5.75				
		2.5	2.75	3.25		4.75		6.5			
				3	3.5		5		6.75		
						4		5.5		7.25	
									7	8.75	
											10.5

From Example 5.15 if we require a central confidence interval with confidence coefficient at least 0.90, then the appropriate values of c_1, c_2, are $c_1 = 2$, $c_2 = 19$ or $c_1 + 1 = 3$, $c_2 - 1 = 18$. Hence in the list of change-points we find the third from the bottom, which has value 2, and the third from the top, which is 7.25. The obtained confidence interval is $(2, 7.25)$.

The method illustrated in Example 5.16 is tedious, though we belatedly note that we need only calculate as far as is strictly necessary and that the second and third lines in the table are redundant. See also Problems 5.12 and 5.13.

A reasonable *point estimate* of the distribution median would yield a realized

Wilcoxon signed-rank statistic close to its expected value. We have seen that if the trial value for the median moves from below the smallest sample value to above the greatest, W_+ decreases by one or more as it passes through each of the Walsh averages. Furthermore, if we subtract any constant a from each sample value, each resultant Walsh average is reduced by a. Hence if we subtract the observed median $\hat{m}$, of the existing Walsh averages from each of the sample values, then about half of the resulting Walsh averages will be positive and W_+ will be near to its expectation. This suggests we take $\hat{m}$ as the point estimate of the distribution median.

5.2.6   *Ties*

Because of the way the ranks are assigned for this test, two sample members are said to be tied if they have the same absolute values. Thus suppose the sample is

$$0.4 \quad -1.2 \quad -2.3 \quad -0.8 \quad +0.8 \quad -2.1 \quad -1.1$$

Here the values -0.8 and $+0.8$ have different signs but the same absolute magnitude. The following alternative procedures have been suggested for dealing with ties.

1. Omit both -0.8, $+0.8$. To drop two observations from a sample of size 7 seems a touch unwise.
2. Omit -0.8. This will be conservative, since it diminishes the sum of the negative and positive signed ranks. The one-sided p-value level of $W_+ = 3$, with six ranks, is $5/2^6 = 0.078125$.
3. Break the ties at random and average the p-values. Thus if $+0.8$ is given rank 2 and -0.8 rank 3, then $W_+ = 3$ with p-value $5/128 = 0.039$. If $+0.8$ has rank 3 and -0.8 has rank 2, then $W_+ = 4$ with p-value $7/128 = 0.055$. The average is $6/128 = 0.047$.
4. Average the ranks that the tied values could assume. In that case, $+0.8$, -0.8 each acquire rank $2\frac{1}{2}$ and $W_+ = 3\frac{1}{2}$. With this method, W_+ no longer has the tabulated Wilcoxon signed-rank distribution and to find the p-value we must resort to listing the sets of possible ranks as described in Example 5.11. We find $\Pr(W_+ \leqslant 3\frac{1}{2}) = 6/128 = 0.047$.

At first sight, the case when the tied sample values have the *same* sign and magnitude seems easier. Suppose we have the sample

$$0.4 \quad -1.2 \quad -2.2 \quad -0.8 \quad +0.9 \quad -2.2 \quad -1.1$$

Since the values tied at -2.2 have the same sign it would seem reasonable to give rank 6 to one and 7 to the other. $W_+ = 4$ and the table of W_+ need be consulted only once and the p-value is evaluated *without regard to the assignment order*. This fortunate facility may not extend to method (4). Thus for the sample

	+0.4	−1.2	+0.9	−0.7	−1.2
signed-rank	1	$-4\frac{1}{2}$	+3	−2	$-4\frac{1}{2}$

$W_+ = 4$. It is clear that the shared rank $4\frac{1}{2}$ is excluded from appearing in a list of achievable values of W_+ less than or equal to 4.

The solution to the problem of ties is basically a choice between breaking the ties at random and sharing the ranks. Note that, as always, the method should be chosen before seeing the data. The following points deserve consideration:

1. Breaking the ties at random restores the original Wilcoxon distribution without ties. To avoid the calculations involved in computing the average *p*-value, one can stick to the most conservative tie break. In spite of the advantage conferred by tabulation, workers in some fields feel disinclined to import any element of randomization into the decision process.
2. Averaging the ranks has the superficial advantage of requiring only one calculation at the cost of using a statistic for which no tables are available. Nevertheless, conclusions drawn from the usual critical levels given by tables are not likely to be seriously misleading.

If the sample size exceeds 20, then a normal approximation may be used, in which case the variance of W_+ is no longer as given in (2). For a full discussion the reader should consult reference 1.

5.2.7   *Ties in the presence of zeros*

When the differences derived from comparing matched pairs have a discrete distribution, the sample data may contain both zeros and ties. We have previously recommended that the zeros be discarded – but should ranks be assigned before or after so doing? The ordinary sign test raises no such question, because the remaining sample values all have equal weight. The motivation for assigning the ranks first is that omitted zeros should not alter the ranks which would otherwise have been applied. This is called the *signed-rank zero* procedure.

———— EXAMPLE 5.17 ————————————————————————————————

Suppose the sample differences are

$$0 \quad 0 \quad -1.5 \quad 1.5 \quad 1.5 \quad -2 \quad -2 \quad 2$$

and in a one-sided test, large values of W_+ are taken to be significant.

1. If the zeros are discarded and the ranks 1 to 6 are shared between the non-zero values, we obtain the signed-ranks $-2, 2, 2, -5, -5, 5$. The obtained $W_+ = 9$, and by a complete enumeration, there are 44 cases for which $W_+ \geqslant 9$. If all $2^6 = 64$ assignments of signs are equally likely, $\Pr(W_+ \geqslant 9) = 44/64 = 0.6875$ (see Problem 5.16).
2. If the ranks 1, 2 are formally allotted to the two zeros and *then* omitted, the ranks 3 to 8 are shared between the non-zero values and we obtain the signed ranks -4,

4, 4, −7, −7, 7. Now $W_+ = 15$ and a complete count shows that $\Pr(W_+ \geqslant 15) = 41/64 = 0.6406$.

There are no tables readily available for either method of scoring the ranks. A normal approximation may be employed, but in that case, modifications to the mean and variance should be applied (see Problems 5.10, 5.11 and 5.14–5.16).

5.2.8 *Use of MINITAB*

The MINITAB command WTEST K C1 tests whether sample data in C1 has been drawn from a distribution with median K and calculates W_+ (reporting it as W), evaluates the *p*-value of the test, and supplies an estimate of the distribution median. To calculate the *p*-value, MINITAB uses the same normal approximation, with a continuity correction, whether or not there are ties. The estimate of the distribution median is calculated as the sample median of the Walsh averages.

Thus, if the data 0.4 0.7 −0.9 −1.1 −1.2 −2.1 −2.3 are set in C1, and we wish to test the hypothesis that the distribution median is zero against the alternative that it is less than zero, the command

        WTEST 0 C1;
        ALT = −1.

provides

N	N FOR TEST	WILCOXON STATISTIC	P-VALUE	ESTIMATED MEDIAN
C1 7	7	3	0.038	−0.975

For this data, we found in Example 5.11, that the exact *p*-value was $5/128 = 0.039$.

If the data 0.4 0.8 −0.8 −1.1 −1.2 −2.1 −2.3 is tested by the same command, the corresponding output is

N	N FOR TEST	WILCOXON STATISTIC	P-VALUE	ESTIMATED MEDIAN
C1 7	7	3.5	0.045	−0.950

MINITAB shares the tied ranks 2,3 and reports W as 3.5. The exact *p*-value for this sample has been shown to be $6/128 = 0.046875$.

If the hypothetical median falls at one or more of the sample values, then MINITAB drops these and assigns ranks to the remainder, starting with rank 1. Thus, for the sample 0 0 0.4 0.7 −0.9 −1.1 −1.2 −2.1 −2.3, the command

        WTEST 0 C1;
        ALT = −1.

yields

N	N FOR TEST	WILCOXON STATISTIC	P-VALUE	ESTIMATED MEDIAN
C1 9	7	3	0.038	−0.700

which gives the same *p*-value as obtained in our first illustration.

Similarly, the command WINTERVAL uses a normal approximation to calculate central confidence intervals for the median. If the data 1   2.5   3   4   7   10.5 (Example 5.16) is set in C1 and we require a central confidence interval with confidence coefficient at least 90%, the command WINTERVAL 0.9 C1 prints

N	ESTIMATED MEDIAN	ACHIEVED CONFIDENCE	CONFIDENCE INTERVAL
6	4	90.7	(2.000, 7.2500)

The estimated median is the eleventh of the 21 Walsh averages. The confidence interval is from the third to the nineteenth Walsh average. Both these results can be checked from the display in Example 5.16.

The Walsh averages can be calculated and displayed using the command WALSH. Thus

WALSH C 1 C2 [C3 C4]

calculates the Walsh averages for the sample values in C1, places them in C2, and in C3 and C4 records the position indices of the C1 values used to obtain the C2 values. The use of [C3, C4] is optional.

5.2.9   *Problems*

In the following problems the samples are supposed to have been drawn from a continuous distribution with median zero. Where a normal approximation is suggested for a small sample, this is intended for comparison with the exact calculation or with the MINITAB method.

5.8   By considering all possible selections of the ranks $1, 2, 3, 4$, find the exact probability distribution for the Wilcoxon signed-rank statistic for a sample of size 4. Verify that the mean and variance of this distribution are 5, $15/2$ respectively.

5.9   Find the Walsh averages for the sample $1, 2, 3, 4$. Subtract 2.25 from each sample value and find the Walsh averages for the residues. Check that the number of positive averages is six. Explain how this result is related to testing that the sample $1, 2, 3, 4$ is drawn from a distribution with median 2.25.

5.10 $U_1, U_2, \ldots, U_n$ are independent random variables each with expectation $1/2$ and variance $1/4$. If

$$W_+^* = \sum_{i=k+1}^{n} iU_i$$

calculate the mean and variance of W_+^*. Explain how this result can be used in connection with the procedure of signed-rank zeros.

5.11 $U_1, U_2, \ldots, U_n$ are independent random variables each with expectation $1/2$ and variance $1/4$. If

$$W_+' = \sum_{i=1}^{s-1} iU_i + \sum_{i=s}^{s+t-1} \left[s + \left(\frac{t-1}{2} \right) \right] U_i + \sum_{i=s+t}^{n} iU_i$$

calculate the mean and variance of W_+'. Explain how this result can be used in connection with the Wilcoxon signed-rank test when t-values are tied in absolute value and these are assigned the average of the ranks they would otherwise occupy. (Hint: let the tied elements share ranks $s, s+1, \ldots, s+t-1$.)

5.12 Suppose $X_1, X_2, \ldots, X_n$ is a random sample from a continuous symmetric distribution with median zero. Show that the statistic W_+ may be calculated as the total number of pairs (X_i, X_j) where $i \leq j$ and $X_i + X_j > 0$. (Hints: (i) Without loss of generality we may take the sample values to be in increasing order of absolute value, so that x_j has signed rank $\pm j$ according to its sign. (ii) Using (i) note $X_i + X_j > 0$ $(i \leq j)$ if and only if X_j is positive.)

5.13 $X_1, X_2, \ldots, X_n$ is a random sample of size n from a continuous symmetric distribution with unknown median m. Let $A_{ij} = (X_i + X_j)/2$, $i \leq j$ and $A_{(1)}, A_{(2)}, \ldots, A_{(n(n+1)/2)}$ be the order statistics for the $n(n+1)/2$ Walsh averages A_{ij}. Show that $A_{(k)} \leq m$ if and only if W_+ for the differences $X_1 - m, X_2 - m, \ldots, X_n - m$ does not exceed $[n(n+1)/2] - k$. Use this result to construct confidence intervals for m.

5.14 Let the obtained sample be   0   4   8   12   16.
(a) If the zero is discarded and ranks 1   2   3   4 assigned to the remaining values, calculate $\Pr(W_+ \geq 10)$ both exactly and using a normal approximation.
(b) If the zero is included but given signed rank zero and ranks 2   3   4   5 assigned to the remaining values, calculate $\Pr(W_+ \geq 14)$ both exactly and using a normal approximation with parameters modified to take account of the zero.
(c) Use MINITAB to test that the median is zero against the alternative that the median is positive. Compare the returned significance level with the results of (a) and (b).

5.15 Let the obtained sample be 4   4   4   4   4.
(a) If the ranks 1   2   3   4   5 are shared, calculate $\Pr(W_+ \geq 15)$ using a normal approximation.
(i) With unmodified parameters
(ii) With parameters modified to take account of the ties.

(b) Use MINITAB to test that the median is zero against the alternative that the median is positive. Compare the returned significance level with the result in (a).

5.16 For the sample 0 0 −1.5 1.5 1.5 −2 −2 +2, find the frequency distribution of W_+ by considering all 64 distinguishable ways of assigning signs when
(a) The zeros are omitted and the ranks of the elements tied in absolute value are shared
(b) The zeros are given signed rank zero and the remaining six values share ranks 3 to 8 inclusive. Calculate $\Pr(W_+ \geqslant 9)$ for (a) and $\Pr(W_+ \geqslant 15)$ for (b).

5.3   Solutions to problems

5.1 $\Pr(S_- \leqslant 3 \mid 13, 1/2) = 0.046$. If $\theta = 3/2$, $\Pr(X < 1/4) = \int_0^{1/4} (3x^{1/2}/2)\,dx = 1/8$ $= 0.125$. $\Pr(S_- \leqslant 3 \mid 13, 0.12) = 0.9391$, $\Pr(S_- \leqslant 3 \mid 13, 0.13) = 0.9224$.

5.2 $\Pr(S_- \leqslant 5 \mid 20, 1/2) = 0.0207$, $\Pr(S_- \geqslant 15 \mid 20, 1/2) = 0.0207$. Place the sample values in increasing order. Then the required confidence interval is $(X_{(6)}, X_{(15)})$.

5.3 $\Pr(X \leqslant 9 \mid p_2) = \Pr(16 - X \geqslant 7 \mid p_2) = \Pr(Y \geqslant 7 \mid q_2) = 0.025$, where $Y = 16 - X$. From tables, $0.19 < q_2 < 0.20$ or $0.80 < p_2 < 0.81$. Similarly, the greatest p_1 such that $\Pr(X \geqslant 9 \mid p_1) \leqslant 0.025$ satisfies $0.29 < p_1 < 0.30$. The interval $(0.29, 0.81)$ is conservative.

5.4 $\Pr(X > 0 \mid X \neq 0) = \Pr(X > 0)/\Pr(X \neq 0) = p_+/(1 - p_0) = p_+/(p_+ + p_-)$. If $p_+ = \lambda p_-$, $p_+/(p_+ + p_-) = \lambda/(1 + \lambda)$, hence result. $\Pr(\text{sample recorded} > 0)$ $= \Pr(X > 0) + \frac{1}{2}\Pr(X = 0) = p_+ + \frac{1}{2}p_0 = p_+ - \frac{1}{2}(1 - p_+ - p_-) = \frac{1}{2} + \frac{1}{2}(p_+ - p_-)$. When $\lambda = 1$, distribution of S_+ given $S_0 = s_0$ is $b(n - s_0, \frac{1}{2})$ and of T_+ is $b(n, \frac{1}{2})$.

5.5 When $\lambda = 1$, the distribution of T_+ is $b(7, \frac{1}{2})$ and $\Pr(T_+ \geqslant 6) = 8/128 = 0.0625$. When $S_0 = 2$, and $\lambda = 3$, the distribution of T_+ is $b(7, \frac{1}{2} + p_-)$, which depends on p_-. However the distribution of S_+ given $s_0 = 2$ is $b(5, 3/4)$.
Here we can calculate

$$\Pr(T_+ \geqslant 6 \mid S_0 = 2) = \Pr(S_- + = 5 \mid S_0 = 2) \times \tfrac{3}{4} + \Pr(S_+ = 4 \mid S_0 = 2) \times \tfrac{1}{4}$$
$$= 0.1780 + 0.0989 = 0.2769$$

5.6   From the properties of the trinomial distribution (1.3)

$$E(S_+ - S_-) = E(S_+) - E(S_-) = np_+ - np_- = n(p_+ - p_-)$$
$$\begin{aligned} V(S_+ - S_-) &= V(S_+) + V(S_-) - 2\,\mathrm{Cov}(S_+, S_-) \\ &= np_+(1 - p_+) + np_-(1 - p_-) + 2np_+p_- \\ &= n(p_+ + p_-) - n(p_+ - p_-)^2 \end{aligned}$$

$$E\left[S_+ + S_- - \frac{(S_+ - S_-)^2}{n}\right] = E(S_+ + S_-) - \frac{1}{n}\left[V(S_+ - S_-) + \{E(S_+ - S_-)\}^2\right]$$

$$= np_+ + np_- - \frac{1}{n}\left[V(S_+ - S_-) + n^2(p_+ - p_-)^2\right]$$

$$= n(p_+ + p_-) - n(p_+ - p_-)^2 - \frac{V(S_+ - S_-)}{n} = \frac{n-1}{n}V(S_+ - S_-)$$

5.7 If the null hypothesis is $p_+ = p_-$, since $n = 30$, $S_+ = 14$, $S_- = 6$, $S_+ - S_- = 8$, the estimate of $V(S_+ - S_-) = 30(20 - 64/30)/29 \approx 18.48$. If Z has the distribution $N(0, 1)$, applying a continuity correction of $\frac{1}{2}$,

$$\Pr(Z \geqslant 7.5/\sqrt{18.48}) = \Pr(Z \geqslant 1.744) \approx 0.04.$$

5.8 There are two selections, for example, which have rank sum 6, namely $\{2, 4\}$, $\{1, 2, 3\}$.

W_+	0	1	2	3	4	5	6	7	8	9	10
Selections	1	1	1	2	2	2	2	2	1	1	1

$E(W_+) = (0 + 1 + 2 + 6 + 8 + 10 + 12 + 14 + 8 + 9 + 10)/16 = 5,$

$E(W_+^2) = (0 + 1 + 4 + 18 + 32 + 50 + 72 + 98 + 64 + 81 + 100)/16 = 520/16$

$V(W_+) = (520/16) - 25 = 15/2$

5.9

	1	2	3	4	Subtract 2.25 from each Walsh average			
1	1	1.5	2	2.5	-1.25	-0.75	-0.25	0.25
2		2	2.5	3		-0.25	0.25	0.75
3			3	3.5			0.75	1.25
4				4				1.75

If 2.25 is subtracted from each sample value, residues are -1.25 -0.25 $+0.75$ $+1.75$. The two positive values have signed ranks 2, 4. After subtracting 2.25, the number of positive Walsh averages is the sum of the positive signed ranks.

5.10 In the method of signed-rank zeros, the absolute values of the non-zero sample values are assigned the ranks $k + 1$, $k + 2$, ..., n. If $\Pr(U_i = 1) = 1/2$, $\Pr(U_i = 0) = 1/2$ then W_+^* is the appropriate modified Wilcoxon statistic in the presence of zeros:

$$E(W_+^*) = \frac{1}{2} \sum_{i=k+1}^{n} i = \frac{1}{2} \left[\sum_{i=1}^{n} i - \sum_{i=1}^{k} i \right] = \frac{1}{4} [n(n + 1) - k(k + 1)]$$

$$V(W_+^*) = \frac{1}{4} \sum_{i=k+1}^{n} i^2 = \frac{1}{4} \left[\sum_{i=1}^{n} i^2 - \sum_{i=1}^{k} i^2 \right] = \frac{1}{24} [n(n + 1)(2n + 1) - k(k + 1)(2k + 1)]$$

These are the mean and variance for the approximating normal distribution for this procedure.

5.11 The average of the ranks $s, s + 1, ..., s + t - 1$ is $s + (t - 1)/2$. Hence W_+' is the modified Wilcoxon signed-rank statistic for *this* treatment of ties:

$$E(W_+') = \frac{1}{2} \left[\sum_{i=1}^{s-1} i + t \left(s + \frac{t-1}{2} \right) + \sum_{i=s+t}^{n} i \right] = \frac{n(n + 1)}{4}$$

$$V(W'_+) = \frac{1}{4}\left[\sum_{i=1}^{s-1} i^2 + t\left(s + \frac{t-1}{2}\right)^2 + \sum_{i=s+t}^{n} i^2\right]$$

$$= \frac{1}{4}\left[\sum_{i=1}^{n} i^2 + t\left(s + \frac{t-1}{2}\right)^2 - \sum_{i=s}^{s+t-1} i^2\right]$$

$$= \frac{1}{24}\left[n(n+1)(2n+1) - \frac{1}{2} t(t-1)(t+1)\right]$$

Note that the result does not depend on s.

5.12 If $X_j < 0$, since $|X_j|$ exceeds the absolute magnitude of all its predecessors, $X_i + X_j < 0$, $i \leqslant j$. If $X_j > 0$, then, similarly, each $X_i + X_j > 0$, $i \leqslant j$ and the number of these is j. Summing over j, we have the sum of the positive signed ranks.

5.13 $A_{(k)} \leqslant m$ if and only if at least k averages satisfy $(X_i + X_j)/2 \leqslant m$ or $[(X_i - m) + (X_j - m)]/2 \leqslant 0$. That is, at most, $[n(n+1)/2] - k$ values satisfy $[(X_i - m) + (X_j - m)]/2 > 0$. By Problem 5.12, this means that the sum of the signed ranks is, at most, $[n(n+1)/2)] - k$. Since $(X - m)$ is continuous, symmetric with median zero, the sum of signed ranks has the distribution of Wilcoxon's W_+. Hence

$$\Pr(A_{(k)} \leqslant m) = \Pr\left(W_+ \leqslant \frac{n(n+1)}{2} - k\right)$$

and, similarly,

$$\Pr(A_{(l)} > m) = \Pr\left(W_+ \geqslant \frac{n(n+1)}{2} - l + 1\right)$$

These results can be used to find confidence coefficients appropriate to a proposed confidence interval $(A_{(k)}, A_{(l)})$, $k < l$. For example, if $n = 6$, $k = 3$, $l = 19$, $(A_{(3)}, A_{(19)})$ fails to contain the median if either $A_{(3)} > m$ or $A_{(19)} < m$.

$$\Pr(A_{(3)} > m) = \Pr(W_+ \geqslant 21 - 3 + 1) = \Pr(W_+ \geqslant 19) = 0.047$$
$$\Pr(A_{(19)} < m) = \Pr(W_+ \leqslant 21 - 19) = \Pr(W_+ \leqslant 2) = 0.047$$

The confidence coefficient is $1 - 2(0.047) = 0.906$.

5.14 (a) There are $2^4 = 16$ ways of allotting signs. $W_+ = 10$ is the maximum possible value and $\Pr(W_+ \geqslant 10) = 1/16 = 0.0625$. For $n = 4$, $E(W_+) = 5$, $V(W_+) = 7.5$. $\Pr(W_+ \geqslant 10) = \Pr(Z \geqslant (9.5 - 5)/\sqrt{7.5}) = 0.05$.

(b) $W_+ = 14$ is the maximum achievable and $\Pr(W_+ \geqslant 14) = 0.0625$. The modified mean is 7, and the modified variance is $54/4$ so that $\Pr(W_+ \geqslant 14) \approx \Pr(Z \geqslant (13.5 - 7)/\sqrt{54/4}) \approx \Pr(Z \geqslant 1.77) \approx 0.038$.

(c) Reported as a sample of four for the test and a one-sided significance level of 0.05 – which agrees with (a) using a normal approximation.

5.15 (a) $W_+ = 15$. If the approximating distribution is taken to be $N(7.5, 55/4)$, $\Pr(W_+ \geqslant 15) \approx \Pr(Z \geqslant (14.5 - 7.5)/\sqrt{54/4}) \approx \Pr(Z \geqslant 1.89) \approx 0.03$. The modified variance, taking account of the tie is

$$\frac{1}{24}(5 \times 6 \times 11) - \frac{1}{48}(5 \times 4 \times 6) = \frac{45}{4}$$

Since W_+ can only assume the values 0 3 6 9 12 15, we use a continuity correction of $3/2$. $\Pr(W_+ \geqslant 15) \approx \Pr(Z \geqslant (13.5 - 7.5)/\sqrt{45/4}) \approx \Pr(Z \geqslant 1.788) \approx 0.0367$.

(b) Reported as a sample of five for the test and a one-sided significance level of 0.03 – which agrees with the unmodified normal distribution.

5.16 (a)

Sample values	−1.5	1.5	1.5	−2	−2	2
Shared ranks	−2	+2	+2	−5	−5	+5

W_+	0	2	4	5	6	7	9	10	11	12	14	15	16	17	19	21
Frequency	1	3	3	3	1	9	9	3	3	9	9	1	3	3	3	1

For example, $W_+ = 7$ if one of the three values with magnitude 1.5 receives a positive sign *and* one of the three values with magnitude 2 receives a positive sign. $\Pr(W_+ \geqslant 9) = 44/64$.

(b)

Sample values	−1.5	1.5	1.5	−2	−2	2
Shared ranks	−4	4	4	−7	−7	7

W_+	0	4	7	8	11	12	14	15	18	19	21	22	25	26	29	33
Frequency	1	3	3	3	9	1	3	9	9	3	1	9	3	3	3	1

For example, $W_+ = 11$, if one of the three values with magnitude 1.5 and one of the three values with magnitude 2 are both assigned a positive sign. $\Pr(W_+ \geqslant 15) = 41/64$.

Reference

1. Pratt, J. and Gibbons, J., *Concepts of Nonparametric Theory*, Springer-Verlag, New York, 1981.

Chapter Six

Tests related to the hypergeometric distribution

6.1   Fisher's test

A graphologist claims to be able to distinguish between the handwriting of men and women. To test this claim, he is invited to classify 10 specimens of handwriting and is informed that five have been written by men and five by women. To focus on the reputed powers of discrimination and reduce corruption of the results from other sources, all specimens are of the same sentence, written with standard black ballpoints on similar paper and are presented in random order. The experimenter believes that the graphologist has no such ability. If this null hypothesis is true and the graphologist accepts the information provided, then he will divide the specimens into two groups of five and all $\binom{10}{5} = 252$ possible divisions of the 10 specimens have equal probability. Suppose the outcome of one such experiment results in the correct allocation of four of the men and, hence, four of the women. Is the outcome sufficiently impressive for the experimenter to reject his null hypothesis?

Now the four specimens written by men can be chosen from the 5 available in $\binom{5}{4} = 5$ ways and the one specimen classified as male but who is actually female can be chosen in $\binom{5}{1} = 5$ ways. Hence such a selection can be made up in $5 \times 5 = 25$ ways and each such selection determines the compositions of the remaining group. Note that the number of male and female specimens correctly classified must be equal. In the light of the graphologist's positive claim, the alternative is one-sided. Since there is just one way in which all the specimens could be correctly assigned, the probability of classifying at least eight correctly is $(25 + 1)/252 = 13/126 = 0.1032$, perhaps not sufficiently small to influence the experimeter's disbelief in the graphologist's claimed power.

The result of such an experiment can be conveniently displayed in a 2×2 table in which the columns relate the actual status and the rows the classification status. Thus, for the present result

		Actual		
		Male	Female	Total
	Male	4	1	5
Classified	Female	1	4	5
	Total	5	5	10

It is now apparent that the number of male specimens correctly classified has a hypergeometric distribution (see Section 1.3) with parameters $N = 10$, $w = 5$, $n = 5$. The observed significance level is $0.0992 + 0.0040 = 0.1032$.

Considerable prominence has been given in the literature to Fisher's discussion of an example of the above type and this kind of analysis has become known as *Fisher's exact test*. The example has become a classic and begins as follows: 'A lady claims that by tasting a cup of tea made with milk she can discriminate whether the milk or the tea infusion was first added to the tea.... Our experiment consists of mixing eight cups of tea, four in one way and four in the other and presenting them to the subject in random order. The subject has been told in advance of what the test will consist ....'

An important feature of both examples is that both the row and column totals are fixed and thus is referred to as the case of both margins fixed. The analysis has, however, rather wider application than its seemingly narrow scope would suggest. Suppose there is a disease, from which patients may recover spontaneously, though the preconditions for this fortunate result are unknown. From a total of N subjects who have developed the symptoms, a random sample of n_1 is chosen to undergo a new treatment which it is hoped will improve the recovery rate. The remaining $n_2 = N - n_1$ are given no treatment and act as controls. The results of a trial are displayed in the 2×2 table

	Treated	Control	Total
Recovered	a	b	m_1
Not recovered	c	d	m_2
Total	n_1	n_2	N

How, then, should we test the null hypothesis that the treatment is ineffective against the alternative that it leads to an improvement? We argue as follows. If the treatment is useless, then m_1 would have recovered in any case and the number a of those treated is classified only as a result of random selection. Originally the column totals were fixed. Arguing conditionally on the m_1 observed to have recovered, the row totals are fixed and the Fisher's exact test may now be applied. The only new feature is that the marginal totals need no longer be equal.

———— EXAMPLE 6.1 ————————————————————————————

Suppose the outcome of the treatment trial was

	Treated	Control	Total
Recovered	4	2	6
Not recovered	0	6	6
Total	4	8	12

From the distribution $h(12, 6, 4)$, the chance that at least four of the assumed six spontaneous recoverers are randomly assigned to the treatment group is 0.0303.

In the case of the graphologist, only better than random allocation would have cast doubt on the null hypothesis, while markedly worse performance would have left it undisturbed. But in a medical trial, especially for a new treatment, we might wish to take account of the negative effects, i.e. that the treatment actually made the patient worse. Thus a possible outcome is:

	Treatment	Control	Total
Recovered	0	6	6
Not recovered	4	2	6
Total	4	8	12

which also has probability 0.0303. Here the *p*-value is $2(0.0303) = 0.0606$ against a two-sided alternative.

For a particular trial which does not correspond to a symmetric distribution, the solution to the two-sided significance may not appear quite as comfortable.

──────── **EXAMPLE 6.2** ────────────────────────

Suppose that the outcome had been

	Treatment	Control	Total
Recovered	0	5	5
Not recovered	4	3	7
Total	4	8	12

The number of recovered patients who undertook treatment now has the $h(12, 5, 4)$ distribution. The probability of 0 is 0.0707. This suggests, against the two-sided alternative, that the treatment has some effect (for better or for worse), that the observed significance level should be taken as 0.1414. However, the distribution $h(12, 5, 4)$ is not symmetric. For the same marginal totals, the treatment effectiveness would be best supported by the outcome

	Treatment	Control	Total
Recovered	4	1	5
Not recovered	0	7	7
Total	4	8	12

and this result, conditional on five recovered, has probability 0.0101. There is no matching set of favourable outcomes with total probability 0.0707 in the other tail of $h(12, 5, 4)$. Indeed the next most favourable case, three of the four treated patients recover, has probability 0.1414.

As noted in Chapter 4, there is some support for the view that, in such a case, the observed *p*-value should be reported as $0.0707 + 0.0101 = 0.0808$. This takes into account only those outcomes favourable to the treatment with probability not exceeding that computed for the unfavourable case observed.

6.1.1 Large-sample approximation – both margins fixed

If the number of subjects involved in the trial is large, then an approximation may be employed. We recall the general lay-out for the observations

	Treatment	Control	Total
Recovered	a	b	m_1
Not recovered	c	d	m_2
Total	n_1	n_2	N

If N is large and neither of the proportions n_1/N, m_1/N is small, then the distribution $h(N, m_1, n_1)$ can be approximated by a normal distribution with mean $n_1 m_1/N$ and variance $(m_1 m_2 n_1 n_2)/(N-1)N^2$ – see Section 1.3. It is usual to incorporate a continuity correction of $1/2$.

─────── EXAMPLE 6.3 ───

	Treatment	Control	Total
Recovered	40	10	50
Not recovered	35	15	50
Total	75	25	100

For the case $N = 100$, $m_1 = 50$, $n_1 = 75$, it is required to calculate the approximate probability that $A \geqslant 40$, where A is the number of treated subjects recovering:

$$E(A) = 75 \times 50/100 = 37.5 \quad \text{and} \quad V(A) = \frac{50 \times 50 \times 75 \times 25}{100 \times 100 \times 99} \approx (2.176)^2$$

so that

$$\Pr(A \geqslant 40) = \Pr\left(\frac{A - 37.5}{2.176} \geqslant \frac{40 - 37.5}{2.176}\right) \approx \Pr\left(Z \geqslant \frac{39.5 - 37.5}{2.176}\right) = 0.179$$

after applying a continuity correction, where Z has the distribution $N(0, 1)$.

Here the approximate observed significance level in a two-sided alternative would be stated as $2(0.179) = 0.358$. Note that applying the normal approximation implicitly places equal probability in both tails.

───

6.1.2 The chi-squared test of fit

The reader may already have some experience of the use of the χ^2 distribution to test the goodness-of-fit in a contingency table. N elements are each assessed as to the possession or absence of two different attributes I, II. Suppose the probability that an element has attribute I is p_1 and that it has attribute II is p_2. If the attributes are *independent* then the probability that an element possesses both attributes is p_1p_2 and the expected number to be found in a sample of N elements is Np_1p_2. However, p_1, p_2 are, in general, unknown and have to be estimated from the sample results, which may be displayed in tabular form as

Attribute I

		Present	Absent	Total
	Present	a	b	m_1
Attribute II	Absent	c	d	m_2
	Total	n_1	n_2	N

The obvious estimators of p_1, p_2 are the observed proportions $\hat{p}_1 = n_1/N$, $\hat{p}_2 = m_1/N$. Hence, for example, the expected number with both attributes, under independence, is estimated by $N\hat{p}_1\hat{p}_2 = n_1m_1/N$, with similar results for the other combinations in the table. Pearson showed that the statistic X^2, which in terms of the observed and expected frequencies, is given by

$$X^2 = \frac{\left(a - \dfrac{m_1 n_1}{N}\right)^2}{\dfrac{m_1 n_1}{N}} + \frac{\left(b - \dfrac{m_1 n_2}{N}\right)^2}{\dfrac{m_1 n_2}{N}} + \frac{\left(c - \dfrac{n_1 m_2}{N}\right)^2}{\dfrac{n_1 m_2}{N}} + \frac{\left(d - \dfrac{n_2 m_2}{N}\right)^2}{\dfrac{n_2 m_2}{N}} \qquad (6.1)$$

has approximately the $\chi^2(1)$ distribution if N is large and none of the estimated expected frequencies is too small. In Problem 6.3 we show that X^2 can be written in the form

$$X^2 = \frac{N(ad - bc)^2}{n_1 n_2 m_1 m_2} \qquad (6.2)$$

The approximation is improved by applying a continuity correction. This consists of reducing the absolute value of the difference between each observed frequency and its estimated expectation by $1/2$. This modifies equation (6.2) to give

$$X^2 = N \frac{[|ad - bc| - N/2]^2}{n_1 n_2 m_1 m_2} \qquad (6.3)$$

X^2 can be used to test the null hypothesis that the attributes are independent against the two-sided alternative hypothesis that they are not independent. The null hypothesis becomes suspect if the differences of the observed frequencies a, b, c, d from their predicted expectation are greater than can be reasonably accepted. The differences will be reflected in the value of X^2, which will be larger than is likely when the null hypothesis is true.

———— **EXAMPLE 6.4** ————

For the 2 ×2 table in Example 6.3

40	10	50
35	15	50
75	25	100

we calculate from equation (6.2) $X^2 = 100(40 \times 15 - 10 \times 35)^2/(75 \times 25 \times 50 \times 50) = 4/3$ with $\Pr(X^2 \geqslant 4/3) \approx 0.2482$ (MINITAB value). Using the continuity corrected formula (6.3),

$$X^2 = 100(250 - 50)^2/(75 \times 25 \times 50 \times 50) = 64/75$$

and $\Pr(X^2 \geqslant 64/75) \approx 0.3556$. This second result agrees closely with the two-sided p-value obtained using a normal approximation.

6.1.3   *Use of MINITAB*

Unless you have stored a macro for calculating hypergeometric probabilities, there is no direct method of evaluating p-values for Fisher's exact test. However, if the rows of a 2×2 table are read into C1, C2, then the command CHISQUARE C1 C2, will provide the value of X^2 as a measure of fit.

———— **EXAMPLE 6.5** ————

For the table

40	10	50
35	15	50
75	25	100

entering 40 35 in C1 and 10 15 in C2, the command CHISQUARE C1 C2 produces

	C1	C2	TOTAL
	40	10	50
	37.50	12.50	
	35	15	50
	37.50	12.50	
TOTAL	75	25	100

CHISQ = 0.167 + 0.500 +
            0.167 + 0.500 = 1.333

DF = 1

Notes:

1. No p-value is supplied, and this may easily be found from the command

   CDF 1.333;
   CHISQUARE 1.

   The p-value is then $1 - \text{CDF}$ value
2. A warning concerning the number of cells with expected frequencies less than five will be provided.

6.1.4   *Problems*

6.1   In a population of N elements, R are of type A, the remainder of type 'not A'. A random sample of n elements is drawn without replacement. If X is the number of elements in the sample of type A, explain why

$$\Pr(X = r \,|\, N, R, n) = \binom{R}{r}\binom{N-R}{n-r} \Big/ \binom{N}{n}$$

Verify that this probability is equal to

$$\binom{n}{r}\binom{N-n}{R-r} \Big/ \binom{N}{R}$$

and briefly describe how this result may be applied.

From 14 patients suffering from a minor disease, seven are selected at random and given a new treatment. Subsequently four of these are judged to have recovered. The remaining seven received no treatment and two recovered during the trial. Test at the 5% level whether the treatment had a significantly good effect.

6.2   An experimenter shuffles 15 cards of which seven are red and eight are black. A clairvoyant is given this information and invited to divide the cards into their correct colours through inspecting the backs of the cards. If the clairvoyant has no special powers, use Fisher's test to calculate the probability that he nominates the correct colour of at least 13 cards. Find the corresponding probability if the clairvoyant disregards the information provided and divides the cards into a pile of six which he calls 'red' and one of nine which he calls 'black'.

6.3   In equation (6.1), show that $(a - n_1 m_1/N)^2 = (ad - bc)^2/N^2$, and verify that all the numerators in the equation have this value. Hence, or otherwise, deduce that the goodness-of-fit statistic X^2 satisfies

$$X^2 = N(ad - bc)^2/(n_1 n_2 m_1 m_1 m_2)$$

(This is the result in equation (6.2).)

6.4 If in a 2×2 table the entry A has the hypergeometric distribution, show that $[A - E(A)]^2/V(A) = (N-1)X^2/N$, where X^2 is the goodness-of-fit statistic.

6.2 Runs test

Suppose a computer is reputed to be generating a sequence of random binary digits – that is, each digit is a 1 with probability p, or a 0 with probability $1-p$, independently of all previous output. Two different attributes might be tested – the independence of successive sequence members and the constancy of the probability p. If there is dependence, each digit might be influenced by its predecessors and then there would be a tendency for longer or shorter sequences of the same digit than would otherwise be expected. Unbroken sequences of symbols of either kind are said to constitute a *run*. The total number of runs will be denoted by R.

——— EXAMPLE 6.6 ———

If six digits are generated then there are $2^6 = 64$ different possible sequences. The number of sequences containing three ones and three zeros is $\binom{6}{3} = 20$, since this is the number of ways of selecting three places from six to be ones. Among these 20, it is possible to obtain 2, 3, 4, 5, 6 runs in all. These are

Two runs	111000	000111			2 cases
Three runs	100011	110001	011100	001110	4 cases
Four runs	101100	100110	110100	110010	
	010011	011001	001011	001101	8 cases
Five runs	101001	100101	010110	011010	4 cases
Six runs	101010	010101			2 cases

The conditional probability that the number of runs is less than 4 is $(2+4)/20 = 3/10$.

6.2.1 *The probability distribution of* R

Suppose we observe N independent random variables, $X_1, X_2, \ldots, X_N$, where the outcome of each X_i is classified as being of Type I or Type II with probabilities p, $1 - p$. Then the conditional probability that any particular collection of n of the X_i are of Type I and the remaining $m = N - n$ are of Type II is

$$1 \Big/ \binom{m+n}{n}$$

That is, each of the possible

$$\binom{m+n}{n}$$

collections has equal probability.

To find the conditional probability of just r runs in all, we distinguish between the cases when r is even or odd. Runs of either type alternate, hence if $r = 2j$, then there must be j runs of each type, while if $r = 2j + 1$, there must be j runs of one type and $j + 1$ runs of the other type. Now any run of type I symbols must contain at least one symbol of that kind. Hence to obtain j runs of Type I, we must arrange n symbols of Type I into j groups so that no group is empty. This can be done in

$$\binom{n-1}{j-1}$$

distinguishable ways. In combination with each of these, there are

$$\binom{m-1}{j-1}$$

distinguishable ways of obtaining j runs of Type II. Since the alternation of runs may commence with one of either type,

$$\Pr(R = 2j) = 2\binom{m-1}{j-1}\binom{n-1}{j-1} \bigg/ \binom{m+n}{n}$$

Similarly, if $r = 2j + 1$

$$\Pr(R = 2j + 1) = \left[\binom{n-1}{j-1}\binom{m-1}{j} + \binom{n-1}{j}\binom{m-1}{j-1}\right] \bigg/ \binom{m+n}{n}$$

─────── EXAMPLE 6.7 ───────

Let $m = 4$, $n = 4$. If $r = 5$, there are

$$\binom{3}{1}\binom{3}{2} + \binom{3}{2}\binom{3}{1} = 18$$

distinguishable ways of obtaining just five runs. If the

$$\binom{4+4}{4} = 70$$

arrangements of the eight symbols are equally likely, the probability of five runs is $18/70$.

─────── EXAMPLE 6.8 ───────

Let $m = 3$, $n = 3$. If $r = 4$, there are

$$2\binom{2}{1}\binom{2}{1} = 8$$

ways of obtaining just four runs. If the

$$\binom{3+3}{3}$$

possible arrangements of the six symbols are equally likely, the probability of four runs is 8/20.

The distribution of R is displayed in Table A3.2 in Appendix 3 for $4 \leqslant m \leqslant n \leqslant 15$. Since the distribution of R is only symmetric when $m = n$, the upper and lower tail probabilities have to be computed separately.

--- **EXAMPLE 6.9** ---

Let $m = 8$, $n = 9$. From Table A3.2, $\Pr(R \leqslant 5) = 0.02028$ but $\Pr(R \geqslant 14) = 0.0415$. However, for $m = 9$, $n = 9$, $\Pr(R \leqslant 6) = \Pr(R \geqslant 13) = 0.04447$. When the distribution of R is not symmetric, there may be some difficulties concerning the reporting of two-sided p-values (see Chapter 4).

6.2.2 *Normal approximation*

If R is the total number of runs of either kind, then it can be shown that

$$E(R) = 1 + 2mn/(m+n)$$
$$V(R) = 2mn(2mn - m - n)/[(m+n)^2(m+n-1)]$$

These results are established in Problem 6.8. Furthermore, if both $m \geqslant 10$ and $n \geqslant 10$, it can be shown that R is approximately normally distributed, that is, $[R - E(R)]/\sqrt{V(R)}$ is approximately $N(0, 1)$.

--- **EXAMPLE 6.10** ---

In a random sequence of 20 symbols of each kind, we calculate the approximate probability of 26 runs or more.

From equation (6.3), $E(R) = 21$, from equation (6.4), $V(R) = 380/39 \approx (3.121)^2$.

$$\Pr(R \geqslant 26) \approx \Pr(Z \geqslant (26 - 21)/3.121) \approx \Pr(Z \geqslant 1.60) = 0.0548.$$

In Example 6.1 the number of digits of a particular type found was a random variable. Tests on the number of runs would then be conditional on the number found of that type. In another type of application, the number of elements of Type I is fixed in advance. Suppose some examination results are arranged in increasing order and the women's and men's marks define elements of Types I and II respectively. In such a case, too few runs suggests clumping – particularly when the marks of one type tend to be lower than those of the other type. We note that a pattern such as MMMMFFFFFFMMM, which corresponds to one misconceived stereotype about examination performance, might fail to appear significant when exposed to the usual *t*-test applied to the original marks.

──────── EXAMPLE 6.11 ────────────────────────────────

Twelve men and eight women take an examination, the marks obtained being

Men	80, 72, 77, 81, 73, 78, 62, 60, 68, 79, 67, 61
Women	69, 66, 65, 63, 70, 64, 58, 59

If the marks are arranged in increasing order and then replaced by M or F, as appropriate, the total number of runs is six. From Table A3.2 the probability of six or fewer runs is 0.02461. There is therefore significant evidence of clumping.

──

6.2.3   Use of MINITAB

MINITAB finds the total number of runs in a sequence of numbers, each member of which can be classified as belonging to one of two well-defined classes. If any sequence of numbers is set in C1, then each number can be so classified according, for given K, as it is $\leq K$ or $>K$. The command RUNS   K   C provides the observed and expected number of runs, the number of values above and not above K, and a two-sided significance level obtained from a normal approximation (without a continuity correction).

──────── EXAMPLE 6.12 ────────────────────────────────

If   −3 2 −8 7 −4 3 −4 5 0 6 −1 6 1 2 −1 9 −4 8
−2 2 −1 −1 −1 2 2 2 0 0 0 3 3 3 1 1 1 1 4 4 4
4 are set in C1 and k is selected to be 1, the command RUNS 1 C1 supplies

    THE OBSERVED NUMBER OF RUNS = 26
    THE EXPECTED NUMBER OF RUNS = 21.000
    20 OBSERVATIONS ABOVE K 20 BELOW
    THE TEST IS SIGNIFICANT AT 0.1095
    CANNOT REJECT AT ALPHA = 0.05

Since the normal approximation is regarded as unsatisfactory when the number of sample values above or below K is less than 10 a warning note will be appended in such cases. If K is not specified, the default value will be the average of C1.

When the sample is numerical, there are other possibilities. We might be interested in detecting whether there is a trend, in which case a natural choice of K would be the sample median. A paucity of runs indicates that either many of the early values are less than the median while the latter exceed it, or conversely.

On the other hand, each element of the sample may belong to one of two exclusive categories, in which case the data may have to be coded before applying the RUNS command. In Example 6.11 the data should be set in C1 and indexed in C2 by 1 if male or by 2 if female. The sequence

>   SORT C1 C2 C3 C4
>   RUNS 1 C4

supplies

>   OBSERVED NO OF RUNS = 6
>   EXPECTED NO OF RUNS = 10.600
>   8 OBSERVATION ABOVE K 12 BELOW
>   THE TEST IS SIGNIFICANT AT 0.0276

Since the programme deals with numbers, any symbols must be suitably coded. Thus in ABABABABABABABABABABABAAABBBAAABBBAAAABBBB, we may replace A by 1 and B by 2. For the sequence of letters above, after coding, the RUNS command with $K = 1$ supplies the same output as in Example 6.12.

The effect of selecting a K is to impose a dichotomy on the sample values. For a sequence of three different letters, coded say with A = 1 B = 2 C = 3, then $K = 2$ will amalgamate A with B, while $K = 1$ will amalgamate B with C.

6.2.4   Problems

6.5   There are three elements of Type I and four elements of Type II. If these are arranged at random in a row, calculate the probability function for the total number R of runs and deduce $E(R)$.

6.6   There are n elements of Type I and m elements of Type II. If these are arranged at random in a row and K is the number of runs of Type I, show that

$$\Pr(K = k) = \binom{n-1}{k-1}\binom{m+1}{k}\Big/\binom{m+n}{n} \qquad 1 \leq k \leq n$$

(Hint: there must be $k-1$, k, or $k+1$ runs of Type II.)

6.7   For Problem 6.6, verify that $\sum_{k=1}^{n} \Pr(K = k) = 1$ and calculate $E(K)$. Deduce the expected total number of runs of either kind.

6.8 n white balls and m black balls are arranged at random in a row. Let $X_i = 1$ if the ith ball starts a run of balls of the same colour and $X_i = 0$ otherwise. Note $X_1 \equiv 1$ and for $i \geq 2$, $X_i = 1$ if and only if balls $i-1$, i have different colours. Show that

$$\Pr(X_i = 1) = 2mn/[(m+n)(m+n-1)]$$

Explain why the total number R of runs can be written

$$R = \sum_{i=1}^{m+n} X_i = 1 + \sum_{i=2}^{m+n} X_i$$

and deduce the mean of R. To calculate the variance of R, show that for $i \geqslant 2$,

$$\Pr(X_i = 1 \text{ and } X_{i+1} = 1) = mn/[(m+n)(m+n-1)],$$
$$\Pr(X_i = 1 \text{ and } X_j = 1)$$
$$= 4mn(m-1)(n-1)/[(m+n)(m+n-1)(m+n-2)(m+n-3)]$$

for $i \neq j-1, j, j+1$. Write

$$R^2 = 1 + 2 \sum_{i=2}^{m+n} X_i + \left(\sum_{i=2}^{m+n} X_i\right)^2$$

Find $E(R^2)$ and deduce $V(R)$. (Hint: $X_i^2 \equiv X_i$.)

6.3   The Wilcoxon, Mann–Whitney rank sum test

The test we now describe is variously termed the *Wilcoxon rank sum test*, the *Mann–Whitney test* or simply the *rank sum test*, equivalent definitions having been given by these authors.

Suppose it is required to test the effect of a new treatment on a particular type of element. To detect a possible influence, it will be necessary to compare the response of the treated elements with other elements which have either received an old treatment or which have not been treated and act as controls. Suppose the members of a random sample of n elements receive the treatment and we record, on some suitable response variable, the values $x_1, x_2, \ldots, x_n$. An independent random sample of m elements act as controls and their responses on the same response variable are $y_1, y_2, \ldots, y_m$. To analyse the results, it is frequently assumed that the effect of the treatment is to add (or subtract) a fixed amount to the neutral (untreated) response of any element. The overall effect is to increase (or decrease) the mean (or median) of the distribution of the responses of the treated elements. That is, the treatment and control distributions have the same shape but differ in location. The natural statistic to study is $\bar{X} - \bar{Y}$, where $\bar{X}, \bar{Y}$ are the means of the samples and $\bar{x}, \bar{y}$ are the obtained means.

We first consider a permutation test. If there is no treatment effect, then $x_1, x_2, \ldots, x_n, y_1, y_2, \ldots, y_m$ are effectively a random sample of size $n + m$ from the same distribution. Hence any one of the

$$\binom{n+m}{n}$$

selections of n elements from these $n + m$ values could have qualified as a *possible* result for the treatment values $x_1, x_2, \ldots, x_n$, with equal probability

$$1 \Big/ \binom{n+m}{n}$$

Now if c is the sum of the $n + m$ sample values, since $c = m\bar{y} + n\bar{x}$

$$\bar{x} - \bar{y} = \frac{1}{m}(m\bar{x} - m\bar{y})$$

$$= \frac{1}{m}[(n+m)\bar{x} - c]$$

$$= \frac{n+m}{mn}\left[\sum_{i=1}^{n} x_i - \frac{nc}{n+m}\right] \qquad (6.4)$$

From equation (6.4), we see that essentially we need a listing of the possible values of $\sum_{i=1}^{n} x_i$ which can arise from the selections, given the data set.

——— EXAMPLE 6.13 ———

A random sample of four values, from a continuous distribution with median m_1, yielded

$$x_1 = 4.2, \ x_2 = 3.1, \ x_3 = 1.0, \ x_4 = 7.9$$

An independent random sample of four values, from a continuous distribution with mean m_2, yielded

$$y_1 = 6.2, \ y_2 = 5.1, \ y_3 = 3.0, \ y_4 = 8.9$$

The null hypothesis is that $m_1 = m_2$, the alternative that $m_1 < m_2$. The pooled sample values, in order of magnitude, are

> *1.0   3.0   3.1   4.2*   5.1   6.2   7.9   8.9

The sum of the italicized x-values is 16.2. We require the probability that the selection has sum less than or equal to 16.2. There are fourteen selections of size four with sum not exceeding 16.2, namely

1.0	3.0	3.1	4.2	1.0	3.0	4.2	5.1	1.0	3.1	4.2	6.2
1.0	3.0	3.1	5.1	1.0	3.0	4.2	6.2	1.0	3.1	4.2	7.9
1.0	3.0	3.1	6.2	1.0	3.0	4.2	7.9	1.0	3.1	5.1	6.2
1.0	3.0	3.1	7.9	1.0	3.0	5.1	6.2	3.0	3.1	4.2	5.1
1.0	3.0	3.1	8.9	1.0	3.1	4.2	5.1				

There are

$$\binom{8}{4} = 70$$

selections of four in all. Hence the required probability is $14/70 = 0.2$.

6.3.1   *Ranking the data*

Such a permutation test, although it enjoys the benefit of being conditionally distribution-free, suffers from the inconvenience that it cannot be tabulated. This defect can be removed by working with the ranks as determined by ordering the pooled sample values, assuming that no values are equal.

───────── EXAMPLE 6.14 ───

For the data in Example 6.13, the ranks run from 1 to 8 and the values in italic have ranks 1, 3, 4, 7 with sum 15. The selections of four which have rank sum not exceeding 15 are

1	2	3	4		1	2	4	5		1	2	5	7	2 3 4 5
1	2	3	5		1	2	4	6		1	3	4	5	2 3 4 6
1	2	3	6		1	2	4	7		1	3	4	6	
1	2	3	7		1	2	4	8		1	3	4	7	
1	2	3	8		1	2	5	6		1	3	5	6	

There are 17 such selections, whence the (one-sided) significance level is $17/70 \approx 0.243$.

───

The following procedures for comparing a treatment with a control will have the same probability structure as that discussed for comparing independent samples. A random sample of $n + m$ is drawn from a population. From this *finite* population, a random sample of n is drawn, *without replacement*, and these elements receive the treatment. The remaining m elements act as controls.

───────── EXAMPLE 6.15 ───

A new drug is thought to extend the lives of patients suffering from a disease. Ten patients are available with approximately equal symptom conditions. Five of these are selected at random and receive the drug. The survival times, in months, were

Treated	3.2	5.5	6.8	12.3	15.5
Control	0.2	2.3	2.9	4.3	5.9

When the 10 values are placed in increasing order of magnitude, it is found that the treated persons have ranks 4, 6, 8, 9, 10 . The sum of these treatment ranks is 37 and looks impressively high. Alternatively the sum of the control ranks is 18 and looks depressingly low. There are seven selections of five ranks from 10 with rank sum less than or equal to 18. In the absence of a treatment effect, the

$$\binom{10}{5} = 252$$

possible selections of ranks have equal probability, and the probability of a sum not exceeding 18 is $7/252 \approx 0.028$.

6.3.2   The rank-sum statistics

In general, if R_i is the rank of the ith treatment element, then $R_t = \sum_{i=1}^{n} R_i$ is the sum of the n treatment ranks, and it is possible to tabulate the critical levels for each n and size m of the control group. Some reduction in the amount of tabulation is effected by noting that R_t must be at least $\frac{1}{2}n(n+1)$, as when it is composed of the first n ranks. Also R_t is a maximum when the control elements, with rank sum R_c, comprise the first m ranks. Since the sum of all $n+m$ ranks is $\frac{1}{2}(n+m)(n+m+1)$, then R_t satisfies

$$\tfrac{1}{2}n(n+1) \leqslant R_t \leqslant \tfrac{1}{2}(n+m)(n+m+1) - \tfrac{1}{2}m(m+1) \tag{6.5}$$

If we define $U_t = R_t - \frac{1}{2}n(n+1)$, then from equation (6.5), after simplification,

$$0 \leqslant U_t \leqslant nm \tag{6.6}$$

Similarly, if $U_c = R_c - \frac{1}{2}m(m+1)$, then

$$0 \leqslant U_c \leqslant nm \tag{6.7}$$

Moreover, $U_t + U_c = nm$.

    U_t, U_c are called the *Mann–Whitney* statistics after the authors who introduced them. The statistics R_t, R_c were originally employed by *Wilcoxon*.

6.3.3   Distributions of U_t, U_c

We shall show that U_c, U_t have the same distribution. Suppose the ranks of the treated elements are $R_1, R_2, \ldots, R_n$. To this sample of n ranks we associate another sample of n ranks, $n+m+1-R_1, n+m+1-R_2, \ldots, n+m+1-R_n$. That is, to the rank R_i from the bottom, we associate the rank $n+m+1-R_i$ which is R_i steps down from the greatest rank $n+m$. The sum of the ranks for this second sample is

$$\sum_{i=1}^{n} (n+m+1-R_i) = n(n+m+1) - \sum_{i=1}^{n} R_i$$

$$= nm + n(n+1) - [U_t + \tfrac{1}{2}n(n+1)]$$

that is,

$$\sum_{i=1}^{n} (n+m+1-R_i) - \tfrac{1}{2}n(n+1) = nm - U_t \tag{6.8}$$

Thus for each sample with $U_t = k$, there is another sample for which $U_t = nm - k$. But all samples of n ranks have equal probability, whence the distribution of U_t is symmetric. Since $U_c = nm - U_t$,

$$\Pr(U_t \leqslant x) = \Pr(U_t \geqslant nm - x) = \Pr(nm - U_t \leqslant x) = \Pr(U_c \leqslant x)$$

and U_t, U_c have the same probability distribution. Table A3.3 in Appendix 3 supplies the common distribution function of U_c and U_t for $2 \leqslant m \leqslant n \leqslant 15$.

――――― EXAMPLE 6.16 ―――――――――――――――――――――――――――

If $n = m = 5$, and $R_c = 18$, then $U_c = 3$. From Table A3.3, $\Pr(U_c \leqslant 3) = 0.02778$. Hence the obtained R_c has approximate p-value 0.028 in a one-sided test.

――――― EXAMPLE 6.17 ―――――――――――――――――――――――――――

In Example 6.15, we required $\Pr(U_t \geqslant 22)$. From the symmetry, this is also $\Pr(U_t \leqslant 3) = 0.02778$.

Furthermore, if we require a two-sided test of the null hypothesis of no treatment effect, then we calculate $U_t, U_c = nm - U_t$ and examine whether the minimum of U_t, U_c is significantly small. This is equivalent to examining whether U_t, say, is either significantly small or large, since $U_c = nm - U_t$ implies that if U_t is large then U_c is small.

6.3.4   *Mean and variance of* U_t, U_c

The expected value of U_c is immediate, since we have seen that its distribution is symmetric. Hence

$$E(U_t) = E(U_c) = \tfrac{1}{2}nm \tag{6.9}$$

Writing $\bar{R}_t = (1/n) \sum_{i=1}^{n} R_i$, we have $V(U_t) = V(R_t - \tfrac{1}{2}n(n+1)) = V(R_t)$ so that

$$V(U_t) = n^2 V(\bar{R}_t) \tag{6.10}$$

To obtain $V(\bar{R}_t)$, we apply the result of Appendix 2, Example A2.2, to the sample mean of n ranks from a finite population of ranks $1, 2, \ldots, n + m$. We have

$$V(\bar{R}_t) = \frac{\sigma^2}{n} \frac{(n + m - n)}{n + m - 1} = \frac{m\sigma^2}{n(n + m - 1)} \tag{6.11}$$

where,

$$\sigma^2 = (n + m + 1)(n + m - 1)/12. \tag{6.12}$$

Substituting equations (6.11) and (6.12) in equation (6.10) we obtain

$$V(U_t) = nm(n + m + 1)/12 \qquad (6.13)$$

These results are required to enable a normal approximation to be used when both n, m are large.

6.3.5 Normal approximation

In the absence of ties, we have $E(R_t) = n(n + m + 1)/2$, $V(R_t) = nm(n + m + 1)/12$ from equations (6.9) and (6.13), and it can be shown that $[R_t - E(R_t)]/\sqrt{V(R_t)}$ is approximately $N(0, 1)$.

────── EXAMPLE 6.18 ──────────────────────────────

In Example 6.15, with $m = n = 5$, we observed that $R_t = 37$ and noted that $\Pr(R_t \geqslant 37) \approx 0.028$. Suppose we use a normal approximation, together with a continuity correction of $1/2$. Then $E(R_t) = 27.5$, $V(R_t) = 275/12 = (4.787)^2$ and

$$\Pr(R_t \geqslant 37) \approx \Pr(Z \geqslant (36.5 - 27.5)/4.787) \approx \Pr(Z \geqslant 1.88) = 0.0301$$

6.3.6 Effect of ties

────── EXAMPLE 6.19 ──────────────────────────────

Suppose the data is

Treatment	3.0	5.5	6.8	12.3	15.5
Control	0.2	3.0	3.0	4.3	6.8

There is one tie of extent three, straddling ranks 2, 3, 4 and another of extent two covering ranks 7, 8 . Such ties can be resolved by any of the ways discussed in some detail in connection with the Wilcoxon signed-rank test (Section 5.2).

We consider the method of shared ranks. The observations in order of magnitude are

0.2 *3.0* 3.0 3.0 4.3 *5.5* 6.8 *6.8* *12.3* *15.5*

with shared ranks

1 *3* 3 3 5 *6* 7.5 *7.5* *9* *10*

The sum of the (italic) treatment ranks is 35.5. Suppose we are interested in whether the treatment sum of ranks is significantly high. Then, for an exact *p*-value, we must list all the selection of five ranks with a sum of ranks of at least 35.5,

$$
\begin{array}{lll}
10 + 9 & + 7.5 + 7.5 + 6 = 40 & \text{one case} \\
10 + 9 & + 7.5 + 7.5 + 5 = 39 & \text{one case} \\
10 + 9 & + 7.5 + 7.5 + 3 = 37 & \text{three cases} \\
10 + 9 & + 7.5 + 6 \ \ + 5 = 37.5 & \text{two cases} \\
10 + 9 & + 7.5 + 6 \ \ + 3 = 35.5 & \text{six cases} \\
10 + 7.5 + 7.5 + 6 \ \ + 5 = 36 & \text{one case}
\end{array}
$$

giving a total of 14 cases. Hence, if all 252 selections of 5 from 10 ranks have equal probability,

$$\Pr(R_t \geqslant 35.5) = 14/252 \approx 0.056$$

In the method of rank sharing, $E(R_t)$ remains unchanged. The variance of R_t is modified since a finite population is sampled in which some ranks appear more than once and others not at all. In Appendix 2, it is shown that if the *i*th tie is of extent t_i and $N = n + m$, then

$$V(R_t) = \frac{mn}{12N(N-1)} \left[N(N^2 - 1) - \sum_i t_i(t_i^2 - 1) \right]$$

——— EXAMPLE 6.20 ———

Thus in Example 6.19, $m = n = 5$, $N = 10$, $t_1 = 3$, $t_2 = 2$.

$$V(R_t) = 25[10 \times 99 - 3(9 - 1) - 2(4 - 1)]/(12 \times 10 \times 9) = 200/9 = (4.714)^2.$$

Hence, using *this* variance for the normal approximation, and a continuity correction

$$\Pr(R_t \geqslant 35.5) \approx \Pr(Z \geqslant (35 - 27.5)/4.714)$$
$$\approx \Pr(Z \geqslant 1.59) = 0.0559$$

where *Z* has the distribution $N(0, 1)$.

6.3.7   *Confidence interval for a location parameter*

Suppose $x_1, x_2, \ldots, x_n$ be the obtained values of a random sample from a continuous distribution and $y_1, y_2, \ldots, y_m$ be an independent random sample from another continuous distribution – these distributions being thought to differ only by a location parameter θ. Now the hypothesis $\theta = \theta_0$ can be tested by pooling the residues $y_i - \theta_0$, $i = 1, 2, \ldots, m$ with the x_j, $j = 1, 2, \ldots, n$ and applying the rank sum test to the resultant ranks. We might prefer to have a confidence interval for θ based on the obtained sample values. It is clear that if a sufficiently large θ_2 is *subtracted* from the y_i, then the $y_i - \theta_2$ will all be below all the x_j and the sum of the ranks, R_t, of the x_j, will be maximal. If a sufficiently large θ_1 is *added* to all the y_i, then all the $y_i + \theta_1$ will be above the x_j and R_t will be minimal. To obtain a confidence level with coefficient at least $1 - \alpha$, we require (θ_1, θ_2) to contain all values of θ which would be accepted at the significance level α.

──────── EXAMPLE 6.21 ────────

When the data of Example 6.13 are placed in increasing order, we obtain

x_1	y_1	x_2	x_3	y_2	y_3	x_4	y_4
1	3.0	*3.1*	*4.2*	5.1	6.2	*7.9*	8.9

The sum R_t of the ranks of the italic values is 15 . It is required to calculate a central confidence interval for θ with confidence coefficient at least 0.90. From tables, $\Pr(R_t \leqslant 11) \leqslant 0.05$ and $\Pr(R_t \geqslant 25) \leqslant 0.05$, so that $\Pr(12 \leqslant R_t \leqslant 24) \geqslant 0.90$. Now if the y values are moved en bloc, either to the left or to the right, it is apparent that changes in R_t do not occur continuously but in jumps when some y_i, x_j change places. Thus the distances between such pairs will determine changes in R_t arising from increasing or decreasing all the y values. Now x_2, with rank 3, *exceeds* y_1, which has rank 2. The distance between their values is $3.1 - 3 = 0.1$, and this is the closest distance between such a pair. Hence if slightly more than 0.1 is added to all the y data then x_2 and y_1 exchange ranks and R_t falls to 14. The next closest are x_1 and y_2, with y_2 *exceeding* x_1 by 0.9. If slightly more than 0.9 is subtracted from each original y value then x_1 and y_2 exchange ranks and R_t increases to 16.

Hence movements of size $d_{ij} = y_i - x_j$ will produce jumps of one (or more) in R_t. We need a table of the sixteen values of d_{ij} – termed the *Walsh differences*.

		x		
y	1	3.1	4.2	7.9
3.0	2.0	−0.1	−1.2	−4.9
5.1	4.1	2.0	0.9	−2.8
6.2	5.2	3.1	2.0	−1.7
8.9	7.9	5.8	4.7	1

The differences, in numerical order, are

$$-4.9 \quad -2.8 \quad -1.7 \quad -1.2 \quad -0.1 \quad 0.9 \quad 1 \quad 2 \quad 2 \quad 2 \quad 3.1 \quad 4.1 \quad 4.7$$
$$5.2 \quad 5.8 \quad 7.9$$

To lower R_t to 12, there must be a fall of $15 - 12 = 3$. The third (in absolute magnitude) is negative difference -1.7. In accordance with the previous discussion, if any number more than 1.7 but less than 2.8 is added to the y values, then $R_t = 12$. Hence $\theta_1 = -2.8$. Similarly, to raise R_t to 24, there must be a rise of $24 - 15 = 9$. If any number between 5.2 and 5.8 is subtracted from the y values, then $R_t = 24$. Hence $\theta_2 = 5.8$. The required confidence interval is the open interval $(-2.8, 5.8)$. To reach the upper end of this confidence interval, we stepped over a triple in the differences of magnitude $+2$. This led to an increase of 3 in R_t. Should such a tie appear at an end of an interval, we have to modify the procedure by adopting a different confidence coefficient.

Suppose we require a *point estimator* for the location parameter θ. One approach is to find an estimator $\hat{\theta}$ which, when used to reduce each member of the y sample, provides a set of translated values, $y_1 - \hat{\theta}, y_2 - \hat{\theta}, \ldots, y_m - \hat{\theta}$, which appears to be a random sample of size m from the same source distribution which generated the x values. How then should $\hat{\theta}$ be chosen so as to make the 'appearance' as close as possible? A common recommendation, also adopted by MINITAB, is to take $\hat{\theta}$ as the median of the original Walsh differences (see reference 1).

It deserves noting that if $\hat{\theta}$ is 'subtracted' from each of the y values then about half the resulting Walsh differences, in the absence of ties, will be positive. In the light of the result of Problem 6.13, this forces the sum of the ranks of the translated values to be close to the expected value for a random sample of size n from $n + m$ ranks.

6.3.8   Use of MINITAB

In this section the reader should bear in mind that MINITAB actually uses the WILCOXON rank sum statistic but describes it as MANN–WHITNEY.

The single command MANN–WHITNEY C1 C2 provides several analyses concerning the medians m_1 and m_2 of two distributions, samples from which are set in C1 and C2. In particular, using ETA1 and ETA2 for m_1 and m_2, MINITAB displays:

1. A point estimate for $m_1 - m_2$
2. A 95% confidence interval for $m_1 - m_2$
3. A p-value for the two-sided test of $m_1 = m_2$ against $m_1 \neq m_2$
4. The sum of the ranks for the values in C1
5. The medians for each sample.

--- EXAMPLE 6.22 ---

Suppose 3 5.1 6.2 8.9 is set in C1 and 1 3.1 4.2 7.9 is set in C2. The printout for MANN–WHITNEY C1 and C2 is

C1 N = 4 MEDIAN = 5.65
C2 N = 4 MEDIAN = 3.65
POINT ESTIMATE FOR ETA1–ETA2 IS 2.00
97.0 PCT C.I. FOR ETA1–ETA2 IS $(-4.901, 7.899)$
W = 21
TEST OF ETA1 = ETA2 VS ETA1 ≠ ETA2 IS SIGNIFICANT AT 0.4705.

MINITAB also allows one-sided tests and the confidence coefficient to be pre-set. The confidence coefficient should appear in the main command and the type of one-sided test is to be specified by the sub-command ALTERNATIVE.
Notes:

1. Any discrepancies between the text and the printout arise from MINITAB's use of a normal approximation. In particular, if the samples are small or there are ties it would be more sensible to use an exact calculation. Thus if C1 contains 0 0 1 and C2 contains 0 1   1 then the significance level is returned by the command MANN–WHITNEY as 0.6625. If the ranks are shared, then $W = 9$ and $\Pr(W \leqslant 9) = 1/2$. Hence the exact, two-sided significance level is 1.
2. MINITAB provides a confidence interval with coefficient as close as possible to the default value of 95%.

The point estimate for $m_1 - m_2$ is the median of the *Walsh differences*, as described in the text. MINITAB finds these differences behind the scenes. The command WDIFF C1 C2 C3 finds all the differences for data stored in C1, C2 and places them in C3. The use of this command on the samples in Example 6.22 confirms that the point estimate is 2.

6.3.9 *Problems*

6.9 A random sample of 5 ranks is chosen at random from 1, 2, 3, 4, 5, 6, 7, 8, 9, 10. By listing the possible selections, find $\Pr(R_t \leqslant 17)$. Deduce $\Pr(R_t \geqslant 38)$ and hence the exact value for $\Pr(18 \leqslant R_t \leqslant 37)$. Compare the exact value with the value obtained using a normal approximation to calculate $\Pr(18 \leqslant R_t \leqslant 37)$.

6.10 For Example 6.15, calculate a central confidence interval for the difference between the medians with confidence coefficient at least 95%. Apply the appropriate MINITAB command to this data and compare the result with that for the normal approximation used in Problem 6.9.

6.11 For the data of Example 6.13

      1.0   3.1   4.2   7.9
      3.0   5.1   6.2   8.9

use the table of the distribution function of U_t and the method of differences described above to find a central confidence interval for the difference in population medians with confidence coefficient at least 95%. Use a normal approximation to evaluate the (two sided) *p*-value of the observed value of R_t.

6.12 If the treatment and control data for Example 6.14 had been

      Treatment   3.0   5.5   6.8   12.3   15.5
      Control   0.2   3.0   3.0   4.3   6.8

use the method of Example 6.21 to calculate a central confidence interval for the difference between the medians with confidence coefficient at least 95%.

Compare with the MINITAB result for the data. Use the normal distribution, modified to take account of the ties, to find an approximation to the confidence coefficient.

6.13 $X_1, X_2, \dots, X_n$ and $Y_1, Y_2, \dots, Y_m$ are two samples. Show that the number of pairs (X_i, Y_j) such that $X_i < Y_j$ is equal to the Mann–Whitney statistic for the y sample.

6.14 Let $\phi(r; m, n)$ denote the number of selections of m from the $m+n$ ranks $1, 2, \dots, m+n$ with rank sum equal to r. Show that

$$\phi(r; m, n) = \phi(r - m - n; m - 1, n) + \phi(r; m, n - 1)$$

Hence deduce that the conditional probability $\Pr(r \mid m, n)$ of obtaining r given m, n satisfies

$$\Pr(r \mid m, n) = \frac{m}{m+n} \Pr(r - m - n \mid m - 1, n) + \frac{n}{m+n} \Pr(r \mid m, n - 1)$$

6.4   Solutions to problems

6.1 Write out the probabilities in terms of factorials and check that they are the same. Allows tables to be used when $n > R$. The question is equivalent to asking whether significantly many of the treated subjects recovered. From tables, the probability of at least 4 is $0.2448 + 0.0490 + 0.0023 = 0.2961$.

6.2 The clairvoyant divides the cards into a pile of seven, which he will call red, the remainder he will call black. The possibilities for at least 13 correct are

		Actual				Actual	
		R	B			R	B
Call	R	7	0		R	6	1
	B	0	8		B	1	7

From the table of $h(15, 7, 7)$, the required probability is $0.0087 + 0.0002 = 0.0089$. In the second case, six actual reds must be classified as red and eight actual blacks be classified as black. There is no entry for $h(15, 6, 7)$, so transpose the rows and columns for $h(15, 7, 6)$; the required probability is 0.0014.

6.3 In $aN - m_1 n_1$ write $N = a + b + c + d$ and $m_1 = a + b$, $n_1 = a + c$.

6.4 $E(A) = n_1 m_1/N$, $V(A) = n_1 m_1 n_2 m_2 / [N^2(N-1)]$. Substitute to show equivalence.

6.5 There are

$$\binom{7}{3} = 35$$

distinguishable arrangements. The number of runs varies from 2 to 7 inclusive. Using formulae (6.1) and (6.2),

r	2	3	4	5	6	7	
$\Pr(R = r)$	2/35	5/35	12/35	9/35	6/35	1/35	$E(R) = 31/7$.

6.6 $\Pr(k \text{ of Type I and } k - 1 \text{ of Type II}) = \binom{n-1}{k-1}\binom{m-1}{k-2} \Big/ \binom{m+n}{n}$

$\Pr(k \text{ of Type I and } k \text{ of Type II}) = 2\binom{n-1}{k-1}\binom{m-1}{k-1} \Big/ \binom{m+n}{n}$

$\Pr(k \text{ of Type I and } k + 1 \text{ of Type II}) = \binom{n-1}{k-1}\binom{m-1}{k} \Big/ \binom{m+n}{n}.$

Add these probabilities and simplify to obtain the result.

6.7 $\displaystyle\sum_{k=1}^{n} \binom{n-1}{k-1}\binom{m+1}{k} = \sum_{k=1}^{n} \binom{n-1}{k-1}\binom{m+1}{m-(k-1)} = \binom{m+n}{m}$

$\displaystyle\sum_{k=1}^{n} k\binom{n-1}{k-1}\binom{m+1}{k} = (m+1)\sum_{k=1}^{n} \binom{n-1}{k-1}\binom{m}{k-1} = (m+1)\binom{m+n-1}{m}$

Hence

$$E(K) = (m+1)\binom{m+n-1}{m} \Big/ \binom{m+n}{m} = \frac{n(m+1)}{m+n}$$

after cancelling.

Interchanging n and m, expected number of runs of Type II is $m(n+1)/(m+n)$. Add to obtain $1 + 2mn/(m+n)$.

6.8 $E(X_i) = \Pr(X_i = 1) = \Pr(\text{balls } i - 1, i \text{ have different colours})$

$\qquad\qquad = 2mn/[(m+n)(m+n-1)]$

$E(R) = 1 + \displaystyle\sum_{i=2}^{m+n} E(X_i) = 1 + 2mn/(m+n)$

Now $X_i = 1$ and $X_{i+1} = 1$ if and only if $i - 1, i + 1$ have the same colour which differs from that of place i. Hence $\Pr(X_i = 1 \text{ and } X_{i+1} = 1) = mn/[(m+n)(m+n-1)]$ and for $j \neq i - 1, i, i + 1$,

$\qquad \Pr(X_i = 1 \text{ and } X_j = 1) = \Pr(X_j = 1 \mid X_i = 1)\Pr(X_i = 1)$

$$= \frac{2(m-1)(n-1)}{(m+n-2)(m+n-3)} \times \frac{2mn}{(m+n)(m+n-1)}$$

$$\left(\sum_{i=2}^{m+n} X_i\right)^2 = \sum_{i=2}^{m+n} X_i^2 + \sum\sum_{i \neq j} X_i X_j$$

Note: the expansion of $(\sum_{i=2}^{m+n} X_i)^2$ contains $(m+n-1)^2$ terms in all; of these $(m+n-1)$ are of type X_i^2, $2(m+n-2)$ of type $X_i X_{i+1}$, and by subtraction, $(m+n-2)(m+n-1)$ of type $X_i X_j$ ($j \neq i - 1, i, i + 1$).

6.9   The qualifying selections are 1, 2, 3, 4, 5   1, 2, 3, 4, 6   1, 2, 3, 4, 7   1, 2, 3, 5, 6. Hence $\Pr(R_t \leqslant 17) = 4/252 \approx 0.016$ since there are 252 equally likely selections of 5 from 10. The maximum value of R_t is 40 and the minimum value is 15. Hence, by symmetry

$$\Pr(R_t \geqslant 38) = \Pr(R_t \leqslant 40 - 38 + 15) = \Pr(R_t \leqslant 17) = 4/252.$$
$$\Pr(18 \leqslant R_t \leqslant 37) = 1 - 8/252 = 244/252 \approx 0.968$$

Since $E(R_t) = 27.5$, $V(R_t) \approx (4.787)^2$, using a continuity correction of $1/2$,

$$\Pr(18 \leqslant R_t \leqslant 37) \approx \Pr((17.5 - 27.5)/4.787 \leqslant Z \leqslant (37.5 - 27.5)/4.787)$$
$$\approx \Pr(-2.089 \leqslant Z \leqslant 2.089) \approx 0.963.$$

6.10   The table of differences for the sample values is

	3.2	5.5	6.8	12.3	15.5
0.2	3.0	5.3	6.6	12.1	15.3
2.3	0.9	3.2	4.5	10.0	13.2
2.9	0.3	2.6	3.9	9.4	12.6
4.3	−1.1	1.2	2.5	8.0	11.2
5.9	−2.7	−0.4	0.9	6.4	9.6

When these differences are put into numerical order then, from Problem 6.9, we require those values at which R_t attains 18 and 37 . These are at the third value from the bottom and top, namely -0.4, 12.6. MINITAB provides a confidence interval $(-0.399, 12.601)$ with coefficient 96.3% which agrees with the normal approximation in Problem 6.9. The point estimate of the difference of the medians is 4.5, which is the middle of the 25 differences.

6.11   The largest integer i such that $\Pr(U_t \leqslant i) \leqslant 0.02$ is, from tables, $i = 0$, corresponding to $R_t = 10$. The table of differences for this sample is displayed earlier. The confidence interval is $(-4.9, 7.9)$. There are 70 selections of 4 ranks from 8. The confidence coefficient is $1 - 2/70 \approx 0.97$, i.e. 97%. The observed value of R_t is 15. Since $E(R_t) = 18$, $V(R_t) = 12$, $\Pr(R_t \leqslant 15) \approx \Pr(Z \leqslant (15.5 - 18)/\sqrt{12}) \approx \Pr(Z \leqslant -0.7217) \approx 0.235$. The two sided *p*-value is thus 0.470, using a continuity correction.

6.12   From Problem 6.9, we require those differences corresponding to $18 \leqslant R_t \leqslant 37$. The third difference from the bottom is now -1.3 and from the top is 12.5. The coefficient is again 96.8%. MINITAB supplies $(-1.297, 12.4999)$ with confidence coefficient 96.3%. If a normal distribution with mean 27.5 and modified variance $200/9$ is used, then with a continuity correction of $1/2$, $\Pr(18 \leqslant R_t \leqslant 37) \approx 0.966$.

6.13   Suppose that $Y_{(j)}$, the jth largest of the y values, has rank R_j (among all $n + m$ values). Then there are $j - 1$ members of the y sample less than $Y_{(j)}$. But $R_j - 1$ values from the pooled samples are less than $Y_{(j)}$. Hence there are $(R_j - 1) - (j - 1)$ values from the x sample less than $Y_{(j)}$. Hence the total number of pairs (X_i, Y_j) with $X_i < Y_j$ is

$$\sum_{j=1}^{m}(R_j - j) = \sum_{j=1}^{m}R_j - \tfrac{1}{2}m(m + 1)$$

as required.

6.14 Each selection either includes the rank $m + n$, or it does not. If it does, then the residue $r - m - n$ must be made up from the remaining $m - 1$ elements. If not, then r must arise from m elements chosen from $m + (n - 1)$ ranks. Since

$$\Pr(r | m, n) = \phi(r; m, n)\bigg/\binom{m + n}{m}$$

divide both sides of the relation by

$$\binom{m + n}{m}$$

and simplify.

Reference

1. Hodges, J. L. and Lehman, E. L., Estimates of location based on rank tests. *Ann. Math. Stat.*, **34**, 1963, 598–611.

Chapter Seven

Rank correlation coefficients

7.1   Introduction

In this chapter we develop distribution-free tests of the hypothesis that the correlation coefficient of a bivariate distribution is zero. We recall, from Section 1.1, that if X, Y have a joint distribution then the *distribution* correlation coefficient ρ, when it exists, is given by

$$\rho = \frac{\mathrm{Cov}(X, Y)}{\sqrt{\mathrm{V}(X)\mathrm{V}(Y)}} \tag{7.1}$$

and satisfies $\rho^2 \leqslant 1$. In fact, $\rho = 0$ only implies that X and Y are uncorrelated, though in the bivariate normal case it further guarantees independence.

If (X_1, Y_1), (X_2, Y_2), ..., (X_n, Y_n) is a random sample of n pairs of values from the distribution of X, Y, then the product moment correlation coefficient R for the sample is

$$R = \frac{\sum\limits_{i=1}^{n} (X_i - \bar{X})(Y_i - \bar{Y})}{\sqrt{\sum\limits_{i=1}^{n} (X_i - \bar{X})^2 \sum\limits_{i=1}^{n} (Y_i - \bar{Y})^2}} \tag{7.2}$$

We first consider a permutation test based on the obtained sample values (x_1, y_1), (x_2, y_2), ..., (x_n, y_n). The value of R yielded from the obtained sample values is

$$
\begin{aligned}
r &= \frac{\sum\limits_{i=1}^{n} (x_i - \bar{x})(y_i - \bar{y})}{\sqrt{\sum\limits_{i=1}^{n} (x_i - \bar{x})^2 \sum\limits_{i=1}^{n} (y_i - \bar{y})^2}} \\[2ex]
&= \frac{\sum\limits_{i=1}^{n} x_i y_i - n\bar{x}\bar{y}}{\sqrt{\left(\sum\limits_{i=1}^{n} x_i^2 - n\bar{x}^2\right)\left(\sum\limits_{i=1}^{n} y_i^2 - n\bar{y}^2\right)}}
\end{aligned}
\tag{7.3}
$$

Now if X, Y are independent, the $n!$ permutations in which the $y_1, y_2, \ldots, y_n$ could be paired off with $x_1, x_2, \ldots, x_n$ have equal probability. Conditional on the obtained sample, the terms $\bar{x}$, $\bar{y}$, $\sum_{i=1}^{n} x_i^2$ and $\sum_{i=1}^{n} y_i^2$ remain fixed. The only term in equation (7.3) which varies under the permutations is $\sum_{i=1}^{n} x_i y_i$. For any such permutation, the obtained correlation coefficient will be denoted by r_p.

——— EXAMPLE 7.1 ———

Suppose the obtained sample is $(1,2)$, $(3,8)$, $(5,4)$, $(7,6)$. Then we calculate

$$\bar{x} = 4, \bar{y} = 5, \sum_{i=1}^{4} x_i^2 = 84, \sum_{i=1}^{4} y_i^2 = 120, \sum_{i=1}^{4} x_i y_i = 88$$

Hence the value obtained for r_p is $[88 - 4 \times 20]/[(84 - 4 \times 16)(120 - 4 \times 25)]^{1/2}$ $= 2/5$. Regarding the x values 1, 3, 5, 7 as fixed, there are $4! = 24$ permutations of the y values 2, 8, 4, 6. We display the sums of products of six of these:

x_i	1	3	5	7	$\Sigma\, x_i y_i$
	4	2	6	8	96
	8	2	6	4	72
y_i	8	2	4	6	76
	6	2	4	8	88
	4	2	8	6	92
	2	6	8	4	88

The 24 permutations yield the frequency table

$\Sigma\, x_i y_i$	60	64	68	72	76	80	84	88	92	96	100
Frequency	1	3	1	4	2	2	2	4	1	3	1

The reader should check:

1. That the average of the 24 values of $\sum_{i=1}^{4} x_i y_i$ is 80; and
2. That since, from equation (7.3),

$$r_p = \left(\sum_{i=1}^{4} x_i y_i - 80\right)\Big/20$$

the average of the 24 values of r_p is zero.

From the frequency table, for example, the conditional probability that r_p is greater than or equal to $2/5$ is also the proportion of values for which $\sum_{i=1}^{4} x_i y_i \geq 88$ and hence is $3/8$.

7.2   Spearman's rank correlation coefficient

The permutation correlation coefficient is conditionally distribution-free. This restriction of conditionality can be removed by replacing the variate values by their ranks. We shall assume that the sample of pairs (x_i, y_i) has been drawn from a continuous distribution and thus ties can be ignored. Suppose the x_i and y_i are ranked separately. Without loss of generality we may relabel the data with $x_1, x_2, \ldots, x_n$ in increasing order and then the pairs of ranks will take the form (i, r_i).

─────── EXAMPLE 7.2 ───────────────────────────────────────

Suppose we have four pairs of variate values $(1, 2)$, $(5, 4)$, $(3, 8)$, $(7, 6)$. If these pairs are rearranged so that the x components are in increasing order, we have $(1, 2)$, $(3, 8)$, $(5, 4)$, $(7, 6)$. If the values are then replaced by their ranks we arrive at $(1, 1)$, $(2, 4)$, $(3, 2)$, $(4, 3)$ and $r_1 = 1$, $r_2 = 4$, $r_3 = 2$, $r_4 = 3$.

───

To apply formula (7.3) we compute the quantities

$$\sum_{i=1}^{n} i = \sum_{j=1}^{n} r_j = \frac{1}{2} n(n+1) \qquad \sum_{i=1}^{n} i^2 = \sum_{j=1}^{n} r_j^2 = \frac{1}{6} n(n+1)(2n+1)$$

and

$$\sum_{i=1}^{n} i^2 - \left(\sum_{i=1}^{n} i\right)^2 \Big/ n = \sum_{i=1}^{n} r_j^2 - \left(\sum_{i=1}^{n} r_j\right)^2 \Big/ n = \frac{1}{12} n(n^2 - 1) \tag{7.4}$$

Substituting these values in equation (7.3), the correlation coefficient between the rankings becomes

$$r_S = \frac{\sum_{i=1}^{n} i r_i - n(n+1)^2/4}{n(n^2 - 1)/12} \tag{7.5}$$

This quantity is called *Spearman's rank correlation coefficient*.

─────── EXAMPLE 7.3 ───────────────────────────────────────

For Example 7.2, $\sum_{i=1}^{4} i r_i = 27$ and from equation (7.2), with $n = 4$, $r_S = 2/5$.

───

Now equation (7.5) is the obtained value of Spearman's rank correlation coefficient. The sequence of ranks $r_1, r_2, \ldots, r_n$ is just one of the $n!$ sequences obtained by permuting the integers $1, 2, \ldots, n$. Under these permutations, r_i is the obtained value of a random variable R_i. We write

$$R_S = \frac{\sum_{i=1}^{n} i R_i - n(n+1)^2/4}{n(n^2 - 1)/12}$$

7.2.1   *Properties of* R_S

We show that the distribution of R_S is symmetric about zero, so that $\Pr(R_S \geq c) = \Pr(R_S \leq -c)$. For each ranking $r_1, r_2, \ldots, r_n$ let $r_i^* = n + 1 - r_i$. Then the correlation coefficient for the pairs (i, r_i^*) $i = 1, 2, \ldots, n$ is, from equation (7.5),

$$r_S^* = \frac{\sum_{i=1}^n i r_i^* - n(n+1)^2/4}{(n^3 - n)/12}$$

$$= \frac{-\sum_{i=1}^n i r_i + (n+1) \sum_{i=1}^n i - n(n+1)^2/4}{(n^3 - n)/12}$$

$$= -r_S, \text{ after simplification}$$

The version of r_S given in equation (7.5) is not the most convenient computationally and r_S is usually expressed in terms of the quantities $d_i = i - r_i$. Now

$$\sum_{i=1}^n (i - r_i)^2 = \sum_{i=1}^n i^2 + \sum_{i=1}^n r_i^2 - 2 \sum_{i=1}^n i r_i$$

so that

$$\sum_{i=1}^n i r_i = \sum_{i=1}^n i^2 - \frac{1}{2} \sum_{i=1}^n d_i^2 \tag{7.6}$$

When equation (7.6) is substituted in equation (7.5) we obtain

$$r_S = 1 - \frac{6}{n^3 - n} \sum_{i=1}^n d_i^2 \tag{7.7}$$

7.2.2   *Spearman's* S

The quantity $\sum_{i=1}^n d_i^2$ is known as *Spearman's S*. It also is used as a measure of association between two sets of rankings, and although S is not a correlation coefficient it has the merit of simplicity.

The properties of S are easily derived from those of R_S. From equation (7.7),

$$S = (1 - R_S)(n^3 - n)/6$$

Hence, since $E(R_S) = 0$,

$$E(S) = (n^3 - n)/6 \tag{7.8}$$

and

$$V(S) = (n^3 - n)^2 V(R_S)/36 = n^2(n-1)(n+1)^2/36 \tag{7.9}$$

since, from Appendix 2, Example A2.4, $V(R_S) = 1/(n-1)$. Moreover, since R_S is symmetric about zero, S is symmetric about $(n^3 - n)/6$.

───────── EXAMPLE 7.4 ─────────────────────────────────────

Suppose that there are eight pairs of rankings (i, r_i), with no ties. There are $8! = 40\,320$ permutations of the integers $1, 2, \ldots, 8$ and each of these provides a value of Spearman's S. For instance, for the permutation 1, 2, 6, 4, 5, 3, 7, 8, $S = (1-1)^2 + (2-2)^2 + (3-6)^2 + (4-4)^2 + (5-5)^2 + (6-3)^2 + (7-7)^2 + (8-8)^2 = 18$. Tables give the lower critical 5% value as 30 when $n = 8$. That is, 30 is the largest integer j such that $\Pr(S \leqslant j) \leqslant 0.05$. To find the upper 5% value, we note that the maximum value of S, being twice its expectation, is $(1/3)8(8^2 - 1) = 168$ and, since the distribution of S is symmetric,

$$\Pr(S \leqslant 30) = \Pr(S \geqslant 168 - 30) = \Pr(S \geqslant 138)$$

and 138 is the least integer j such that $\Pr(S \geqslant j) \leqslant 0.05$.

The exact value of $\Pr(S \leqslant 30)$ would require a complete enumeration of the values of S when $n = 8$. Table A3.4 in Appendix 3 provides the cumulative distribution function of S for $4 \leqslant n \leqslant 10$. This table gives $\Pr(S \leqslant 30)$ as 0.04809.

───

7.2.3   *Normal approximation*

Most tables are so extensive in terms of critical values that a normal approximation is required only when the sample size exceeds 40. From equations (7.8) and (7.9) we obtain the appropriate mean and variance. As an illustration, when $n = 8$, $E(S) = 84$, $\sqrt{V(S)} = 12\sqrt{7}$. Since $[S - E(S)]/\sqrt{V(S)}$ is distributed approximately N(0, 1),

$$\Pr(S \leqslant 30) \approx \Pr(Z \leqslant (31 - 84)/12\sqrt{7}) \approx \Pr(Z \leqslant -1.67) = 0.0475$$

including a continuity correction of 1, because S increases in steps of 2.

7.2.4   *Effect of ties*

Ten pairs of matched observations (x, y) yielded the result

x	68	44	24	82	58	32	75	66	67	46
y	0.86	0.90	1.23	0.80	0.99	1.45	1.05	1.05	1.04	0.93

When the y values are ranked, there is a tie at $y = 1.05$, competing for ranks 7 and 8:

x ranks	1	2	3	4	5	6	7	8	9	10
y ranks	9	10	3	4	5	?	6	2	?	1.

We consider the effect on S of some techniques of resolving the tie where ranks 7, 8 would (normally) appear. It will be convenient to denote $\sum_{i=1}^{n} d_i^2$ by S even though d_i may no longer be expressible in the form $(i - r_i)$ when there are ties in the x-data.

1. Omit one of the tied pairs. This may have a dramatic effect, depending on the magnitude of the x-value discarded. The reader should check that if $(75, 1.05)$ is discarded $S = 198$, while if $(66, 1.05)$ is discarded then $S = 188$.
2. Calculate S for each of the possible ways of assigning ranks to the tied observations and average the results. The tied values compete for ranks 7 and 8. The possible pairings are

x	9	6		x	9	6
y	7	8		y	8	7

   The corresponding values of S are 254 and 248 with average 251 – not comparable with method (1) since it is based on ten observations.
3. The most common recommendation is to share the possible ranks, assigning 7.5 to each. If (s_i, r_i) are the rankings for the ith pair and $S = \sum_{i=1}^{n} (s_i - r_i)^2$, then the reader should check that the calculated value of S is 250.5.

7.2.5   *Evaluating significance in the presence of ties*

If the ranks are shared, we cannot use a table of the distribution of S to evaluate the p-value. This requires an enumeration of all possible values of S.

———— **EXAMPLE 7.5** ————

Suppose for y, three elements share ranks $1, 2, 3$, with average 2 and three elements share ranks $4, 5, 6$ with average 5, in the table

x rank	1	2	3	4	5	6
y rank	2	5	2	2	5	5

There are no longer $6! = 720$ distinct permutations for the y rankings. For six elements there are

$$\binom{6}{3} = 20 \text{ ways}$$

of selecting three to have rank 2 and the others to have rank 5. The twenty distinguishable arrangements yield values of S increasing in steps of 6. The smallest value, 4, arises from the arrangement 2 2 2 5 5 5, the largest, 58, from 5 5 5 2 2 2. Since four of the twenty have values of S less than or equal to 16, if all assignments are equally likely, then, conditional on the ties, $\Pr(S \leqslant 16) = 4/20 = 0.2$.

For large samples, the method of Example 7.5 would be intolerable. The use of the normal approximation is much facilitated by the observation that the conditional variance of R_S remains $1/(n-1)$ provided equation (7.3) is used to compute R_S.

In case there are ties, the mean and variance of S are no longer as given in equations (7.8) and (7.9). More generally, whether there are ties or not, if $S = \sum_{i=1}^{n}(s_i - R_i)^2$ then, applying Appendix 2, Example A2.5, to the pairs (s_i, R_i)

$$E(S) = \sum_{i=1}^{n} s_i^2 + \sum_{i=1}^{n} r_i^2 - 2n\bar{s}\bar{r} \tag{7.10}$$

$$V(S) = 4\left(\sum_{i=1}^{n} s_i^2 - n\bar{s}^2\right)\left(\sum_{i=1}^{n} r_i^2 - n\bar{r}^2\right)\bigg/(n-1) \tag{7.11}$$

——— EXAMPLE 7.6 ———

For Example 7.5, to apply a normal approximation, we calculate

$$\sum_{i=1}^{6} s_i^2 = 91, \quad \bar{s} = 3.5, \quad \sum_{i=1}^{6} r_i^2 = 87, \quad \bar{r} = 3.5$$

and substitute in equations (7.10) and (7.11) to obtain $E(S) = 31$, $V(S) = 189$. Since S increases by steps of 6, we use a continuity correction of 3. $\Pr(S \leq 16) \approx \Pr(Z \leq (19 - 31)/\sqrt{189}) \approx \Pr(Z \leq -0.8729) \approx 0.19$.

7.2.6   *Use of MINITAB*

MIN1TAB does not, as yet, have an instruction which directly calculates the Spearman rank correlation coefficient. However, if the data is ranked, the CORRELATION command may be used. If there are ties, then the corresponding ranks are shared.

Thus if the 10 pairs of observations (x, y)

x	68	44	24	82	58	32	75	66	67	46
y	0.86	0.90	1.23	0.80	0.99	1.45	1.05	1.05	1.04	0.93

are set in C1, C2 then the sequence

    RANK C1 C3
    RANK C2 C4
    CORRELATION C3 C4

returns r_S in the form

    CORRELATION OF C3 AND C4 = −0.523.

If Spearman's S is required, then we must find the sum of the squares of the differences in the pairs of components in C3 and C4, via the command SSQ(C3 − C4), which prints 250.500 for our example.

7.2.7   Problems

7.1   For 10 elements, the obtained values of two variables X, Y were

x	68	44	14	82	58	32	75	66	67	46
y	88	90	123	80	99	145	105	106	104	98

Calculate Spearman's S for this sample. Test the hypothesis that X, Y are independent against the alternative that they are not independent by finding the p-value using
(a) Tables
(b) A normal approximation.

7.2   Calculate Spearman's S for the twelve pairs of values

x	1	2	3	4	5	6	7	8	9	10	11	12
y	1	−2	3	−4	5	−6	7	−8	9	−10	11	−12

and test the null hypothesis that X, Y are independent against the alternative that they are dependent. Comment on your conclusion.

7.3   For the pair of rankings

x	1	2	3	4	5	6
y	2	5	2	2	5	5

show that the obtained value of $S = 16$. Find the frequency distribution of Spearman's S for the 20 distinguishable arrangements of the y ranks. Calculate $E(S)$, $V(S)$ directly and compare the results with Example 7.6.

7.4   For the years 1913 to 1924 inclusive, the average yield of wheat in kilograms per 10^4 square metres and the average maximum temperature for the winter previous to the harvest were as follows:

YEAR	1913	1914	1915	1916	1917	1918	1919	1920	1921	1922	1923	1924
YIELD	1990	1950	1630	1720	1560	1680	1980	2180	2370	1790	2400	1400
TEMPERATURE	2.7	3.1	1.9	1.3	1.0	1.5	2.3	1.7	3.0	1.1	1.6	0.1

Calculate Spearman's S between yield and temperature and show that it is less than the lower $2\frac{1}{2}\%$ point of the distribution of S. Examine also whether there is a significant trend in yield over time.

7.3   Kendall's rank correlation coefficient

A correlation coefficient is one way of measuring the association between two random variables X, Y. We now describe an alternative measure which depends on probabilities rather than expected values.

Let (X_1, Y_1), (X_2, Y_2) be a random pair of observations from the distribution of X, Y. We assume that $X_1 \neq X_2$ and $Y_1 \neq Y_2$. The pairs are said to be *concordant* if

$$\text{either } X_1 < X_2 \text{ and } Y_1 < Y_2 \qquad \text{or} \qquad X_1 > X_2 \text{ and } Y_1 > Y_2$$

that is, $(X_1 - X_2)(Y_1 - Y_2) > 0$
The pairs are said to be *discordant* if

$$\text{either } X_1 < X_2 \text{ and } Y_1 > Y_2 \qquad \text{or} \qquad X_1 > X_2 \text{ and } Y_1 < Y_2$$

that is, $(X_1 - X_2)(Y_1 - Y_2) < 0$

The properties as defined do not depend on location or scale and embody the intuitive notion of X, Y going up or down together – or the converse. Suitable measures of concordance and discordance are

$$p_c = \text{probability of concordance} = \Pr((X_1 - X_2)(Y_1 - Y_2) > 0) \text{ and}$$
$$p_d = \text{probability of discordance} = \Pr((X_1 - X_2)(Y_1 - Y_2) < 0)$$

Kendall's coefficient of association τ is defined by

$$\tau = p_c - p_d$$

It is clear that $p_c + p_d \leqslant 1$, with equality if and only if ties have zero probability. It follows that $-1 \leqslant \tau \leqslant +1$.

If X, Y have a *continuous* joint distribution then $\Pr(X_1 = X_2 \text{ or } Y_1 = Y_2) = 0$ and $p_c + p_d = 1$. In that case τ may also be written $\tau = 2p_c - 1 = 1 - 2p_d$.

When X, Y are not only continuous but independent then

$$\begin{aligned}
p_c &= \Pr(X_1 < X_2 \text{ and } Y_1 < Y_2) + \Pr(X_1 > X_2 \text{ and } Y_1 > Y_2) \\
&= \Pr(X_1 < X_2)\Pr(Y_1 < Y_2) + \Pr(X_1 > X_2)\Pr(Y_1 > Y_2) \\
&= \Pr(X_1 > X_2)\Pr(Y_1 < Y_2) + \Pr(X_1 < X_2)\Pr(Y_1 > Y_2)
\end{aligned}$$

since X_1, X_2 are independent and have the same distribution. Thus $p_c = p_d$, and in that case each is $1/2$ and $\tau = 0$. Hence a necessary (but not sufficient) condition for X, Y to be independent is that $\tau = 0$.

7.3.1   *Testing that $\tau = 0$*

The obvious statistic is based on the difference between the number of concordances C and discordances D. Thus if (X_1, Y_1), (X_2, Y_2), ..., (X_n, Y_n) is a random sample of n pairs from the distribution of X, Y, then Kendall's (sample) coefficient of correlation is

$$R_K = (C - D)/[\tfrac{1}{2} n(n - 1)] = K/[\tfrac{1}{2} n(n - 1)]$$

where $K = C - D$ is called *Kendall's score*.

It is to be noted that, to compute r_K,

$$\binom{n}{2} = \frac{1}{2}n(n-1)$$

comparisons have to be made. In the case when there are no ties in the data, the number of comparisons will be $C + D$, so that

$$R_K = \{2C/[\tfrac{1}{2}n(n-1)]\} - 1$$

Since the property of concordance depends only on the *relative* values of the sample data, we may, without loss of generality, replace the variate values by their ranks. For this reason, R_K is more usually called *Kendall's rank correlation coefficient*. Table A3.5 in Appendix 3 provides the upper tail probabilities for R_K, i.e. $\Pr(R_K \geqslant x)$, for $4 \leqslant n \leqslant 20$.

The labour of making the comparisons is much reduced by placing one of the rankings in increasing order. In the case of data drawn from a continuous distribution we may restrict attention to the value of C. We then need only find the total *number* of ranks to the right which exceed each particular value in the second ranking. In the next example these numbers are displayed as components of the observed value of C.

——— **EXAMPLE 7.7** ———

Twelve elements are each measured on two variables. The values are then ranked, yielding the result

Element	C	B	A	E	K	F	D	L	H	I	G	J
X ranking	1	2	3	4	5	6	7	8	9	10	11	12
Y ranking	3	1	2	5	11	4	6	9	7	8	10	12

The number of y ranks to the right of 3 which exceed 3 is 9, and so on:

$$c = 9 + 10 + 9 + 7 + 1 + 6 + 5 + 2 + 3 + 2 + 1 = 55,$$
$$r_K = \{2 \times 55/[\tfrac{1}{2}12(12-1)]\} - 1 = 2/3$$

From Table A3.5, $\Pr(R_K \geqslant 0.6667) = 0.0009$, strongly indicating that the rankings are not independent.

7.3.2   *The distributional properties of C; no ties*

C involves $\tfrac{1}{2}n(n-1)$ comparisons of pairs. If every comparison is concordant, then C assumes this maximum value of $\tfrac{1}{2}n(n-1)$. When every comparison is discordant, then C assumes its minimum value of 0. Moreover, the distribution of C is symmetric, since for each permutation of the ranks $1, 2, ..., n$ with concordance c

there is a *conjugate* ranking in which rank i is replaced by $(n + 1) - i$ and which has concordance $\frac{1}{2}n(n - 1) - c$.

──────── EXAMPLE 7.8 ────────────────────────────────

From Example 7.7, we had the two rankings

X	1	2	3	4	5	6	7	8	9	10	11	12
Y	3	1	2	5	11	4	6	9	7	8	10	12

The conjugate rankings for Y are

$$10 \quad 12 \quad 11 \quad 8 \quad 2 \quad 9 \quad 7 \quad 4 \quad 6 \quad 5 \quad 3 \quad 1$$

where the sum of each rank and its conjugate is 13. The number of higher ranks to the right of 5 in the ranking for Y is 7. The corresponding rank in the conjugate permutation is 8 and there is only one conjugate higher rank to the right.

──

Consider element i which has rank r_i. If $i < j$ and $r_i < r_j$ then in the conjugate permutation the positive concordance corresponds to a negative concordance (since $n + 1 - r_i > n + 1 - r_j$). Hence if c_i is the number of positive concordances to the right of element i, then this number of positive concordances for the conjugate element is $n - i - c_i$. Hence the total number of positive concordances for the conjugate permutation is

$$\sum_{i=1}^{n} (n - i - c_i) = \frac{1}{2} n(n - 1) - C$$

This is the same amount below the maximum as is C above the minimum. Hence the distribution of C is symmetric.

7.3.3   *Normal approximation; no ties*

It can be shown that if n is large then C has approximately a normal distribution. We show in Problem (7.10) that, for all n, if the rankings are independent,

$$E(C) = n(n - 1)/4, \qquad V(C) = n(n - 1)(2n + 5)/72 \tag{7.12}$$

──────── EXAMPLE 7.9 ────────────────────────────────

Suppose 10 pairs of values (x_i, y_i) are ranked as

x ranking	1	2	3	4	5	6	7	8	9	10
y ranking	3	1	8	5	10	4	6	9	7	2

It is easily verified that $C = 26$. Now $E(C) = 22.5$ and $V(C) = 31.25$, from equation (7.12). Hence, for instance,

$$\Pr(C \geqslant 26) \approx \Pr(Z \geqslant (25.5 - 22.5)/\sqrt{31.25}) \approx \Pr(Z \geqslant 0.5367) = 0.296$$

Notice that as C increases in steps of 1, a continuity correction of $1/2$ is used. From Table A3.5, $\Pr(C \geqslant 26) = \Pr(R_K \geqslant 14/90) \approx 0.300$.

When there are no ties, we can just as easily use Kendall's K. In this case,

$$K = C - D = 2C - n(n-1)/2$$

Hence $E(K) = 0$, $V(K) = 4V(C) = n(n-1)(2n+5)/18$. However, when there *are* ties, K should be preferred, as discussed below.

7.3.4   *Kendall's score in the presence of ties*

If there are ties, in either or both rankings, then it is usually recommended that tied elements be assigned the average of the ranks they would otherwise occupy and K then be calculated in the usual way. Some of the comparisons will now be neither concordant nor discordant and hence yield zeros. Thus the significance of an obtained value of K can no longer be found from a table of the distribution of R_K. This can be assessed from the resulting conditional distribution of K.

———— EXAMPLE 7.10 ————————————————————————————————

Suppose that, after the ranks are shared, we have

> x ranks   1   2   3   4   5   6
> y ranks   2   5   2   2   5   5.

Comparing each y rank with its successors,

$$k = (3 - 0) + (0 - 2) + (2 - 0) + (2 - 0) + (0 - 0) = 5$$

There are 20 distinguishable arrangements of the shared ranks 2   5   2   2   5   5. The frequency distribution of the possible values of K is

k	−9	−7	−5	−3	−1	1	3	5	7	9
Frequency	1	1	2	3	3	3	3	2	1	1

If all arrangements are equally likely, it is easily checked that this distribution has mean 0 and variance 21. We can also calculate results such as $\Pr(K \geqslant 5) = 4/20 = 0.2$. For $2 \leqslant n \leqslant 6$. Burr (see reference 1) has displayed the distribution of K for all possible sets of ties and provided an algorithm for $n \geqslant 7$.

7.3.5   *Normal approximation*

If the number of pairs is sufficiently large, the distribution of K can be approximated. We restrict attention to the case when there are ties in only one of the rankings. Kendall (see reference 2) has proved that if $n > 10$, and the ith tie has extent t_i, then K is approximately normal with mean zero and variance

$$\left\{ n(n-1)(2n+5) - \sum_i t_i(t_i - 1)(2t_i + 5) \right\} \Big/ 18 \tag{7.13}$$

(See Problem 7.10.)

──────── EXAMPLE 7.11 ──────────────────────────────────────

$n = 6$, $t_1 = t_2 = 3$ then from equation (7.13), $V(K) = 21$, and, using a continuity correction of 1 since K increases in steps of 2, $\Pr(K \geqslant 5) \approx \Pr(Z \geqslant 4\sqrt{21}) \approx 0.19$, which should be compared with Example 7.10.

───

7.3.6   *Use of MINITAB*

This coefficient must also be obtained indirectly. Consider again the pairs of rankings in Example 7.7:

y ranking	1	2	3	4	5	6	7	8	9	10	11	12
x ranking	3	1	2	5	11	4	6	9	7	8	10	12

When the pair $(3, 1)$ is compared with $(1, 2)$, there is a contribution of -1 to Kendall's K. Now if $(3, 1)$ and $(1, 2)$ are regarded as two points in the (x, y) plane, then to say this pair is discordant is equivalent to the slope of the line joining these points being negative. The MINITAB command WSLOPE calculates the slopes for all $n(n-1)/2$ lines defined by n points. Ties in the y ranking will yield a zero slope and ties in the x ranking will be reported as $*$. Thus, WSLOPE C1 C2 C3 treats the sample in C1 as the y coordinate, the sample in C2 as the x coordinate and records the slopes in C3. The command K1 = SUM (SIGN (C3)) finds the score = $C - D$, treating $*$ as zero.

──────── EXAMPLE 7.12 ──────────────────────────────────────

If, after ranking the data, we obtain

y ranking	1	2	3	4	5	6	7	8	9	10	11	12
x ranking	2	2	2	5	5	5	8	8	8	11	11	11

the above sequence of commands supplies K1 = 54. A print of SIGN (C3) reveals 54 ones with a positive sign and 12 stars in the display of 66 elements.

───

7.3.7 Problems

7.5 For the pair of rankings

      x ranking 1 2 3 4 5 6
      y ranking 2 5 2 2 5 5

Kendall's K is 5. Suppose instead of sharing ranks we randomly assign 1, 2, 3, to the places occupied by 2 and 4, 5, 6 to those occupied by 5. Find the frequency distribution of the 36 values of Kendall's K. Calculate the mean and variance of the distribution.

7.6 For the pairs of values (x_i, y_i), $i = 1, 2, \ldots, n$, the x_i are all different but the y_i take only two values l, u, where $l < u$. There are s values equal to l and $t = n - s$ values equal to u. Let the ranks of the x values for those pairs for which the y value is l be $r_1, r_2, \ldots, r_s$. Show that the number of positive concordances is

$$st - \left(\sum_{i=1}^{s} r_i - \sum_{i=1}^{s} i \right)$$

Relate this result to a Mann–Whitney statistic.

7.7 Suppose the sample of n pairs (x_i, y_i) consists of a pairs of type $(0,0)$, b of type $(1,0)$, c of type $(0,1)$ and d of type $(1,1)$. Show that Kendall's $K = ad - bc$.

7.8 Suppose we have two rankings for n elements and there are no ties. The first ranking is fixed in the order $1, 2, 3, \ldots, n$ and $\phi_n(c)$ is the number of different permutations of the second ranking which yield a total concordance of value c. By considering the possible ranks of the nth element, show that

$$\phi_n(c) = \sum_{r=1}^{\min(c+1, n)} \phi_{n-1}(c - r + 1)$$

Hence, or otherwise, deduce that the generating function

$$G_n(t) = \sum_{c=0}^{n(n-1)/2} t^c \phi_n(c)$$

satisfies

$$G_n(t) = (1 + t + t^2 + \cdots + t^{n-1}) G_{n-1}(t)$$

Deduce that if all permutations have equal probability, then the probability generating function $P_n(t)$ for C satisfies

$$P_n(t) = \left(\frac{t^n - 1}{t - 1} \right) \frac{P_{n-1}(t)}{n}$$

7.9 By writing the probability generating function for C which was obtained in Problem 7.8 in the form

$$n(t-1)P_n(t) = (t^n - 1)P_{n-1}(t)$$

evaluate successive derivatives at $t = 1$ to show that

$$2P'_n(1) = (n-1) + 2P'_{n-1}(1)$$
$$3P''_n(1) = (n-1)(n-2) + 3(n-1)P'_{n-1}(1) + 3P''_{n-1}(1)$$

Hence show that $E(C) = n(n-1)/4$, $V(C) = n(n-1)(2n+5)/72$. (Hint: use the method of induction.)

7.10 Let a first ranking be the integers $1, 2, ..., n$ arranged in increasing order. The second ranking is a random permutation of the integers $1, 2, ..., n$ and K is Kendall's score for the two rankings. Consider any t successive ranks in the second ranking and suppose K_t is their contribution to K, arising from comparison of members of this set with each other. Let K_0 be the score obtained by replacing such a set by their average. Show that $E(K_0) = 0$, $V(K_0) = [n(n-1)(2n+5) - t(t-1)(2t+5)]/18$. (Hint: $K_0 = K - K_t$.)

7.4   Solutions to problems

7.1 $S = 254$. The test is two-sided. The maximum value possible when $n = 10$ is 330, whence $\Pr(S \geqslant 254) = \Pr(S \leqslant 76) = 0.05693$ from Table A3.4. The (two-sided) p-value of the observed S is ≈ 0.114. When $n = 10$, $E(S) = 165$, $V(S) = 3045$.

$$\Pr(S \leqslant 76) \approx \Pr(Z \leqslant (77 - 165)/55.18) \approx \Pr(Z \leqslant -1.59) = 0.0559$$

Hence the two-sided p-value is about 0.112.

7.2 The ranks of the y values are 7   6   8   5   9   4   10   3   11   2   12   1, and the value of S is 322. This exceeds the lower 5% point and is less than the upper 5% point. We would therefore fail to reject independence, even though the y-values are a function of the x-values.

7.3

S	4	10	16	22	28	34	40	46	52	58
freq	1	1	2	3	3	3	3	2	1	1

$$E(S) = (1 \times 4 + 1 \times 10 + 2 \times 16 + \cdots + 1 \times 58)/20 = 31$$
$$E(S^2) = (1 \times 4^2 + 1 \times 10^2 + 2 \times 16^2 + \cdots + 1 \times 58^2)/20 = 1150$$
$$V(S) = 1150 - 31^2 = 189$$

7.4 The value of S between temperature and yield is 108. The lower $2\frac{1}{2}\%$ point for $n = 12$ is 118. The value of S between yield and year is 245 and is not significant.

7.5

K	−9	−7	−5	−3	−1	1	3	5	7	9
freq	1	1	2	3	3	3	3	2	1	1

(a) $E(K) = 0$ from the symmetry and $V(K) = [1 \times (-9)^2 + 1 \times (-7)^2 + \cdots + 1 \times 9^2]/20 = 21$.   From   equation   (7.13), $V(K) = [n(n-1)(2n+5) - \sum t_i(t_i - 1)(2t_i + 5)]/18$. There are two ties of extent 3 and $n = 6$ so that $V(K) = (6 \times 6 \times 17 - 2 \times 3 \times 2 \times 11)/18 + 21$.

(b) $\Pr(K \geqslant 5) = 4/20 = 0.2$. K increases by steps of 2. Hence using a continuity correction of 1, $\Pr(K \geqslant 5) \approx \Pr(Z \geqslant (4-0)/\sqrt{21}) \approx \Pr(Z \geqslant 0.873) = 0.19$.

7.6   Rearrange the pairs in order of increasing x-value. In the y-values, there are $r_i - i$ values to the left of the pair (r_i, l) equal to u, and $t - (r_i - i)$ values to its right equal to u. Hence the number of concordances is

$$\sum_{i=1}^{s} [t - (r_i - i)] = st - \sum_{i=1}^{s} (r_i - i)$$

Note that

$$\sum_{i=1}^{s} (r_i - i) = \sum_{i=1}^{s} r_i - \frac{1}{2} s(s+1)$$

is the Mann–Whitney statistic for those pairs for which the y-value equals l.

7.7   Any pair $(0,0)$ compared to $(1,1)$ scores $+1$, and there are *ad* such comparisons. Any pair $(0,1)$ compared to $(1,0)$ scores -1, and there are *bc* such comparisons. All the other comparisons score 0. $K = ad - bc$. Note that if the description of the sample is displayed in a 2×2 table,

		First component	
		0	1
Second component	0	a	b
	1	c	d

the relationship to the usual test of goodness of fit becomes apparent.

7.8   Consider those permutations for which the nth element has rank r. There are $r - 1$ elements to its left, each of which contributes $+1$ to c. Hence, if the first $n - 1$ elements yield a concordance sum $c - (r - 1)$, the total for all n will be c, so that

$$\phi_n(c) = \sum_r \phi_{n-1}(c - r + 1)$$

Equate coefficients of t^c, bearing in mind that there are two cases to consider according as $c \geqslant n - 1$ or $c < n - 1$. For the last part, note that $P_n(t) = G_n(t)/n!$.

7.9   If $E(C) = \mu_n$, $V(C) = \sigma_n^2$, for n rankings, the first equation gives $\mu_n = (n-1)/2 + \mu_{n-1}$. Since $\mu_2 = \frac{1}{2}$, $\mu_n = n(n-1)/4$. Using $\sigma_n^2 = P''_n(1) + \mu_n - \mu_n^2$, we have $\sigma_n^2 = \sigma_{n-1}^2 + \frac{1}{12}(n^2 - 1)$. But $\sigma_2^2 = \frac{1}{4}$ satisfies $\sigma_n^2 = n(n-1)(2n+5)/72$, whence the result by induction.

7.10   The places for the t successive ranks can be chosen in $\binom{n}{t}$ ways. The set of t can be permuted in $t!$ ways and the remaining $n - t$ independently permuted in

$(n-t)!$ ways. The product of these is $n!$ and accounts for all the $n!$ permutations of the n ranks:

$$V(K_t) = t(t-1)(2t+5)/18, \quad \text{and} \quad V(K) = n(n-1)(2n+5)/18$$

If each member of the set is replaced by their average, the contribution to K from them is zero. However, comparisons between members of the t successive ranks and the remaining $n-t$ ranks are unchanged. Hence $K_0 = K - K_t$, that is, $K = K_0 + K_t$. But the obtained score and K_t are independent, whence $V(K) = V(K_0) + V(K_t)$. The result gives the modified variance when there is a tie of extent t in one ranking.

References

1. Burr, E. J., The distribution of Kendall's score s for a pair of tied rankings, *Biometrika* **47**, 1960, 151–71.
2. Kendall, M. G., *Rank Correlation Methods*, Griffin, London, 1955.

Chapter Eight

Distribution-free tests for many samples

8.1 Multiple testing

So far, we have been concerned with one- or two-sample tests. We shall shortly discuss some tests which are also applicable to three or more samples. These will typically test the null hypothesis that all the treatment effects are equal against the very general alternative that at least one is different from the others. Such a test could reasonably be called an overall test of significance. A new difficulty now arises. If such an overall test *does* suggest significant differences, what subsequent hypotheses should be tested? The more common procedures include (1) testing pairs of treatments against each other; (2) testing a control against several treatments; (3) testing a prescribed order of merit for the treatments.

We illustrate some of the difficulties with a relatively simple example, which, however, does not specifically relate to the tests discussed in this chapter. Suppose that X_1 is a random value from the distribution $N(\mu_1, 1)$ and X_2 is an independent random value from the distribution $N(\mu_2, 1)$. The overall null hypothesis is that $\mu_1 = \mu_2 = 0$, while the (rather broad) alternative is that at least one of μ_1, μ_2 is not zero. One way of tackling the problem is to test each part of the null hypothesis separately. The 'usual' tests, at the 5% level, of each statement are to reject $\mu_1 = 0$ if $|X_1| > 1.96$ and $\mu_2 = 0$ if $|X_2| > 1.96$.

When the null hypothesis is true, we might

- Correctly *accept* both $\mu_1 = 0$ and $\mu_2 = 0$, with probability $(0.95)^2$ or
- Incorrectly *reject* $\mu_1 = 0$ and *accept* $\mu_2 = 0$, with probability $(0.05)(0.95)$ or
- Incorrectly *reject* $\mu_2 = 0$ and *accept* $\mu_1 = 0$, with probability $(0.05)(0.95)$ or
- Incorrectly *reject* both $\mu_1 = 0$ and $\mu_2 = 0$, with probability $(0.05)^2$.

How, then, should the performance of this procedure be assessed? A plausible measure, which mimics our broad alternative, is to calculate the *family error rate*, that is, the probability of at least one error. Here it is $[1 - (0.95)^2] = 0.0975$. If this seems too high then, as the reader can easily check, it can be reduced to 0.05, by choosing $(1 - \sqrt{0.95})$ as the significance level of the individual tests.

More elaborate approaches usually involve an attempt to decide whether there is evidence that the null hypothesis is false. Only when such an overall rejection is suggested by the data is any attempt made to discover which components might be deemed suspect.

A well-known test of the above type is now applied to our example. It exploits the fact that if the null hypothesis is true, then $X_1^2 + X_2^2$ has the $\chi^2(2)$ distribution for which the upper 5% point is 5.99.

- *Stage 1* If $X_1^2 + X_2^2 < 5.99$ we accept the complete null hypothesis and stop.
- *Stage 2* If $X_1^2 + X_2^2 > 5.99$, then if $|X_1| > 1.96$, reject $\mu_1 = 0$, and if $|X_2| > 1.96$, reject $\mu_2 = 0$.

The application of this test appears to be quite straightforward but turns out to have two unexpected features. Figure 8.1 shows how these features appear.

1. The four black zones in Figure 8.1 consist of pairs of values, such as $X_1 = 1.75$, $X_2 = 1.75$, which satisfy $X_1^2 + X_2^2 > 5.99$ but fail to reject the null hypothesis at the second stage! The probability of the sample values falling in these zones is clearly small.
2. The shaded region consists of those pairs of values which reject $\mu_1 = 0$. The probability of the sample values falling in this region is difficult to calculate but is clearly less than 5%.

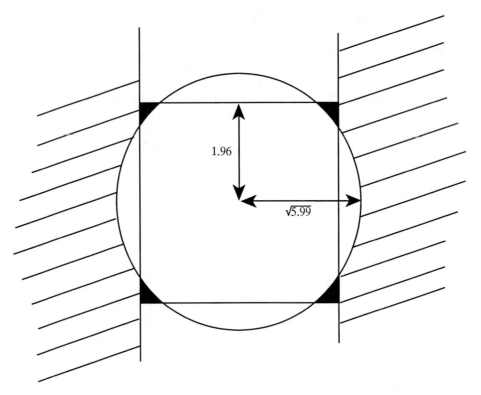

Figure 8.1

If we are prepared to accept upper bounds on the family error rate, then a basis for constructing such bounds can be derived from ideas discussed by Scheffé, through a useful inequality.

Lemma   For all random variables $X_1, X_2, \ldots, X_k$ and real numbers $a_1, a_2, \ldots, a_k$,

$$\Pr\left(\left| \sum_{i=1}^{k} a_i X_i \right| \leq \left(c(\alpha) \sum_{i=1}^{k} a_i^2 \right)^{\frac{1}{2}} \right) \geq 1 - \alpha$$

where $c(\alpha)$ is the upper $100\alpha\%$ point of the distribution of $\sum_{i=1}^{k} X_i^2$.

Proof   By Cauchy's inequality,

$$(\textstyle\sum_{i=1}^{k} a_i X_i)^2 \leq (\sum_{i=1}^{k} a_i^2)(\sum_{i=1}^{k} X_i^2)$$

Hence if $c(\alpha)$ is the upper $100\alpha\%$ point of the distribution of $\sum_{i=1}^{k} X_i^2$, then $\sum_{i=1}^{k} X_i^2 < c(\alpha)$ implies that any expression of the form $(\sum_{i=1}^{k} a_i X_i)^2 / \sum_{i=1}^{k} a_i^2$ is less than $c(\alpha)$ and therefore

$$\Pr(| \textstyle\sum_{i=1}^{k} a_i X_i | \leq (c(\alpha)\sum_{i=1}^{k} a_i^2)^{1/2}) \geq 1 - \alpha$$

We conclude that, if we have a collection of quantities of the form $| \sum_{i=1}^{k} a_i X_i | / (\sum_{i=1}^{k} a_i^2)^{1/2}$, the probability that at least one of them exceeds $c(\alpha)$ is not more than α.

Remarks

(i)   A judicious choice leads to useful probability bounds for families of linear comparisons. For example, if $a_i = 1$, $a_j = -1$, $a_l = 0$ for $l \neq i$ or j, then we have such bounds for expressions of the form $| X_i - X_j |$.

(ii) It may be that $\sum_{i=1}^{k} X_i^2 > c(\alpha)$ but $(\sum_{i=1}^{k} a_i X_i)^2 / \sum_{i=1}^{k} a_i^2 < c(\alpha)$, for some $\{a_i\}$. This explains the phenomenon of an overall test yielding a statistic which exceeds some critical level while none of the subsequent comparisons are judged to be significant.

──────── **EXAMPLE 8.1** ────────────────────────────────────

X_1, X_2, X_3 are independent $N(0, 1)$, hence the distribution of $X_1^2 + X_2^2 + X_3^2$ has the $\chi^2(3)$ distribution. The upper 5% point is 7.815. By the lemma, the probability that at least one of $| X_1 - X_2 |$, $| X_2 - X_3 |$, $| X_1 - X_2 |$ exceeds $\sqrt{15.63}$ is less than 5%.

──

In each of the three tests discussed in this chapter, multiple testing of the treatment differences will be discussed when the overall hypothesis of equality of treatments is rejected.

8.2   Testing equality of several medians: the Mood test

We now give a brief account of an extension of the sign test. Suppose we have independent samples from several distributions of the same form but possibly differing in location. We wish to test whether these distributions are the same – that is, they have the same median. This is not the same as testing whether they have a common, but particular, median. In such a case, we could use the sign test for each sample. When the reputed common median is unknown, we estimate it with the median of the pooled sample values and refer to it as the overall median. For each sample, we count the number of observations above and below the overall median and apply a chi-squared test of fit to the observed frequencies. The sample sizes need not be equal.

──────── EXAMPLE 8.2 ──

Sample 1	65	41	20	33	17
Sample 2	37	25	41	26	45
Sample 3	57	62	23	71	44
Sample 4	64	83	45	36	46

If all 20 observations are put in increasing order, then the overall median is 42.5 – the average of the tenth and eleventh in magnitude. The numbers above or below 42.5 in each sample are

Sample	Above	Below
1	1	4
2	1	4
3	4	1
4	4	1

The expected number above and below the overall median for each sample is 2.5 and the value of the goodness-of-fit statistic is $8(3/2)^2/(5/2) = 7.2$, with a *p*-value of 0.0658 based on the chi-squared distribution with index 3.

───

8.2.1   *Use of MINITAB*

The MINITAB command MOOD computes the value of chi-squared, evaluates its *p*-value, and provides a 95% confidence interval for each distribution median separately. Thus MOOD C1 C2 will apply the test we have just described to data set in C1 and indexed by sample number in C2.

──────── EXAMPLE 8.3 ──

For the samples in Example 8.2, MINITAB supplies
Mood median test of C1

Chisquare = 7.20   df = 5   p = 0.066

```
                                    Individual 95.0% CI's
C2  N< =  N>   Median   Q3 – Q1   --+---------+---------+---------+--
1    4    1    33.0     34.5      (-------+---------------)
2    4    1    37.0     17.5          (-----+---)
3    1    4    57.0     33.0      (---------------+------)
4    1    4    46.0     33.0              (----+------------------)
                                  --+---------+---------+---------+--
                                    20        40        60        80
```

Overall median = 42.5

* NOTE * Levels with < 6 obs. have confidence < 95.0%

Comments on the printout

1. If an observation falls at the overall median, then MIN1TAB includes it in the classification ≤.
2. The 95% confidence intervals are derived from the non-linear computation employed by SINTERVAL.

8.2.2   *Mood test for two samples*

A glance at the output for Example 8.3 might suggest combining sample 1 with 2 and testing against sample 3 combined with 4. To do this, we only need to change the indices in C2. Thus if C2 contains 1 1 1 1 1 1 1 1 1 1 2 2 2 2 2 2 2 2 2 2, then MOOD C1 C2 returns

Mood median test of C1

Chisquare = 7.20   df = 1   p = 0.007

```
                                    Individual 95.0% CI's
C3  N< =  N>   Median   Q3 – Q1   -+---------+---------+---------+-
1    8    2    35.0     18.2      (---------+-----)
2    2    8    51.5     23.8               (--------+--------
                                  -+---------+---------+---------+-
                                   24        36        48        60
```

Overall median = 42.5

A 95.0% C.I. for median(1) – median(2): (−39.0, −3.0)

As usual, the results of such a second test, performed on the same data should be treated with some reserve.

8.2.3   *Confidence intervals for the difference between two medians*

For two distributions, comparing the numbers below and above the overall median is equivalent to testing for independence in a 2 × 2 table. Thus for samples of sizes

m, n such that $m + n = N$, where N is even, we can summarize the information as follows:

	Above median	Below median	Total
Sample I	a	b	m
Sample II	c	d	n
Total	$N/2$	$N/2$	N

Thus in this case, we can use tables of the hypergeometric distribution to test whether the samples have been drawn from distributions with the same median. This correspondence forms the basis for constructing exact confidence intervals for the difference θ between the two unknown medians.

Suppose the second sample is translated by subtracting a value k from each of its members. For each choice of k, we find the new overall median of the pooled sample of members of the first sample and the translated values from the second. There will be choices of k which will produce an unacceptable number (either too few or too many) of members from the first sample which fall below the corresponding new overall median, and the borderline values determine the confidence interval. We illustrate the process rather informally in the next example.

———— **EXAMPLE 8.4** ————————————————————————————

Sample 1	1	3.1	4.2	7.9
Sample 2	3	5.1	6.2	8.9

When the two samples are pooled and placed in rank order, we have

$$1 \quad 3 \quad 3.1 \quad 4.2 \quad 5.1 \quad 6.2 \quad 7.9 \quad 8.9$$

where the members of the second sample are in italic. The overall median is 4.65. Suppose we decide not to accept any choice of k which leads to either none or all of the members of the first sample falling below the resultant overall median. A glance at the above order shows that if we subtract a little more than 7.9 from each italicised number, then none of the members of the first sample are below the resultant overall median. For example, if 8 is subtracted, we have

$$-5 \quad -2.9 \quad -1.8 \quad 0.9 \quad 1 \quad 3.1 \quad 4.2 \quad 7.9 \quad \text{(the overall median is now 0.95)}$$

Similarly if we add a little more than 4.9 to all the italic numbers then all the members of the first sample are below the new overall median. For example, if 5 is added we have

$$1 \quad 3.1 \quad 4.2 \quad 7.9 \quad 8 \quad 10.1 \quad 11.2 \quad 13.9 \quad \text{(the overall median is now 7.95)}$$

To add 4.9 is to subtract -4.9 so we take the open interval $(-4.9, 7.9)$ as the corresponding confidence interval for θ, although its confidence coefficient is yet to be determined. Now if a random sample of four is drawn from a finite population of eight without replacement, the probability that it contains either none or all of a specified four is, using the $h(8,4,4)$ distribution, $0.0143 + 0.0143 = 0.0286$. Hence the required confidence coefficient is $1 - 0.0286 \approx 0.97$.

8.2.4   Problems

8.1   Three independent samples of size 4 are drawn from the same continuous distribution and there are no ties. Calculate the numbers of distinguishable ways in which some sample contributes i, another j, and the remaining sample k, where $i + j + k = 6$, to the six values above the common median of all 12 values.

8.2   Use the results of Problem 8.1 to find the exact distribution of the goodness-of-fit statistic X^2.

8.3   In Problem 8.1, let m_i be the number of values from sample i above the common median. Find the probability distribution of

$$\max_{i \neq j} |m_i - m_j|$$

8.4   Independent random samples of sizes n_1, n_2 are drawn from the same continuous distribution. Let M_1, M_2 be the corresponding numbers in each sample which exceed the common median of the pooled sample of size $n_1 + n_2 = 2N + 1$. Show that the probability that $M_1 = m_1$ and $M_2 = m_2$ is

$$\binom{n_1}{m_1}\binom{n_2}{m_2}\bigg/\binom{2N+1}{N}$$

(The result complements the discussion when $n_1 + n_2$ is even.)

8.2.5   Multiple comparisons

If the null hypothesis that all the samples are from populations with the same median is rejected, then the 'obvious' statistic for testing a pair is the difference between the numbers above the pooled median. In Example 8.2 we noted that the four samples contained $1, 1, 4, 4$ respectively above the common median. The (positive) values of these differences are $|4 - 4|$, $|4 - 1|$, $|4 - 1|$, $|4 - 1|$, $|4 - 1|$, $|1 - 1|$, i.e. $0, 3, 3, 3, 3, 0$ for the six possible pairs of samples. The maximum of these is 3, and, in the spirit of previous endeavours, we should attempt to calculate the probability of a difference at least as large as 3 when the null hypothesis is true. The configuration displayed in Example 8.2 could be arrived at in

$$\binom{5}{4}\binom{5}{4}\binom{5}{1}\binom{5}{1} = 625 \text{ ways}$$

But the rows can be exchanged in

$$\binom{4}{2} = 6 \text{ ways}$$

Hence, since there must always be 10 of the 20 values above the common median, the probability of a pattern similar to the one displayed is

$$(6 \times 625) \bigg/ \binom{20}{10}$$

But there is another configuration with $4, 3, 2, 1$ members above the common median which also yields a maximum difference of 3. The extent of the labour involved is now clear. However, a more fully worked out example is pursued in Problems 8.1 and 8.3.

A statistic equivalent to the one discussed above is

$$\max_{i \neq j} |M_i - M_j|/(n/2)^{\frac{1}{2}}$$

the critical values for which have been published (see reference 1). As an alternative, it is, of course, possible to carry out the Mood test on any pair of samples. In that case, we should have to find the pooled median for each of the pairs of samples concerned. Thus in Example 8.2, if sample 1 is compared with sample 2, the 10 observations have a median of 35. The numbers above and below this common median are

	Above	Below
Sample 1	2	3
Sample 2	3	2

which, while not showing such an exact match as before, does not suggest that the samples are significantly different. When sample 1 is similarly compared with sample 3, we obtain

	Above	Below
Sample 1	1	4
Sample 3	4	1

From the $h(10, 5, 5)$ distribution, the two-sided p-value is $2(0.0040 + 0.0992) \approx 0.21$. These are two of the $\binom{4}{2}$ possible comparisons of pairs that can be made.

8.3   The Kruskal–Wallis test

There is an analog of the Mann–Whitney test when there are more than two treatments. Suppose there are k treatments and n_i elements are assigned to treatment i. The values, in the absence of ties, of all $\sum_{i=1}^{k} n_i = N$ responses are placed in rank order. If there are no treatment differences, then those ranks assigned to treatment i

are a random sample of size n_i from the integers $1, 2, \ldots, N$. Hence the expectation of T_i, the sum of the ranks for treatment i, is $n_i(N+1)/2$ and the variance of T_i is $n_i(N - n_i)(N+1)/12$. The *Kruskal–Wallis statistic H* is defined as

$$H = \frac{12}{N(N+1)} \sum_{i=1}^{k} \{[T_i - n_i(N+1)/2]^2/n_i\}$$

$$= \frac{12}{N(N+1)} \left[\sum_{i=1}^{k} \left(\frac{T_i^2}{n_i} \right) - \frac{N(N+1)^2}{4} \right] \tag{8.1}$$

and is one measure of the discrepancies between the observed T_i and their expectations. If $k = 2$, it is shown in Problem 8.7 that H is equivalent to the Mann–Whitney statistic.

———— EXAMPLE 8.5 ————

Let $k = 3$, $n_1 = n_2 = n_3 = 5$ and suppose ranking all $N = 15$ observations yields

						T_i
Sample 1	1	2	3	4	5	15
Sample 2	6	7	8	9	10	40
Sample 3	11	12	13	14	15	65
						120

$$H = \frac{12}{15 \times 16} \left[\frac{15^2 + 40^2 + 65^2}{5} - \frac{15(16)^2}{4} \right] = 12.5$$

If the treatment distributions are different, then the T_i will tend to be further from their expectations than might reasonably arise from a random assignment of ranks and hence large values of H will be significant. To assess the significance of any obtained H, we may:

1. For small values of k and n_i either consult a table of the distribution function of H (see reference 2) or the critical values in more widely available tables (see reference 3).
2. For large N, and n_i/N tending to a positive number, approximate the distribution of H with a $\chi^2(k-1)$ distribution.

———— EXAMPLE 8.6 ————

Let $k = 3$, $n_1 = n_2 = n_3 = 5$, and suppose the computed value of H is 5.78.

1. From a table of the distribution function, $\Pr(H \geqslant 5.78) = 0.049$.
2. From a table of critical values, 5.78 is the smallest x, such that $\Pr(H \geqslant x) \leqslant 0.05$.
3. The probability that a random variable with a $\chi^2(2)$ distribution exceeds 5.78 is about 5.6%.

8.3.1   Use of MINITAB

MINITAB has a command KRUSKAL which computes H for data set in C1 and indexed in C2.

──────── EXAMPLE 8.7 ────────────────────────────────────

If	3.0	5.1	6.2	8.9	1.0	3.1	4.2	7.9	is set in C1
and	1	1	1	1	2	2	2	2	is set in C2

then KRUSKAL C1 C2 outputs

LEVEL	NOBS	MEDIAN	AVERAGE RANK	Z VALUE
1	4	5.65	5.3	0.87
2	4	3.65	3.8	−0.87
OVERALL	8		4.5	

$H = 0.7500$ d.f. $= 1$ $p = 0.387$

together with a warning note if one or more of the sample sizes is small. The Z values are $[T_i - E(T_i)]/\sqrt{V(T_i)}]$, for the observed T_i, and give some idea as to which of the samples is contributing most to the value of H. Note the Z values sum to zero when the sample sizes are equal.

──

8.3.2   Effect of ties

By considering samples of size n_i from the finite population of ranks $1, 2, \ldots, N$ it can be shown (see Appendix 2 Example A2.2) that $E(H) = k - 1$, when there are no ties, matching the mean of the approximating chi-squared distribution. If there are tied ranks, and these are *shared*, then when the extent of the jth tie is d_j we find that, using Appendix 2, Example A2.3,

$$E(H) = (k - 1)\left[1 - \sum_j (d_j^3 - d_j)/(N^3 - N)\right] \tag{8.2}$$

In the light of equation (8.2), if the computed H, using the shared ranks, is divided by $1 - \sum_j (d_j^3 - d_j)/(N^3 - N)$ to give an adjusted value $\text{Adj}(H)$, then $E[\text{Adj}(H)]$ is again $k - 1$.

──────── EXAMPLE 8.8 ────────────────────────────────────

Let $k = 3$, $n_1 = n_2 = n_3 = 5$, and suppose after sharing ranks of the 15 observations, we have

		Rankings				T_i
Sample 1	3	3	3	3	3	15
Sample 2	8	8	8	8	8	40
Sample 3	13	13	13	13	13	65

Since the T_i are the same as in Example 8.5, $H = 12.5$. But there are three ties, each of extent five, and using equation (8.2), then $\text{Adj}(H) = 14$.

8.3.3   *MINITAB and ties*

MINITAB will calculate both an unadjusted and an adjusted value of H, together with the associated p-value, using a chi-squared approximation in all cases. We illustrate with an artificial example where there is a tie in each of three samples.

——— EXAMPLE 8.9 ———

If

Cl contains   3 3 3 3 3 4 4 4 4 4 5 5 5 5 5   and
C2 contains   1 1 1 1 1 2 2 2 2 2 3 3 3 3 3,

then KRUSKAL Cl C2 returns

LEVEL	NOBS	MEDIAN	MEAN RANK	Z
1	5	3	3	−3.06
2	5	4	8	0.00
3	5	5	13	+3.06
OVERALL	15		8	

$H = 12.50$ d.f. $= 2$ $p = 0.002$
$H = 14.00$ d.f. $= 2$ $p = 0.001$ (Adj for ties)

8.3.4   *Problems*

8.5   For Example 8.7, use formula (8.1) to confirm that $H = 0.75$.

8.6   If there are k treatment groups and n_i are assigned to treatment i, show that each selection of the ranks $1, 2, ..., N$ to the groups has probability $(n_1! \, n_2! .... n_k!) \, / N!$ if all the treatment responses have identical distributions. If $n_1 = 1$, $n_2 = 1$, $n_3 = 2$, $k = 3$ find the probability function of H.

8.7   Prove that if $k = 2$, then the Kruskal–Wallis statistic is equivalent to the Mann–Whitney statistic. For the samples 3.0 5.1 6.2 8.9 and 1.0 3.1 4.2 7.9, compare the significance levels reported by MINITAB for a two-sided test of equality of medians and comment on any discrepancy observed between the Kruskal–Wallis and Mann–Whitney statistics.

8.8   Use the variance of T_j/n_j to confirm the Z values reported in Example 8.7.

8.3.5   Multiple comparisons

When the Kruskal–Wallis statistic suggests that there are significant differences between treatments, then we may wish to examine which pairs of treatments have markedly different effects. An obvious statistic to use is the difference between the sums of ranks for the members of each pair. We illustrate the calculation of exact two-sided *p*-values.

─────── EXAMPLE 8.10 ───────────────────────────────────

Suppose we have three observations for each of three treatments and there are no ties. It is required to calculate the *p*-value of a difference in sums of ranks of 17 or more. The only possible patterns are all permutations of the rows of

Trt	ranks	sum	ranks	sum	ranks	sum
1	1 2 4	7	1 2 3	6	1 2 3	6
2	3 5 6	14	4 5 7	16	4 5 6	15
3	7 8 9	24	6 8 9	23	7 8 9	24

We observe $|T_3 - T_1| = 17, 17, 18$ in turn.

The rows of each pattern can be permuted in $3! = 6$ ways, but the arrangement of the ranks within each row is not relevant. Hence there are $6 \times 3 = 18$ patterns for which the maximum difference between sums of ranks has modulus at least 17. But there are

$$\binom{9}{3}\binom{6}{3} = 1680 \text{ ways}$$

of assigning nine ranks to three treatments, hence the required probability is $18/1680 \approx 0.0107$.

──

More generally, if $T_1, T_2, \ldots, T_k$ are the sums of ranks for *k* treatments based on *n* observations per treatment, suppose it be required to test all possible $k(k-1)/2$ pairs of treatments. Then, if $c = c(\alpha, k, n)$ satisfies

$$\Pr(\max_{i \neq j} |T_i - T_j| < c) = 1 - \alpha$$

when all the treatment effects are equal then, for any i, j with $|T_i - T_j| \geq c$, we reject the hypothesis that the effects of treatments *i* and *j* are the same. For selected values of *k* and *n*, some useful *c* values are reported in reference 2. The condition is equivalent to the probability that at least one $|T_i - T_j|$ is greater than or equal to *c* is α.

─────── EXAMPLE 8.11 ───────────────────────────────────

(i) $k = 3$, $n = 3$, $c(0.011, 3, 3) = 17$. This should be compared with the calculation in Example 8.10.

(ii) $k = 12$, $n = 2$. Suppose the 24 observations, after ranking, yielded:

Treatment	1	2	...	11	12
Ranks	(1, 2)	(3, 4)	...	(21, 22)	(23, 24)
Sum	3	7	...	43	47

We see that $|T_1 - T_{11}| = 40$, $|T_2 - T_{12}| = 40$, $|T_1 - T_{12}| = 44$. If the 12 treatments are equally effective, then by chance alone the probability that the magnitude of at least one pairwise comparison reaches 40 or more is, from reference 2, about 0.062.

The reader may have observed that the multiple comparisons studied here retain the ranks imposed by *all* the treatment observations – not just those implied by the pair of treatments under consideration. That is, such comparisons can be influenced by variations in other responses – an effect which some authors have thought rather undesirable (see reference 4). An alternative approach is discussed later.

If we are prepared to accept conservative bounds, then a method of quite general application can be supplied using Scheffé's technique. It can be shown (see Problem 8.9) that if there are k treatments and the number of elements receiving treatment i is n_i, while $\bar{T}_i$ is the average of their ranks, then

$$\Pr\left(|\bar{T}_i - \bar{T}_j| \leq h_k^{1/2}(\alpha)\left[\frac{N(N+1)}{12}\left(\frac{1}{n_i} + \frac{1}{n_j}\right)\right]^{1/2} \text{ for all } i \neq j\right) \geq 1 - \alpha \qquad (8.3)$$

where $h_k(\alpha)$ is the $100\alpha\%$ point of the Kruskal–Wallis statistic for k treatments and $\sum_{i=1}^{k} n_i = N$. If all the $n_i = n$, then equation (8.3) can be written

$$\Pr\left(\max_{i \neq j} |\bar{T}_i - \bar{T}_j| \leq h_k^{1/2}(\alpha)\left(\frac{nN(N+1)}{6}\right)^{1/2}\right) \geq 1 - \alpha \qquad (8.4)$$

——— EXAMPLE 8.12 ———

Suppose we have $k = 3$, $n_1 = n_2 = n_3 = 3$. The data when ranked yields

				Sum
A	1	2	4	7
B	3	5	6	14
C	7	8	9	24

The computed value of H is 6.489. The exact probability that H exceeds 6.489 is 0.011. Hence, by the inequality (8.4), with $k = 3$, $n = 3$

$$\Pr\left(\max_{i \neq j} |\bar{T}_i - \bar{T}_j| \leq 17.088\right) \geq 1 - 0.011 = 0.989$$

We were not obliged to use the *p*-value corresponding to the computed value of *H*. We have done this so that the result can be compared with the exact probability in Example 8.10.

8.3.6   *Problem*

8.9   Suppose n_i elements are assigned to treatment *i* and T_i is the sum of the associated ranks, $i = 1, 2, \ldots, k$. Let

$$\bar{T} = \sum_{i=1}^{k} T_i/N \quad \text{and} \quad N = \sum_{i=1}^{k} n_i$$

Confirm inequality (8.3) in the lemma (Section 8.1) by choosing

$$X_i = \sqrt{n_i}\left(\frac{T_i}{n_i} - \bar{T}\right) \quad i = 1, 2, \ldots, k$$

$$a_j = \frac{1}{\sqrt{n_j}}, \quad a_l = -\frac{1}{\sqrt{n_l}}, \quad a_i = 0 \quad \text{for } i \neq j \text{ or } l$$

If the criticism noted in our previous discussion makes the method there employed seem unacceptable, then there is an alternative. For any pair of treatments, we could carry out a Mann–Whitney test. We would still be faced with the problem of the overall significance level when many such comparisons are made. We discuss the underlying calculation for an exact test when a control is compared with two other treatments, and there are no ties.

--------- EXAMPLE 8.13 ---------

$k = 3$, $n = 2$. There are 90 (unordered) ways of allocating two ranks from 6 to each of three treatments. One of these ways is

Control	Trt A	Trt B
1	2	5
3	4	6

It will be equivalent to the Mann–Whitney test if we compare the sum of the ranks of the control against the treatment sums in turn, reranking if necessary. For the displayed allocation, treatment A against the control is $(2 + 4) - (1 + 3) = 2$, and no reranking was required. Note that if the columns were interchanged we obtain a difference of -2. Against B, the relative rankings transpose to

Control	Trt B
1	3
2	4

and the corresponding difference is $7 - 3 = 4$. If the columns were interchanged, we would obtain -4. It is clear that the only differences appearing are -4, -2, 0, 2, 4. Since a pattern such as

Control	Trt B
1	2
4	3

yields 0, both as it stands and when the columns of ranks are interchanged, the possible differences appear in the ratios 1:1:2:1:1. Hence if all 90 selections of ranks have equal probability, then the probability that the sum of ranks for treatment B exceeds that for the control by $+4$ is $1/6$ – similarly for treatment A. The probability that both exceed the control by $+4$ is $6/90$, since in that case the controls must have ranks $1, 2$ and there are six ways of allocating the remaining four ranks to the two treatments and all of these satisfy the desired property. Hence the probability that the sum of the ranks of at least one treatment exceeds the control by $+4$ is

$$\frac{1}{6} + \frac{1}{6} - \frac{6}{90} = \frac{4}{15}$$

For corresponding probabilities for small values of k, n see reference 5.

8.3.7 *Problems*

8.10 For Example 8.10, calculate

$$\Pr\left(\max_{i \neq j} |T_i - T_j| > 15\right)$$

8.11 Suppose one of the treatments is designated as a control and the only contrasts to be made will be with this control. When $k = 3$, $n = 2$, calculate the probability that the difference between the sums of ranks for at least one of the treatments and the control *exceeds* 6.

8.4 Friedman's test

The sign test for equality of two treatments commonly compares the responses on matched pairs of elements. To detect differences between k treatments, we require sets of k matched elements, one for each treatment. Such a set is called a *block* and we suppose there are b blocks. The matching may be based on similarity with respect to variables thought to be related to the response variable of interest, or constitute experiments carried out at different places or times.

———— **EXAMPLE 8.14** ————

With $k = 4$, $b = 5$, the following data was obtained:

TREATMENTS

A	B	C	D
65	37	57	64
41	25	62	83
20	41	23	45
33	26	71	36
17	45	44	46

Each row of the table constitutes a block of four treatment responses.

Let X_{ij} be the observation in block i on treatment j, $i = 1, 2, ..., b$, $j = 1, 2, ..., k$. The differences between blocks may be eliminated by replacing X_{ij} by its rank R_{ij}, *within that block*.

———— **EXAMPLE 8.15** ————

The data for Example 8.14 ranked within blocks gives

TREATMENTS

	A	B	C	D
	4	1	2	3
	2	1	3	4
	1	3	2	4
	2	1	4	3
	1	3	2	4
SUM	10	9	13	18

Let $T_j = \sum_{i=1}^{b} R_{ij}$. Then if there is no difference between the treatment distributions, T_j is the sum of b independent random variables, each of which, in the absence of ties within rows, has mean $(k+1)/2$ and variance $(k^2 - 1)/12$. *Friedman's statistic* Q for detecting differences is defined as

$$Q = \frac{12}{bk(k+1)} \sum_{j=1}^{k} [T_j - b(k+1)/2]^2 \tag{8.5}$$

Since $\mathrm{E}([T_j - b(k+1)/2]^2) = \mathrm{V}(T_j) = b(k^2 - 1)/12$,
$\mathrm{E}(Q) = 12bk(k^2 - 1)/[12bk(k+1)] = k - 1$.

If there are differences in location between the treatments then the T_j will tend to be further from their expectations than might reasonably arise by chance. Thus large values of Q discredit the hypothesis that the treatment effects are equal. There are $(k!)^b$ distinguishable ways of assigning the ranks $1, 2, ..., k$ independently in each of b blocks and hence in principle we can calculate the distribution of Q. Extensive tables of the distribution are available in Hollander and Wolfe (see reference 2), critical values of b and k are provided in Lindley and Scott (see reference 3), while for large b, the statistic Q may be taken as having approximately a $\chi^2(k-1)$ distribution. For Example 8.15, the calculated value of Q is 5.88, and from reference 2 $\Pr(Q \geqslant 5.88) = 0.123$. The approximate probability, using a $\chi^2(3)$ distribution, is about 0.118.

8.4.1 Use of MINITAB

Each of the bk observations can be located by its block and the treatment applied. If the data is in C1, the corresponding treatment numbers in C2 and the block numbers in C3, then the MINITAB command

   FRIEDMAN   C1   C2   C3

will calculate Q and an approximate p-value, using a $\chi^2(k-1)$ distribution. It will also provide estimates of the medians of the treatment responses (for a discussion of which see the end of this section). Note that MINITAB uses S where we have used Q.

───────── EXAMPLE 8.16 ─────────

For Example 8.14 MINITAB returns

   $S = 5.88$ d.f. $= 3$ $p = 0.118$

C2	N	EST MED	SUM OF RANKS
1	5	29.25	10.0
2	5	29.75	9.0
3	5	47.50	13.0
4	5	49.50	18.0

   Grand median $= 39.00$

8.4.2 Effect of ties

If there are ties within the blocks and the corresponding ranks are *shared*, then $E(Q)$, under all permutations of assigned ranks within the blocks, will no longer be $(k-1)$. We now derive an appropriate scaling factor and an adjusted value $\mathrm{Adj}(Q)$ of Q such that $E(\mathrm{Adj}(Q)) = k-1$, conditionally on the ties. In the presence of ties, T_j is still the sum of b independent random variables, each with mean $(k+1)/2$ but with variances σ_i^2, $i = 1, 2, \ldots, b$ which take account of the ties. If d_{il} is the extent of the lth tie in the ith block, then

$$\sigma_i^2 = \frac{1}{12k}\left[k^3 - k - \sum_l (d_{il}^3 - d_{il})\right]$$

We obtain

$$E((T_j - b(k+1)/2)^2) = V(T_j) = \sum_{i=1}^{b} \sigma_i^2 = \left[b(k^3 - k) - \sum_i \sum_l (d_{il}^3 - d_{il})\right]\Big/ 12k$$

so that

$$E(Q) = (k-1)\left[1 - \sum_i \sum_l (d_{il}^3 - d_{il})\right]\Big/ [b(k^3 - k)] = (k-1)a, \text{ say} \qquad (8.6)$$

Hence $Adj(Q) = Q/a$ will have expectation $(k-1)$. That is, Q is computed from equation (8.5) on the basis of sharing the tied ranks, and the result is adjusted by the factor a.

─────── EXAMPLE 8.17 ───────────────────────────────

If in Example 8.14, the data in the fifth block had been 17 44 44 44, then there is one tie of extent 3 and from equation (8.6) $a = 92/100$. The value of Q, on the basis that the ranks in this block are $1, 3, 3, 3$ is now 4.92 and the adjusted value is $4.92/0.92 \approx 5.35$. In the case of ties, MINITAB will report both the unadjusted and adjusted values of Q with a p-value for each as follows

$S = 4.92$ d.f. $= 3$ $p = 0.118$
$S = 5.35$ d.f. $= 3$ $p = 0.149$ (ADJUSTED FOR TIES)

──

8.4.3   Agreement between blocks

Friedman's Q is not entirely suitable as an overall measure of agreement between blocks. It is intuitively obvious that the rankings show the greatest agreement when the ranking of a treatment is the same in each block, giving block totals $b, 2b, \ldots, kb$ in some order and $Q = b(k-1)$. When there are no ties, it can be shown that $b(k-1)$ is also the maximum value of Q (see Problem 8.14). Perfect agreement yields $Q = b(k-1)$ but the meaning of perfect disagreement is not so clear. For $b = 2$ and $k = 3$, having $1, 2, 3$ in block I and $3, 2, 1$ in block II is one kind of disagreement. Even stronger would be the case $b = 6$, $k = 3$ and each block contains one of the $3! = 6$ permutations of the integers $1, 2, 3$. Both examples give $Q = 0$. However, for $b = 4$ and $k = 3$, if blocks I, II each contain $1, 2, 3$ while blocks III, IV contain $3, 2, 1$ then the overall Q for all four blocks is zero although two pairs of blocks 'agree perfectly'.

8.4.4   Problems

8.12 If $k = 3$, $b = 2$, find the probability function of Friedman's Q and calculate $E(Q)$.

8.13 Assuming that Q is a maximum when every treatment receives the same rank in each block, show that for $b = 3$, $k = 3$, $\Pr(Q = 6) = 1/36$.

8.14 Show that Q can be written $(k - 1)[1 + (b - 1)\bar{\rho}]$, where $\bar{\rho}$ is the average of all possible Spearman's rank correlation coefficients between blocks. When is Q a maximum? (Assume no ties.)

8.15 In the case $k = 2$, show that Friedman's test is equivalent to a sign test.

8.4.5 *Multiple testing*

If Friedman's test suggests that there are significant differences between the treatments, then we may wish to compare pairs of treatments. An exact test has been proposed which compares the sums of ranks for treatments over blocks. We indicate how the *p*-values might be calculated.

──────── EXAMPLE 8.18 ────────────────────────────────

Let $k = 3$, $b = 4$, no ties. There are $(3!)^4$ ways of assigning the ranks $1, 2, 3$ to the three treatments over four blocks. The maximum achievable difference is 8, as illustrated in Table 1.

	Table 1 Treatment				Table 2 Treatment				Table 3 Treatment		
Block	A	B	C	Block	A	B	C	Block	A	B	C
I	1	2	3	I	1	2	3	I	1	2	3
II	1	2	3	II	1	2	3	II	1	2	3
III	1	2	3	III	1	2	3	III	1	2	3
IV	1	2	3	IV	2	1	3	IV	1	3	2
Total	4	8	12	Total	5	7	12	Total	4	9	11

There are 3! ways of obtaining such a difference by interchanging the columns of ranks. The probability of obtaining any difference with *magnitude* 8 is $3!/(3!)^4 = 1/216 \approx 0.00463$. There are two basic patterns which yield a difference of 7 as in Tables 2 and 3. In each of these patterns there are also four ways of exchanging the rows and maintaining the same column totals. Hence the probability of obtaining a difference with magnitude 7 *or more* is $2 \times 3! \times 4/(3!)^4 + 1/216 \approx 0.0417$.

───

More generally, when there are k treatments then there are $k(k-1)/2$ comparisons of pairs to be made. If T_i is the sum of the treatment ranks over blocks $i = 1, 2, ..., k$, then for the differences $|T_i - T_j|$ for $i < j$, we need the distribution of the *maximum*. For selected values of k and b the key tail probabilities are reprinted in reference 2.

─────── **EXAMPLE 8.19** ───────────────────────────

Let $k = 3$, $b = 4$. From Example 8.18 we find

$$\Pr\left(\max_{i \neq j} |T_i - T_j| \leq 6\right) = 0.9583$$

That is, the probability that all three differences $|T_1 - T_2|$, $|T_1 - T_3|$, $|T_2 - T_3|$ are simultaneously less than 7 is 0.9583. In other words the probability that at least one of the differences has value at least 7 is $1 - 0.9583 = 0.0417$ and hence such a difference would be regarded as significant. The *p*-value applies when we have not designated the pair to be tested in advance of seeing the data. If we *had* intended to compare treatment A with treatment C and obtained Table 2, then the two-sided *p*-value would be recorded as $2(4 + 4 + 1)/(3!)^4 = 0.0139$.

───

If we are prepared to accept conservative bounds for the probabilities associated with multiple comparisons then we can apply the method already discussed in connection with the Kruskal–Wallis test. If there are k treatments in each of b blocks and T_i is the sum of the ranks for the *i*th treatments over the blocks, then Scheffé's method gives us

$$\Pr\left(\max_{i \neq j} |T_i - T_j| \leq [q(\alpha)]^{1/2} \, [bk(k + 1)/6]^{1/2}\right) \geq 1 - \alpha$$

where $q(\alpha)$ is the upper $100\alpha\%$ point for the corresponding Friedman statistic.

─────── **EXAMPLE 8.20** ───────────────────────────

Let $k = 3$, $b = 4$ and the rankings be

A	B	C
1	2	3
1	2	3
1	2	3
2	1	3
5	7	12

We calculate $Q = 6.5$, and from tables obtain $\Pr(Q \geq 6.5) = 0.042$. Hence

$$\Pr\left(\max_{i \neq j} |T_i - T_j| \leq [(6.5)(4)(3)(4)/6]^{1/2}\right) \geq 1 - 0.042$$

that is

$$\Pr\left(\max_{i \neq j} |T_i - T_j| \leq 7.21\right) \geq 0.958$$

───

8.4.6   *Relative rankings*

The multiple comparison test just discussed has the same feature noted in connection with the Kruskal–Wallis test – namely that the ranks included in any particular comparison depend on those assigned to other treatments. An alternative analysis is now presented which uses only the relative rankings, derived from the data within the comparisons. We have seen that Friedman's test for two treatments is equivalent to a sign test. More particularly it mimics the matched pairs test discussed in Chapter 5. The extra feature is that, for multiple testing, overall probability levels are required.

──────── EXAMPLE 8.21 ────────────────────────────

(i) If there are two treatments and four pairs then the (one-sided) probability that all the differences are in favour of a particular one of the treatments is $1/16$.
(ii) Suppose there are two treatments and a control assigned to four blocks. What is now the one-sided probability that at least one treatment appears superior to the control over all blocks? The probability that either is superior to the control over all blocks is still $1/16$. It remains to find the corresponding probability for both treatments simultaneously. After ranking the data, the pattern of ranks must be one of the $2^4 = 16$ of the type

Control	Trt A	Trt B
1	2	3
1	3	2
1	2	3
1	3	2

in which each of the treatment ranks 2 and 3 may be chosen in two ways. But there are $(3!)^4 = 1296$ ways of assigning the ranks altogether. Hence the probability that both treatments have higher ranks than the control is $16/1296$, and the probability that at least one has higher rank than the control over all blocks is $2/16 - 16/1296 = 146/1296 \approx 0.113$. Some values for small k, b are tabled in reference 6.

──

8.4.7   *Estimating median treatment responses*

For a full discussion of the method employed by MINITAB to estimate the median responses to the treatments, the reader should consult the specialist literature (see reference 2). We provide a rough outline of the method. The underlying assumption is that X_{ij}, the observation on the jth treatment in the ith block, satisfies a linear model of the form $X_{ij} = \mu + t_j + b_i + e_{ij}$, where t_j, b_i are treatment and block effects and the e_{ij} are random errors. Thus $X_{ij} - X_{il}$ does not depend on b_i and is an observation on $t_j - t_l$. We obtain an estimate for each treatment effect based on these differences. We illustrate by finding the treatment effect for A. The table below shows these differences for treatment A compared to B, C, D derived from Example 8.15.

A–B	28	16	−21	7	−28
A–C	8	−21	−3	−38	−27
A–D	1	−42	−25	−3	−29

Summary estimates of the differences of effects between A and each of the other treatments are provided by the medians of these sample differences and are 7, −21, −25 respectively. The estimate of the effect of A compared to the average of all the four treatments is taken to be $(0 + 7 - 21 - 25)/4 = -9.75$ (the zero arises from comparing A with itself). The corresponding values for B, C, D are $-9.25, 8.5, 10.5$ (note: the sum of these four values is zero – giving a check). Next, these additional treatment effects are subtracted from the data. Thus −9.75 is subtracted from each observation in column A, and so on. These residues are estimates of the block responses *adjusted* for treatments. For Example 8.14 the adjusted block responses are:

A	B	C	D	MEDIAN
74.75	46.25	48.5	53.5	51
50.75	34.25	53.5	72.5	52.125
29.75	50.25	14.5	34.5	32.125
42.75	35.25	62.5	25.5	39
26.75	54.25	35.5	35.5	35.5

The medians of the adjusted block responses are as shown. The median of *these* is 39 and provides an estimate of the overall block effect. The median responses are found by adding this grand median to the additional treatment effects and are $39 - 9.75 = 29.25$, $39 - 9.25 = 29.75$, $39 + 8.5 = 47.5$, $39 + 10.5 = 49.5$ respectively.

8.5   Solutions to problems

8.1   There are five possible patterns:
   (a) Each sample contributes 2 values, in

$$\binom{4}{2}\binom{4}{2}\binom{4}{2} = 216 \text{ ways}$$

   (b) The samples contribute 1, 2, 3 values, in

$$\binom{4}{1}\binom{4}{3}\binom{4}{3} \times 3! = 576 \text{ ways}$$

   (c) The samples contribute 0, 2, 4 values, in

$$\binom{4}{0}\binom{4}{2}\binom{4}{4} \times 3! = 36 \text{ ways}$$

(d) The samples contribute 0, 3, 3 values, in

$$\binom{4}{0}\binom{4}{3}\binom{4}{3} \times 3 = 48 \text{ ways}$$

(e) The samples contribute 1, 1, 4 values, in

$$\binom{4}{1}\binom{4}{1}\binom{4}{4} \times 3 = 48 \text{ ways}$$

The total is 924, which is also the number of selections of 6 from 12.

8.2 The values of X^2 for (a), (b), (c), (d), (e) in the solution of Problem 8.1 are 0, 2, 8, 6, 6, respectively. Hence, $\Pr(X^2 = 8) = 36/924$ and so on.

8.3 The maximum values of $|m_i - m_j|$, $i \neq j$, for (a), (b), (c), (d), (e) are 0, 2, 4, 3, 3, respectively. Hence the probability that the maximum modulus is 3, for example, is $(48 + 48)/924 = 8/77$

8.4 When $n_1 + n_2$ is odd, then the common median must be one of the sample members. It comes from the first sample with probability $n_1/(2N + 1)$, and then the conditional probability that $M_1 = m_1$

is $$\binom{n_1 - 1}{m_1}\binom{n_2}{m_2} \Big/ \binom{2N}{N}$$

It comes from the second sample with probability $n_2/(2N + 1)$ and then the conditional publicity that $M_1 = m_1$

is $$\binom{n_1}{m_1}\binom{n_2 - 1}{m_2} \Big/ \binom{2N}{N}$$

Hence the unconditional probability that $M_1 = m_1$ is

$$\left[\frac{n_1}{2N + 1}\binom{n_1 - 1}{m_1}\binom{n_2}{m_2} + \frac{n_2}{2N + 1}\binom{n_1}{m_1}\binom{n_2 - 1}{m_2}\right] \Big/ \binom{2N}{N}$$

$$= \frac{N + 1}{2N + 1}\binom{n_1}{m_1}\binom{n_2}{m_2} \Big/ \binom{2N}{N} = \binom{n_1}{m_1}\binom{n_2}{m_2} \Big/ \binom{2N + 1}{N}$$

8.5 $T_1 = 2 + 5 + 6 + 8 = 21$, $T_2 = 1 + 3 + 4 + 7 = 15$
$H = 12[21^2 + 15^2 - 8 \times 9^2]/8 \times 9 = 0.75$

8.6 The n_1 ranks for treatment 1 can be selected in

$$\binom{N}{n_1} \text{ ways}$$

the n_2 ranks for treatment 2 from the remaining $N - n_1$ ranks in

$$\binom{N - n_1}{n_2} \text{ ways}$$

and so on. Note

$$\binom{N}{n_1}\binom{N - n_1}{n_2} \cdots \binom{N - n_1 - n_2 \ldots - n_k - 1}{n_k} = \frac{N!}{n_1! n_2! \ldots n_k!}$$

If $N = 4$, there are $4!/1!1!2! = 12$ ways of selecting the ranks, and we list all the possibilities:

Trt 1	Trt 2	Trt 3	H				
1	2	3,4	2.7				
1	3	2,4	1.8				
1	4	2,3	2.7				
2	1	3,4	2.7	H	0.3	1.8	2.7
2	3	1,4	0.3	Pr	2/12	4/12	6/12
2	4	1,3	1.8				
3	1	2,4	1.8				
3	2	1,4	0.3				
3	4	2,1	2.7				
4	1	2,3	2.7				
4	2	1,3	1.8				
4	3	1,2	2.7				

8.7 $n_1 + n_2 = N$ since sum of all ranks is $N(N + 1)/2$, $T_2 = N(N + 1)/2 - T_1$. Substitute for T_2 in H and collect terms to obtain

$$12[T_1 - n_1(N + 1)/2]^2/[n_1 n_2(N + 1)]$$

and hence result. The command KRUSKAL uses a $\chi^2(1)$ approximation without continuity correction and gives $p = 0.387$. The command MANN–WHITNEY uses a normal approximation with a continuity correction and gives $p = 0.4705$.

8.8 $N = 8$, $n_1 = n_2 = 4$, $E(T_1/n_1) = 9/2$, $V(T_1/n_1) = (N - n_1)(N + 1)/(12n_1)$, $V(T_1/4), = 3/4$. $Z = [(21/4) - (9/2]/\sqrt{(3/4)} \approx 0.87$.

8.9 $$\sum_{i=1}^{k} a_i X_i = \left[\left(\frac{T_i}{n_j} - \bar{T}\right) - \left(\frac{T_l}{n_l} - \bar{T}\right)\right] = \frac{T_j}{n_j} - \frac{T_l}{n_l}$$

$$\sum_{i=1}^{k} a_i^2 = \left(\frac{1}{n_j} + \frac{1}{n_l}\right), \quad \sum_{i=1}^{k} X_i^2 = \sum_{i=1}^{k} n_i \left(\frac{T_i}{n_i} - \bar{T}\right)^2 = \frac{N(N + 1)H}{12}$$

where H is the Kruskal–Wallis statistic as defined in equation (8.1) and $T = (N + 1)/2$.

8.10 The basic patterns which show a difference of 16 are:

	Total		Total		Total		Total		Total
125	8	123	6	134	8	123	6	124	7
346	13	467	17	256	13	468	17	357	15
789	24	589	22	789	24	679	22	689	23

Each of these is one of $3! = 6$ patterns giving some difference with magnitude 16. Using the results of Example 8.10 the probability is $(5 \times 6 + 18)/1680 \approx 0.0286$.

8.11 Regarding the first row as the control, the following basic patterns show a sum of ranks exceeding that of the control by more than 6, duplicated by interchanging the bottom two rows

		Total		Total		Total
Control	1 2	3	1 3	4	1 2	3
	3 4	7	2 4	6	3 5	8
	5 6	11	5 6	11	4 6	10

Hence probability is $6/90 \approx 0.067$.

8.12 If $k = 3$, $b = 2$, there are 36 distinguishable arrangements. If the ranks in the first block are in the order $1, 2, 3$ then there are six arrangements in the second block as shown below, giving $E(Q) = 2$.

BLOCK II			T_1	T_2	T_3	Q
1	2	3	2	4	6	4
2	3	1	3	5	4	1
3	1	2	4	3	5	1
2	1	3	3	3	6	3
1	3	2	2	5	5	3
3	2	1	4	4	4	0

The ranks in the first blocks can be interchanged in six ways and each of these leads to the same values of Q. $Q = 0, 1, 3, 4$ with probabilities $1/6, 2/6, 2/6, 1/6$ and $E(Q) = (0 + 2 + 6 + 4)/6 = 2$.

8.13 There are $(3!)^3 = 216$ different arrangements of the ranks. There are six arrangements which give the maximum value, which is $[(3 - 6)^2 + (6 - 6)^2 + (9 - 6)^2]/3 = 6$. $\Pr(Q = 6) = 6/216 = 1/36$.

8.14 Let R_{ij} be the rank of the jth treatment in block i. Then

$$\sum_{j=1}^{k} (T_j - \tfrac{1}{2} b(k + 1))^2 = \sum_{j=1}^{k} \left[\sum_{i=1}^{b} \left\{ R_{ij} - \frac{(k+1)}{2} \right\} \right]^2 = \sum_{j=1}^{k} \left(\sum_{i=1}^{b} S_{ij} \right)^2 \text{ say}$$

$$= \sum_{j=1}^{k} \left[\sum_{i=1}^{b} S_{ij}^2 + \sum_{l \neq m} \sum S_{lj} S_{mj} \right] = \sum_{i=1}^{b} \left(\sum_{j=1}^{k} S_{ij}^2 \right) + \sum_{l \neq m} \sum_{l \neq m} \left(\sum_{j=1}^{k} S_{lm} S_{mj} \right)$$

$$= \sum_{i=1}^{b} k\sigma^2 + \sum_{l \neq m} \sum k\rho_{lm}\sigma^2 = bk\sigma^2 + kb(b - 1)\bar{\rho}\sigma^2$$

where $\sigma^2 = (k^2 - 1)/12$. Hence result. Maximum of Q occurs if $\bar{\rho} = 1$, when Q is $b(k - 1)$.

8.15 $Q = 12[(T_1 - 3b/2)^2 + (T_2 - 3b/2)^2]/6b$. Since only ranks 1, 2 are employed, $T_1 + T_2 = 3b$, hence $(T_1 - 3b/2)^2 + (T_2 - 3b/2)^2 = (T_1 - T_2)^2/2$. But $T_1 - T_2$ is the sum of differences $(R_{i1} - R_{i2})$, $i = 1, 2, ..., b$ and each of these is $+1$, -1, or, in the case of tied ranks, zero.

References

1. Nemenyi, P., Some distribution-free multiple comparison procedures in the asymptotic case (abstract). *Ann. Math. Stat.*, **32**, 1961, 921–2.
2. Hollander, M. and Wolfe, D., *Nonparametric Statistical Methods*, John Wiley, Chichester, 1973.
3. Lindley, D. V. and Scott, W. F., *New Cambridge Elementary Statistical Tables*, Cambridge University Press, Cambridge, 1984.
4. Miller, R., *Simultaneous Statistical Inference*, Springer-Verlag, New York, 1984.
5. Steel, R. D. G., A multiple comparison rank sum test: treatment versus control. *Biometrika*, **15**, 1959, 560–72.
6. Steel, R. G. D., A multiple comparison sign test: treatment versus control. *J. Amer. Statist. Ass.*, **54**, 1959, 767–75.

Chapter Nine

Theory of hypothesis testing

9.1 Introduction

In the earlier chapters, the approach to a test of hypothesis has been to conjure up a test statistic and critical region, and to rely on intuition as the motivation and justification of the method. The reader was not expected to enquire whether an alternative choice of test statistic and critical region might, in some sense, have been better, or whether there existed a best test procedure. We now provide a response to the question in terms of the Neyman–Pearson theory of hypothesis testing. As indicated fleetingly in Section 4.3, the criterion to be used for distinguishing between two test statistics (and related critical regions) is the size of the probability of Type II error – the test statistic having the smaller probability of Type II error is preferred.

In the following introductory example note that the calculation of the probabilities is effected by integrating the joint p.d.f. of the sample variables over appropriate regions in two dimensions.

——— EXAMPLE 9.1 ———

Let X_1, X_2 be a sample of size 2 from a population whose distribution has the p.d.f. $f(x \mid \theta) = \theta x^{\theta-1}$ for $0 \leqslant x \leqslant 1$. We compare two tests of the null hypothesis H_0 that $\theta = 2$ against the alternative hypothesis H_1 that $\theta = 1$, using the level of significance α:

1. Test statistic $S_1 = X_1 + X_2$ and critical region $R_1 = \{s : 0 \leqslant s \leqslant a\}$ and
2. Test statistic $S_2 = \max(X_1, X_2)$ and critical region $R_2 = \{s : 0 \leqslant s \leqslant b\}$.

By the independence of the sample random variables, the joint p.d.f. of X_1 and X_2 is

$$f(x, y \mid \theta) = \theta^2 x^{\theta-1} y^{\theta-1} \quad \text{for} \quad 0 \leqslant x \leqslant 1, 0 \leqslant y \leqslant 1$$

reducing to $f(x, y \mid 2) = 4xy$ under H_0 and $f(x, y \mid 1) = 1$ under H_1.

For the statistic $S_1 = X_1 + X_2$, the value a is determined by

$$\alpha = \Pr(S_1 \in R_1 \mid H_0) = \Pr(X_1 + X_2 \leqslant a \mid H_0)$$

$$= \int_0^a \int_0^{a-y} 4xy \, dx \, dy$$

$$= a^4/6$$

so that $a^4 = 6\alpha$ and the probability of Type II error is given by

$$\Pr(S_1 \notin R_1 \mid H_1) = 1 - \Pr(X_1 + X_2 \leqslant a \mid H_1)$$

$$= 1 - \int_0^a \int_0^{a-y} 1 \, dx \, dy = 1 - \tfrac{1}{2}a^2$$

$$= 1 - \tfrac{1}{2}\sqrt{6\alpha}$$

Similarly, for the statistic $S_2 = \max(X_1, X_2)$, we calculate

$$\alpha = \Pr(S_2 \in R_2 \mid H_0) = \Pr(\max(X_1, X_2) \leqslant b \mid H_0)$$

$$= \Pr((X_1 \leqslant b \mid H_0) \Pr(X_2 \leqslant b \mid H_0) \quad \text{(by independence)}$$

$$= \left(\int_0^b 2x \, dx \right)^2 = b^4$$

and the probability of Type II error is

$$\Pr(S_2 \in R_2 \mid H_1) = 1 - \Pr(\max(X_1, X_2) \leqslant b \mid H_1)$$

$$= 1 - \Pr(X_1 \leqslant b \mid H_1) \Pr(X_2 \leqslant b \mid H_1) = 1 - b^2$$

$$= 1 - \sqrt{\alpha}$$

Since $\tfrac{1}{2}\sqrt{6\alpha} > \sqrt{\alpha}$ for $0 < \alpha < 1$, we find that the probability of a Type II error is larger for the statistic S_2 than for S_1, and, by the criterion of smaller Type II error probability, we prefer the statistic S_1. See Problem 9.1 of Section 9.2 for the 'best' choice of test statistic and critical region.

The method of calculating the probabilities in Example 9.1 anticipates the change of emphasis in hypothesis testing as described in the Neyman–Pearson theory. This change involves moving away from the test statistic and one-dimensional critical region to a critical region as a subset of n-dimensional space (in Example 9.1, $n = 2$).

Writing $\Delta = \{ (x_1, x_2) : (x_1 + x_2) \leqslant a \}$ and $\square = \{ (x_1, x_2) : \max(x_1, x_2) \leqslant b \}$ – a sketch will help – we have:

$S_1 \in R_1$ is equivalent to $(X_1, X_2) \in \Delta$ and

$S_2 \in R_2$ is equivalent to $(X_1, X_2) \in \square$.

It follows that the two tests in Example 9.1 could be defined purely in terms of the critical regions Δ and $\square$, with the null hypothesis rejected if the observed value of (X_1, X_2) lies in the respective critical region. This provides a more general view of a statistical test than that provided by specifying, *ab initio*, a test statistic and a one-dimensional critical region. However, in practice, the general procedure usually reduces to the one-dimensional model.

In the Neyman–Pearson structure, a test of a null hypothesis H_0 against an alternative hypothesis H_1 at the exact level of significance α consists of: sample random variables $X_1, X_2, \ldots, X_n$ with observed sample data values $x_1, x_2, \ldots, x_n$; an n-dimensional critical region C satisfying $\Pr((X_1, X_2, \ldots, X_n) \in C \,|\, H_0) = \alpha$; and the decision procedure that the null hypothesis is rejected at the level of significance α, if, and only if, the observed value $(x_1, x_2, \ldots, x_n) \in C$.

Further, given two critical regions C_1 and C_2 of size α – that is, $\Pr((X_1, X_2, \ldots, X_n) \in C_i \,|\, H_0) = \alpha$ for $i = 1, 2$ – the region C_1 is preferred to C_2 if $\Pr((X_1, X_2, \ldots, X_n) \notin C_1 \,|\, H_1) \leqslant \Pr((X_1, X_2, \ldots, X_n) \notin C_2 \,|\, H_1)$, or, in words, if there is a smaller probability of Type II error for the region C_1.

The criterion for choosing between two critical regions is commonly expressed in a complementary fashion. For a critical region C the *power* of the test defined by C is given by $\Pr((X_1, X_2, \ldots, X_n) \in C \,|\, H_1) = 1 - \Pr(\text{Type II error})$. Our search for critical regions with small Type II error probability can be rephrased as looking for critical regions with large (close to 1) power. The statement that the critical region C_1 is preferred to the critical region C_2 may be put succinctly as 'C_1 is more powerful than C_2'.

Explanations of the precise interpretation of these general statements will unfold gradually in Sections 9.2 and 9.3. The reader is referred to the discussion in Section 4.3 regarding the achievability of prescribed significance levels.

9.1.1   Notation

The use of vector notation in n-dimensional space is an aid to clarity in both the writing and interpretation of formulae. We write $\mathbf{x} = (x_1, x_2, \ldots, x_n)$ for a general point and $\mathbf{X} = (X_1, X_2, \ldots, X_n)$ for the random point defined by the sample random variables $X_1, X_2, \ldots, X_n$. If, for example, the set C is given by $\{ \mathbf{x} : \sum_{i=1}^{n} x_i \leqslant a \}$ then $\mathbf{x} \in C$ simply means $\sum_{i=1}^{n} x_i \leqslant a$, and $\mathbf{X} \in C$ signifies that $\sum_{i=1}^{n} X_i \leqslant a$, or $\bar{X} \leqslant a/n$. Further, when $X_1, X_2, \ldots, X_n$ have the p.d.f. $f(\mathbf{x})$ we have $\Pr(\mathbf{X} \in C) = \int \ldots \int_C f(\mathbf{x}) \, d\mathbf{x}$, where $d\mathbf{x} = dx_1 \, dx_2 \ldots dx_n$.

9.2   The Neyman–Pearson theory

Throughout this section, let $X_1, X_2, \ldots, X_n$ denote the sample random variables, with their joint p.f./p.d.f. $f(\mathbf{x} \,|\, \theta)$ showing the unknown parameter θ. For most of the applications we consider, the random variables will be independent with a common p.f./p.d.f. $f(\mathbf{x} \,|\, \theta)$ and, hence, $f(\mathbf{x} \,|\, \theta) = f(x_1 \,|\, \theta) f(x_2 \,|\, \theta) \ldots f(x_n \,|\, \theta)$. The parameter θ will usually be identified with a point in some r-dimensional space, and we will be concerned with statistical hypotheses which relate to the numerical value of the coordinates of θ.

In each test problem we write Ω as the set of values of θ under discussion – Ω is termed the set of *admissible* values for the problem. The null and alternative hypotheses, H_0 and H_1, define a decomposition of $\Omega = \Omega_0 \cup \Omega_1$ into two disjoint sets, where Ω_0 is the set of values of θ specified in H_0 and Ω_1 the values corresponding to H_1 – this is shown by writing $H_0(\theta \in \Omega_0)$ and $H_1(\theta \in \Omega_1)$. If $\Omega_0 = \{\theta_0\}$ consists of a single element we write $H_0(\theta = \theta_0)$ for $H_0(\theta \in \Omega_0)$, and similarly, if $\Omega_1 = \{\theta_1\}$ we write $H_1(\theta = \theta_1)$. We recall from Section 4.2 that a hypothesis is termed simple if it is specified by a single element, and composite otherwise.

──────── EXAMPLE 9.2 ──

1. Let $X_1, X_2, \ldots, X_n$ have the binomial distribution $b(1, \theta)$.
   (a) For the simple null hypothesis H_0 that $\theta = \frac{1}{2}$ against the simple alternative hypothesis H_1 that $\theta = \frac{2}{3}$, the above notation sets $\Omega = \{\frac{1}{2}, \frac{2}{3}\}$, $\Omega_0 = \{\frac{1}{2}\}$ and $\Omega_1 = \{\frac{2}{3}\}$, and we write $H_0(\theta = \frac{1}{2})$ and $H_1(\theta = \frac{2}{3})$.
   (b) For the simple null hypothesis H_0 that $\theta = \frac{1}{3}$ against the composite alternative hypothesis H_1 that $\theta \geq \frac{2}{3}$, we write $\Omega = \{\theta : \theta = \frac{1}{3} \text{ or } \theta \geq \frac{2}{3}\}$, $\Omega_0 = \{\frac{1}{3}\}$, $\Omega_1 = \{\theta : \theta \geq \frac{2}{3}\}$, with $H_0(\theta = \frac{1}{3})$ and $H_1(\theta \geq \frac{2}{3})$ as the descriptions of the hypotheses.
2. Let $X_1, X_2, \ldots, X_n$ have the normal distribution $N(\mu, \sigma^2)$ with μ and σ unknown. To test the composite null hypothesis H_0 that $\mu = 0$ against the composite alternative hypothesis H_1 that $\mu \neq 0$, we set $\Omega = \{\theta = (\mu, \sigma^2) : -\infty < \mu < \infty, \sigma > 0\}$, $\Omega_0 = \{\theta = (0, \sigma^2) : \sigma > 0\}$ and $\Omega_1 = \{\theta = (\mu, \sigma^2) : \mu \neq 0, \sigma > 0\}$ and write the hypotheses as $H_0(\mu = 0)$ and $H_1(\mu \neq 0)$.

9.2.1   *Simple null and alternative hypotheses*

Suppose $\Omega = \{\theta_0, \theta_1\}$ with $H_0(\theta = \theta_0)$ and $H_1(\theta = \theta_1)$ as the simple null and alternative hypotheses respectively. Then, a test C – that is, a test with critical region C – with the exact (achievable) level of significance α satisfies

$$\alpha = \Pr(X \in C \mid \theta_0)$$

and has power given by

$$\Pr(X \in C \mid \theta_1)$$

Our aim, which is realized in Theorem 1 below, is to find a test C, whose power is at least as large as that of any other test at the same level α. Such a test is called a *most powerful (MP)* test of the simple null hypothesis H_0 against the simple alternative hypothesis H_1.

We require one further piece of terminology to describe a desirable property for a test C. It seems reasonable to suggest that the test should be less likely to reject H_0 when it is true than when it is false – that is, we require

$$\alpha = \Pr(\mathbf{X} \in C \mid \theta_0) \leqslant \Pr(\mathbf{X} \in C \mid \theta_1)$$

A test C satisfying this inequality is termed *unbiased*.

In Theorem 1 we consider a critical region C of the form

$$C = \{\mathbf{x}; \; f(\mathbf{x} \mid \theta_1) \geqslant k \, f(\mathbf{x} \mid \theta_0)\} \tag{9.1}$$

where $k > 0$ is a constant.

─────── **EXAMPLE 9.3** ───────

Suppose the sample variables $X_1, X_2, \ldots, X_n$ are independent and have the distribution $N(\theta, 1)$. Then the joint p.d.f. of $X_1, X_2, \ldots, X_n$ is

$$f(\mathbf{x} \mid \theta) = \frac{1}{(\sqrt{2\pi})^n} \exp\left[-\frac{1}{2} \sum_{i=1}^{n} (x_i - \theta)^2 \right]$$

and, for the null hypothesis $H_0 (\theta = 0)$ and alternative hypothesis $H_1 (\theta = 1)$, the above inequality defining the critical region C is

$$\frac{1}{(\sqrt{2\pi})^n} \exp\left[-\frac{1}{2} \sum_{i=1}^{n} (x_i - 1)^2 \right] \geqslant k \frac{1}{(\sqrt{2\pi})^n} \exp\left(-\frac{1}{2} \sum_{i=1}^{n} x_i^2 \right)$$

After some simplification, we obtain $\exp(n\bar{x}) \geqslant k \exp(n/2)$, or $\bar{x} \geqslant k_1$, where, although the formula is of no consequence, $k_1 = \frac{1}{2} + (\log k)/n$.

Thus, the shape of the critical region $C = \{x : \bar{x} \geqslant k_1\}$ is determined by the inequality, and the value k_1 is obtained from the significance level α in the usual way – that is,

$$\begin{aligned} \alpha &= \Pr(\mathbf{X} \in C \mid \theta = 0) = \Pr(\bar{X} > k_1 \mid \theta = 0) \\ &= \Pr(Z \geqslant \sqrt{n}\, k_1) \end{aligned}$$

where Z is a random variable having the standard distribution $N(0, 1)$.

─────────────────────────────────────

For convenience of notation, we write the following discussion for the case when the sample variables have a continuous distribution with joint p.d.f. $f(\mathbf{x} \mid \theta)$ (all calculations given in terms of integration translate, somewhat cumbersomely, into summation notation for a discrete joint distribution). Then, for example, for a test C at level α, we write

$$\alpha = \Pr(\mathbf{X} \in C \mid \theta_0) = \int \ldots \int_C f(\mathbf{x} \mid \theta_0) \, d\mathbf{x}$$

To simplify the proof of Theorem 1, we isolate some probability calculations. Let C be the critical region defined by equation (9.1), and suppose k is determined so that C is a test at level α. Now, for any other test D at level α, write C_1 for those points of C not in D and D_1 for those points of D not in C. Then, since $f(\mathbf{x} \mid \theta_1) \geqslant k \, f(\mathbf{x} \mid \theta_0)$ for $\mathbf{x} \in C_1$, we have

$$\Pr(\mathbf{X} \in C_1 \mid \theta_1) = \int \ldots \int_{C_1} f(\mathbf{x} \mid \theta_1) \, d\mathbf{x}$$

$$\geqslant k \int \ldots \int_{C_1} f(\mathbf{x} \mid \theta_0) \, d\mathbf{x}$$

which implies

$$\Pr(\mathbf{X} \in C_1 \mid \theta_1) \geqslant k \Pr(\mathbf{X} \in C_1 \mid \theta_0) \tag{9.2}$$

Similarly, for $\mathbf{x} \in D_1$ we have $f(\mathbf{x} \mid \theta_1) < k f(\mathbf{x} \mid \theta_0)$, and integrating the inequality over D_1,

$$\Pr(\mathbf{X} \in D_1 \mid \theta_1) < k \Pr(\mathbf{X} \in D_1 \mid \theta_0) \tag{9.3}$$

Combining equations (9.2) and (9.3),

$$\Pr(\mathbf{X} \in C_1 \mid \theta_1) - \Pr(\mathbf{X} \in D_1 \mid \theta_1) \geqslant k[\Pr(\mathbf{X} \in C_1 \mid \theta_0) - \Pr(\mathbf{X} \in D_1 \mid \theta_0)]$$
$$\tag{9.4}$$

Theorem 1   The Neyman–Pearson fundamental lemma
If there exists a positive constant k such that the region

$$C = \{\mathbf{x}: f(\mathbf{x} \mid \theta_1) \geqslant k \, f(\mathbf{x} \mid \theta_0)\}$$

satisfies $\Pr(\mathbf{X} \in C \mid \theta_0) = \alpha$, then C is a most powerful (MP) and unbiased test at level α for the simple null hypothesis $H_0(\theta = \theta_0)$ against the simple alternative hypothesis $H_1(\theta = \theta_1)$.

Proof   By definition, C is a test at level α. Let D be any other test at level α – that is, $\Pr(\mathbf{X} \in D \mid \theta_0) = \alpha$. To show that C is MP we require

$$\Pr(\mathbf{X} \in C \mid \theta_1) \geqslant \Pr(\mathbf{X} \in D \mid \theta_1) \tag{9.5}$$

Writing $C \cap D$ for the common part of C and D, subtracting $\Pr(\mathbf{X} \in C \cap D \mid \theta_1)$ from both sides in equation (9.5) means that we require, with C_1 and D_1 as above,

$$\Pr(\mathbf{X} \in C_1 \mid \theta_1) \geqslant \Pr(\mathbf{X} \in D_1 \mid \theta_1)$$

But, from equation (9.4)

$$\Pr(\mathbf{X} \in C_1 \mid \theta_1) - \Pr(\mathbf{X} \in D_1 \mid \theta_1) \geqslant k[\Pr(\mathbf{X} \in C_1 \mid \theta_0) - \Pr(\mathbf{X} \in D_1 \mid \theta_0)]$$
$$= k[\Pr(\mathbf{X} \in C_1 \mid \theta_0) + \Pr(\mathbf{X} \in C \cap D \mid \theta_0)$$
$$- \Pr(\mathbf{X} \in C \cap D \mid \theta_0) - \Pr(\mathbf{X} \in D_1 \mid \theta_0)]$$
$$= k[\Pr(\mathbf{X} \in C \mid \theta_0) - \Pr(\mathbf{X} \in D \mid \theta_0)]$$
$$= 0$$

Thus $\Pr(\mathbf{X} \in C_1 \mid \theta_1) \geqslant \Pr(\mathbf{X} \in D_1 \mid \theta_1)$, as required.

Finally, we show that C is an unbiased test – that is, $\Pr(\mathbf{X} \in C \mid \theta_1) \geqslant \alpha$. There are two cases:

1. $k \geqslant 1$, then
$$\Pr(\mathbf{X} \in C \mid \theta_1) \geqslant k\Pr(\mathbf{X} \in C \mid \theta_0) = k\alpha \geqslant \alpha \text{ and}$$

2. $k < 1$, when
$$1 - \Pr(\mathbf{X} \in C \mid \theta_1) = \Pr(\mathbf{X} \notin C \mid \theta_1)$$
$$< k\Pr(\mathbf{X} \notin C \mid \theta_0) = k(1 - \alpha) < 1 - \alpha$$

and, again, $\Pr(\mathbf{X} \in C \mid \theta_1) \geqslant \alpha$.

Before commenting on the lemma, we illustrate its use with three examples.

———— **EXAMPLE 9.4** ————

Suppose the sample variables $X_1, X_2, \ldots, X_n$ are independent with the normal distribution $N(\theta, 1)$. We apply the Neyman–Pearson lemma to test the null hypothesis $H_0(\theta = \theta_0)$ against the alternative hypothesis $H_1(\theta = \theta_1)$, where $\theta_1 > \theta_0$.

The inequality defining the critical region C is

$$\frac{1}{(\sqrt{2\pi})^n} \exp\left[-\frac{1}{2} \sum_{i=1}^{n} (x_i - \theta_1)^2\right] \geqslant k \frac{1}{(\sqrt{2\pi})^n} \exp\left[-\frac{1}{2} \sum_{i=1}^{n} (x_i - \theta_0)^2\right]$$

which simplifies to

$$\exp\left[(\theta_1 - \theta_0) \sum_{i=1}^{n} x_i\right] \geqslant k \exp n(\theta_1^2 - \theta_0^2)/2 \tag{9.6}$$

Since $\theta_1 - \theta_0 > 0$, taking logarithms, gives

$$\sum_{i=1}^{n} x_i \geqslant k_1 = (\log k + n(\theta_1^2 - \theta_0^2)/2)/(\theta_1 - \theta_0)$$

and, therefore,

$$C = \left\{\mathbf{x}: \sum_{i=1}^{n} x_i \geqslant k_1\right\}.$$

The value k_1 is determined by the level α in the usual way. Under the null hypothesis $H_0(\theta = \theta_0)$, the random variable $Z = \sqrt{n}(\bar{X} - \theta_0)$ is distributed $N(0, 1)$, and we have

$$\alpha = \Pr(\mathbf{X} \in C \mid \theta_0) = \Pr\left(\sum_{i=1}^{n} X_i \geqslant k_1 \mid \theta_0\right)$$

$$= \Pr\left(Z \geqslant \sqrt{n}\left(\frac{k_1}{n} - \theta_0\right)\right) = 1 - \Phi\left(\sqrt{n}\left(\frac{k_1}{n} - \theta_0\right)\right)$$

and a table of the normal distribution may be consulted.

To calculate the power of this MP test, note that, under the alternative hypothesis $H_1(\theta = \theta_1)$, the random variable $Z = \sqrt{n}(\bar{X} - \theta_1)$ has the normal distribution $N(0, 1)$. Then, the power is given by

$$\Pr(\mathbf{X} \in C | \theta_1) = \Pr\left(\sum_{i=1}^{n} X_i \geq k_1 | \theta_1\right)$$

$$= \Pr\left(Z \geq \sqrt{n}\left(\frac{k_1}{n} - \theta_1\right)\right)$$

$$= 1 - \Phi\left(\sqrt{n}\left(\frac{k_1}{n} - \theta_1\right)\right)$$

For a numerical illustration, let $n = 16$, $\theta_0 = 0$, $\theta_1 = 1$ and $\alpha = 0.05$. Then tables supply

$$\sqrt{n}\left(\frac{k_1}{n} - \theta_0\right) = 1.64$$

or $k_1 = 6.56$, and the power is

$$1 - \Phi\left(4\left(\frac{6.56}{16} - 1\right)\right) \approx 1 - \Phi(0.64) = 1 - 0.7389 = 0.2611$$

The reader will have noted that the critical region provided by the Neyman–Pearson lemma is equivalent to the use of the text statistic $\bar{X}$ suggested in Section 4.2. Note also that, for the null hypothesis $H_0(\theta = \theta_0)$ against the alternative hypothesis $H_1(\theta = \theta_1)$, where $\theta_1 < \theta_0$, the inequality (9.6) is equivalent to $\Sigma x_i \leq k_1$ (since $\theta_1 - \theta_0 < 0$) and the most powerful critical region has the shape $\{\mathbf{x}: \Sigma_{i=1}^{n} x_i \leq k_1\}$.

———— EXAMPLE 9.5 ————

Suppose the sample random variables are independent with the binomial distribution $b(1, \theta)$, and consider the hypotheses $H_0(\theta = \frac{1}{2})$ and $H_1(\theta = \frac{3}{4})$. Now, the common p.f. of the sample variables is $f(x | \theta) = \theta^x (1 - \theta)^{1-x}$ for $x = 0$ or 1, and the joint p.f. of $X_1, X_2, \ldots, X_n$ is therefore $f(\mathbf{x} | \theta) = \theta^t (1 - \theta)^{n-t}$, where $t = \Sigma_{i=1}^{n} x_i$. The inequality defining the critical region C of the Neyman–Pearson lemma is

$$(\tfrac{3}{4})^t (1 - \tfrac{3}{4})^{n-t} \geq k (\tfrac{1}{2})^n$$

or

$$3^t \geq 2^n k$$

reducing, by logarithms, to $t = \Sigma_{i=1}^{n} x_i \geq k_1$.

For a test at level α, the value k_1 must satisfy

$$\alpha = \Pr(\mathbf{X} \in C \mid \theta = \tfrac{1}{2}) = \Pr\left(\sum_{i=1}^{n} X_i \geq k_1 \mid \theta = \tfrac{1}{2}\right)$$

and, since $\sum_{i=1}^{n} X_i$ has the binomial distribution $b(n, \tfrac{1}{2})$ under H_0, we confront the problem of achievability discussed in Section 4.3. For example, with $n = 10$ and $\alpha = 0.05$, we see from tables that there is no value k_1 satisfying $\Pr(\sum_{i=1}^{n} X_i \geq k_1 \mid \theta = \tfrac{1}{2}) = 0.05$. Adopting the conservative solution suggested in Section 4.3, we take $k_1 = 9$ since

$$0.0107 = \Pr\left(\sum_{i=1}^{10} X_i \geq 9 \mid \theta = \tfrac{1}{2}\right) < 0.05 < \Pr\left(\sum_{i=1}^{10} X_i \geq 8 \mid \theta = \tfrac{1}{2}\right) = 0.0547$$

and obtain a most powerful test at level 0.0107.

Calculating the power of this test requires

$$\Pr\left(\sum_{i=1}^{10} X_i \geq 9 \mid \theta = \tfrac{3}{4}\right) = \Pr(Y \leq 1)$$

where $Y = 10 - \sum_{i=1}^{10} X_i$ has the binomial distribution $b(10, \tfrac{1}{4})$ under the alternative hypothesis $H_1 (\theta = \tfrac{3}{4})$. Tables supply 0.2440.

──────── EXAMPLE 9.6 ────────────────────────────────────

Suppose the sample random variables are independent with the common exponential p.d.f. $f(x, \theta) = \theta e^{-\theta x}$, where $x \geq 0$ and $\theta > 0$. We seek a test of the null hypothesis $H_0 (\theta = 1)$ against the alternative hypothesis $H_1 (\theta = 2)$.

Since the joint p.d.f. of the sample variables is $f(\mathbf{x} \mid \theta) = \theta^n e^{-\theta t}$, where $t = \sum_{i=1}^{n} x_i$ and $x_i \geq 0$ for $i = 1, 2, \ldots, n$, the inequality defining the critical region C of the Neyman–Pearson lemma becomes

$$2^n e^{-2t} \geq k e^{-t}$$

or, equivalently, $t = \sum_{i=1}^{n} x_i \leq k_1$

To determine the value k_1 which gives a test at level α – that is, $\Pr(\sum_{i=1}^{n} X_i \leq k_1 \mid \theta = 1) = \alpha$, we require the distribution of $\sum_{i=1}^{n} X_i$ under the null hypothesis. From Problem 9.2 we see that $W = 2 \sum_{i=1}^{n} X_i$ has the chi-squared distribution $\chi^2 (2n)$, and chi-squared tables will provide k_1 from $\Pr(W \leq 2k_1) = \alpha$.

There is an intuitive argument for using the test statistic $S = \bar{X}$ in this problem. Observing that $E(\bar{X}) = 1/\theta$, and recalling that $1 = \theta_0 < \theta_1 = 2$, a small observed value of $\bar{X}$ may be taken as indicating the alternative hypothesis, and the one-dimensional critical region $\{s : 0 \leq s \leq a\}$ is suggested.

───

In the statement of the fundamental lemma, the inequality defining the critical region C may be said to specify the shape of C as a subset of n-dimensional space, and the level α as determining the location of C. As the above examples show, in order to make use of the lemma we must be able to simplify the inequality and then solve the resulting distributional problem. Example 9.5 shows why the statement is made conditional on the existence of a positive constant k – it is the problem of achievability.

One further observation follows from an examination of the proof. The requirement that the region D provides a test at level α is not required – the proof carries through provided D is a test at level $\alpha_1 \leqslant \alpha$. Hence, the critical region C may be described as being more powerful than any region at level $\leqslant \alpha$.

9.2.2   *Simple null hypothesis, composite alternative hypothesis*

Suppose the set Ω of admissible values of the parameter is written as $\Omega = \Omega_0 \cup \Omega_1$ where $\Omega_0 = \{\theta_0\}$ consists of a single element and Ω_1 of more than one element. We extend the language of the Neyman–Pearson theory to cover a test with an n-dimensional critical region C for the simple null hypothesis $H_0(\theta = \theta_0)$ against the composite alternative $H_1(\theta \in \Omega_1)$.

The test is at the exact level of significance α if, as before, $\Pr(X \in C \mid \theta_0) = \alpha$, but now the power of the test must be calculated for each $\theta \in \Omega_1$. We introduce the power function $p(\theta, C)$ defined by

$$p(\theta, C) = \Pr(X \in C \mid \theta)$$

for all $\theta \in \Omega$. Note, by definition, that $p(\theta_0, C) = \alpha$. The test C is unbiased if C provides an unbiased test of the simple hypothesis $H_0(\theta = \theta_0)$ against each simple alternative hypothesis $\theta = \theta_1$ for $\theta_1 \in \Omega_1$. This may be expressed in terms of the power function by saying that $p(\theta, C) \geqslant \alpha$ for all $\theta \in \Omega$.

Comparing two critical regions C and C_1, both at level α, follows on similar lines by comparing the power for each $\theta \in \Omega_1$. The region C is said to be *uniformly more powerful* than the region C_1 if

$$p(\theta, C_1) \leqslant p(\theta, C)$$

for all $\theta \in \Omega$ (note: $p(\theta_0, C_1) = p(\theta_0, C) = \alpha$), and C is said to be *uniformly most powerful (UMP)* if it is uniformly more powerful than any other region C_1 at level α.

A UMP critical region C at level α, if such a region exists, must, by definition, provide a MP test for the simple hypothesis $H_0(\theta = \theta_0)$ against each alternative hypothesis that $\theta = \theta_1$ for every choice of $\theta_1 \in \Omega_1$. This suggests that, in the search for a UMP region, we apply the fundamental lemma to these simple hypotheses, obtaining a critical region $C(\theta_1)$ which, in principle, depends on the choice of $\theta_1 \in \Omega_1$. Should it happen that $C = C(\theta_1)$ is the same for all $\theta_1 \in \Omega_1$, then C must be a UMP test for the null hypothesis $H_0(\theta = \theta_0)$ against the alternative hypothesis $H_1(\theta \in \Omega_1)$.

─────── **EXAMPLE 9.7** ───────

Suppose the sample variables are independent with the normal distribution $N(\theta, 1)$. Let $\Omega = \{\theta: \theta \geq \theta_0\}$, $\Omega_0 = \{\theta_0\}$ and $\Omega_1 = \{\theta: \theta > \theta_0\}$. Then we require a test of the null hypothesis $H_0(\theta = \theta_0)$ against the alternative hypothesis $H_1(\theta > \theta_0)$.

Referring to Example 9.4 for a test of $\theta = \theta_0$ against $\theta = \theta_1$, where $\theta_1 > \theta_0$, the critical region $C(\theta_1)$ is given by $C(\theta_1) = \{\mathbf{x}: \sum_{i=1}^{n} xi \geq k_1\}$ where, apparently, $k_1 = (\log k + n(\theta_1^2 - \theta_0^2)/2)/(\theta_1 - \theta_0)$ is a function of θ_1. However, using $\Pr(\mathbf{X} \leq C(\theta_1) \mid \theta_0) = \alpha$, the value of k_1 is obtained from

$$1 - \Phi\left(\sqrt{n}\left(\frac{k_1}{n} - \theta_0\right)\right) = \alpha$$

which shows that k_1, and hence $C(\theta_1)$, is independent of θ_1.

The critical region $C = \{\mathbf{x}: \sum_{i=1}^{n} x_i \geq k_1\}$ therefore provides a UMP test at level α for the simple null hypothesis $H_0(\theta = \theta_0)$ against the composite alternative hypothesis $H_1(\theta \in \Omega_1)$. We may also read-off, from Example 9.4, the power function

$$p(\theta, C) = 1 - \Phi\left(\sqrt{n}\left(\frac{k_1}{n} - \theta\right)\right)$$
$$= 1 - \Phi(z - \sqrt{n}(\theta - \theta_0))$$

where $z = \sqrt{n}[(k_1/n) - \theta_0]$ is fixed by $\Phi(z) = 1 - \alpha$.

Since $\Phi(x)$ increases, as x increases we see that $\Phi(z - \sqrt{n}(\theta - \theta_0)) \leq \Phi(z)$ for $\theta > \theta_0$, so that $p(\theta, C) \geq \alpha$ for all $\theta \in \Omega$. Further, since $\Phi(x) \to 0$ as x decreases to $-\infty$, we note that $p(\theta_1, C) \to 1$ for each $\theta_1 > \theta_0$ as $n \to +\infty$ and $p(\theta_1, C) \to 1$ as θ_1 increases to $+\infty$, confirming that the test is unbiased, and showing that it has large power if either the sample size is large or if θ_1 is much greater than θ_0.

Similarly (see the comment after Example 9.4) the critical region $C = \{\mathbf{x}: \sum_{i=1}^{n} x_i \leq a\}$ will provide a UMP test for the null hypothesis $H_0(\theta = \theta_0)$ against the alternative hypothesis $H_1(\theta < \theta_0)$. The procedure that is successful in Example 9.7, has, however, a limited range of application.

─────── **EXAMPLE 9.8** ───────

Suppose the sample variables are independent with the normal distribution $N(\theta, 1)$, and we wish to test the null hypothesis $H_0(\theta = \theta_0)$ against the alternative hypothesis $H_1(\theta \neq \theta_0)$. Following the method of Example 9.7, we have for all $\theta_1 > \theta_0$ the common critical region C_1 with shape $C_1 = \{\mathbf{x}: \sum_{i=1}^{n} x_i \geq b\}$. Similarly, for all $\theta_1 < \theta_0$, we obtain the common critical region C_2 with shape $C_2 = \{\mathbf{x}: \sum_{i=1}^{n} x_i \leq a\}$. It follows that there is no UMP critical region for these hypotheses H_0 and H_1. For, if

C is a UMP critical region, we must have

$$\Pr(\mathbf{X} \in C \mid \theta_1) \geq \Pr(\mathbf{X} \in C_1 \mid \theta_1) \text{ for } \theta_1 > \theta_0$$

and

$$\Pr(\mathbf{X} \in C \mid \theta_1) \geq \Pr(\mathbf{X} \in C_2 \mid \theta_1) \text{ for } \theta_1 < \theta_0$$

However, we have shown that C_1 and C_2 are each UMP for the alternatives $\theta > \theta_0$ and $\theta < \theta_0$, respectively, and this implies the inequalities are reversed. These contradictions show that a UMP critical region cannot exist for $H_0(\theta = \theta_0)$ against $H_1(\theta \neq \theta_0)$.

In Chapter 4 we suggested the test statistic $\bar{X}$ and the one-dimensional critical region $R = \{x : |x - \theta_0| \geq c\}$ for this two-tailed problem. Translating into the n-dimensional framework we obtain the critical region $C = \{\mathbf{x} : |\bar{x} - \theta_0| \geq c\}$, which, of course, is the symmetric combination of $C_1 = \{\mathbf{x} : \bar{x} - \theta_0 \geq b\}$ and $C_2 = \{\mathbf{x} : \bar{x} - \theta_0 \leq a\}$ with $b = -a = c$. We now show that this symmetric choice is necessary if we wish to retain the property that the critical region is unbiased.

──────── **EXAMPLE 9.9** ────────────

Suppose the sample variables are independent with the normal distribution $N(\theta, 1)$, and we test the null hypothesis $H_0(\theta = 0)$ against the alternative hypothesis $H_1(\theta \neq 0)$ using the critical region $C = \{\mathbf{x} : \bar{x} \leq a \text{ or } \bar{x} \geq b\}$, and the level of significance α.

Let Z be a random variable with the normal distribution $N(0, 1)$. Then the values a and b must satisfy $\Pr(\bar{X} \in C \mid \theta = 0) = \alpha$ or $\Pr(a\sqrt{n} < \bar{X}\sqrt{n} < b\sqrt{n} \mid \theta = 0) = 1 - \alpha$, which gives

$$\Pr(a\sqrt{n} < Z < b\sqrt{n}) = 1 - \alpha \qquad (9.7)$$

For the region C to be unbiased we require, for all values of θ,

$$\Pr(a < \bar{X} < b \mid \theta) \leq 1 - \alpha$$

or

$$\Pr((a - \theta)\sqrt{n} < \sqrt{n}(\bar{X} - \theta) < (b - \theta)\sqrt{n} \mid \theta) \leq 1 - \alpha$$

which, since $\sqrt{n}(\bar{X} - \theta)$ has the normal distribution $N(0, 1)$, implies that, for all values of θ,

$$\Pr(a\sqrt{n} < Z + \theta\sqrt{n} < b\sqrt{n}) \leq 1 - \alpha \qquad (9.8)$$

However, for a and b fixed by equation (9.7) we see from Problem 9.3 that the probability in equation (9.8) is maximized when $\theta = \theta_1 = (a + b)/2$, and then

$$\Pr(a\sqrt{n} < Z + \theta_1\sqrt{n} < b\sqrt{n}) \geq \Pr(a\sqrt{n} < Z < b\sqrt{n}) = 1 - \alpha$$

This gives a contradiction unless $\theta_1 = 0$, and $a = -b$.

Examples 9.8 and 9.9 provide contrasting views on the theory of hypothesis testing. Example 9.8 shows that the criterion of seeking a uniformly most powerful test is too demanding, while Example 9.9 indicates that the property of unbiasedness is preserved for the equi-tail region. It can be shown that the equi-tail region above is a uniformly most powerful unbiased (UMPU) test – uniformly more powerful than any other unbiased test at the same level of significance. For a readable account of the theory of *unbiased* tests see reference 1, or, for a more advanced and comprehensive treatment, consult reference 2.

9.2.3   *Composite hypotheses*

We close this section by showing how the Neyman–Pearson lemma may be applied to obtain a UMP test for certain types of composite null and alternative hypotheses.

Suppose the decomposition $\Omega = \Omega_0 \cup \Omega_1$ of the set of admissible values of the parameter produces a composite null hypothesis $H_0(\theta \in \Omega_0)$. A critical region C is a test of $H_0(\theta \in \Omega_0)$ against $H_1(\theta \in \Omega_1)$ at the exact level of significance α if

$$\max_{\theta \in \Omega_0} \Pr(\mathbf{X} \in C \,|\, \theta) = \alpha$$

For sets Ω_0 of the type $\{\theta : \theta \leqslant \theta_0\}$ or $\{\theta : \theta \geqslant \theta_0\}$, where $\Pr(\mathbf{X} \in C \,|\, \theta_0) = \alpha$, a UMP test of $H_0(\theta \in \Omega_0)$ against $H_1(\theta \in \Omega_1)$ may sometimes be found by the method of the previous sub-section.

─────── EXAMPLE 9.10 ───────

Suppose the sample random variables are independent with the common p.d.f. $f(x \,|\, \theta) = \theta e^{-\theta x}$ for $x \geqslant 0$ and $\theta > 0$, and that the null and alternative hypotheses are $H_0(\theta \leqslant 1)$ and $H_1(\theta \geqslant 2)$, respectively. Choosing any values $\theta_0 \leqslant 1$ and $\theta_1 \geqslant 2$, the fundamental lemma applied to the null hypothesis $\theta = \theta_0$ against the alternative $\theta = \theta_1$, produces a critical region C of the form $C = \{\mathbf{x} : \sum_{i=1}^{n} x_1 \leqslant c\}$ – see Example 9.6. Now we require, for the test to be at level α,

$$\max_{\theta \leqslant 1} \Pr\left(\sum_{i=1}^{n} X_i \leqslant c \,\Big|\, \theta \right) = \alpha$$

or

$$\max_{\theta \leqslant 1} \Pr\left(2\theta \sum_{i=1}^{n} X_i \leqslant 2c\theta \,\Big|\, \theta \right) = \alpha$$

Writing $W = 2\theta \sum_{i=1}^{n} X_i$, we see from Problem 9.2 that W has the chi-squared distribution $\chi^2(2n)$, and that we require

$$\max_{\theta \leqslant 1} \Pr(W \leqslant 2c\theta) = \alpha$$

Since $\Pr(W \leqslant 2c\theta) \leqslant \Pr(W \leqslant 2c)$ for $\theta \leqslant 1$ we obtain the required bound by choosing c to satisfy $\Pr(W \leqslant 2c) = \alpha$. Hence $C = \{\mathbf{x}: \sum_{i=1}^{n} x_i \leqslant c\}$ is a UMP test for $H_0(\theta \leqslant 1)$ against $H_1(\theta \geqslant 2)$, simply because it is most powerful for the null hypothesis that $\theta = 1$ against the alternative hypothesis $\theta = \theta_1$, where $\theta_1 \geqslant 2$.

Similarly, the reader may show that:

1. When the sample variables are independent with the normal distribution $N(\theta, 1)$ the region $\{\mathbf{x}: \bar{x} \geqslant c\}$ is UMP for the hypotheses $H_0(\theta \leqslant \theta_0)$ and $H_1(\theta \geqslant \theta_1)$, where $\theta_1 \geqslant \theta_0$, and the region $\{\mathbf{x}: \bar{x} \leqslant c\}$ is UMP for the hypotheses $H_0(\theta \geqslant \theta_0)$ and $H_1(\theta \leqslant \theta_1)$, where $\theta_1 \leqslant \theta_0$; and
2. When the sample variables are independent with the binomial distribution $b(1, p)$, the region $\{\mathbf{x}: \bar{x} \geqslant c\}$ is UMP for $H_0(p \leqslant p_0)$ and $H_1(p \geqslant p_1)$ where $0 < p_0 \leqslant p_1 \leqslant 1$.

As a general guide, the approach outlined will provide a UMP critical region whenever, for any critical region C of the required shape, the function $\Pr(\mathbf{X} \in C \mid \theta)$ is a monotonic function of the parameter θ.

9.2.4   Problems

9.1 Let X_1, X_2 be a sample of size 2 from a population whose distribution has the p.d.f. $f(x \mid \theta) = \theta x^{\theta-1}$ for $0 \leqslant x \leqslant 1$. Show that the Neyman–Pearson fundamental lemma, applied to test the null hypothesis $H_0(\theta = 2)$ against the alternative hypothesis $H_1(\theta = 1)$ at the level of significance α provides the critical region $C = \{(x_1, x_2): x_1 x_2 \leqslant c\}$, where the value c satisfies

$$c^2(1 - 2 \log c) = \alpha$$

Given that $c \approx 0.0935$ when $\alpha = 0.05$, find the probability of Type II error, and compare this with the probabilities of Type II error for the critical regions $\square$ and $\triangle$ in Section 9.1, Example 9.1.

9.2 If the independent random variables $X_1, X_2, \ldots, X_n$ have the common p.d.f. $f(x \mid \theta) = \theta e^{-\theta x}$ for $x > 0$ and $\theta > 0$, find the moment-generating function of the random variable $2\theta X_1$, and deduce the distribution of $W = 2\theta \sum_{i=1}^{n} X_i$ (see Example 9.10).

9.3 If the independent random variables $X_1, X_2, \ldots, X_n$ each have the uniform distribution $U(0, 1)$ find the p.d.f. of $-2 \log X_1$, and deduce the distribution of $W = -2 \sum_{i=1}^{n} \log X_i$ (Hint: Use Problem 9.2.)

9.4 If $X_1, X_2, \ldots, X_n$ is a random sample from the distribution $N(0, \sigma^2)$ find the shape of the most powerful critical region for a test of the null hypothesis $H_0(\sigma = \sigma_0)$ against the alternative hypothesis $H_1(\sigma = \sigma_1)$, where $\sigma_0 < \sigma_1$. Specify the region when $n = 16$, $\sigma_0 = 1$, $\sigma_1 = 2$, and $\alpha = 0.05$, and find the power. Show also that this region provides a UMP test for the hypotheses $H_0(\sigma \leqslant 1)$ and $H_1(\sigma \geqslant 2)$.

9.5 Let $X_1, X_2, \ldots, X_n$ be a random sample from the binomial distribution $b(1, p)$. Use the normal approximation to find the sample size required for testing the hypotheses $H_0(p = \frac{1}{2})$ and $H_1(p = \frac{3}{4})$ at the 1% level of significance, and having power 0.99.

9.3 The likelihood ratio test

The notation of the previous section continues to apply: the sample random variables $X_1, X_2, \ldots, X_n$ have the joint p.f./p.d.f. $f(\mathbf{x} \mid \theta)$; the set Ω of admissible values of the parameter θ is decomposed as the disjoint union $\Omega_0 \cup \Omega_1$; and the null and alternative hypotheses are $H_0(\theta \in \Omega_0)$ and $H_1(\theta \in \Omega_1)$, respectively.

When both the null and alternative hypotheses are simple, the search for a best test has been dealt with, in an unequivocal way, by the fundamental lemma. Further, the examples in Section 9.2 show how the idea of the lemma may be extended, in suitable cases, to provide best tests for some composite hypothesis situations. The likelihood ratio test (LRT) is a general method for defining a critical region should, as in Example 9.9 of Section 9.2, the hypotheses be such that the considerations of that section fail to provide a best test. Although intuitively appealing, as a combination of the idea of the fundamental lemma and the principle of maximum likelihood, the critical region defined by the LRT is not supported by any general theorem outlining desirable properties of a test. However, as our examples will show, the LRT does produce what are regarded as natural tests in the familiar situations.

To introduce the formal definition of the LRT recall that:

1. The fundamental lemma rejects $\theta = \theta_0$ in favour of $\theta = \theta_1$ if the ratio $f(\mathbf{x} \mid \theta_0)/f(\mathbf{x} \mid \theta_1)$ is too small; and
2. Regarding the values $x_1, x_2, \ldots, x_n$ as fixed, the function $f(\mathbf{x} \mid \theta)$ is treated as a function of θ – called the likelihood function (see Section 3.2) – and the principle of maximum likelihood estimates θ by maximizing $f(\mathbf{x} \mid \theta)$.

Combining these concepts, the LRT procedure is to compute the ratio

$$\Lambda(\mathbf{x}) = \frac{\max_{\theta \in \Omega_0} f(\mathbf{x} \mid \theta)}{\max_{\theta \in \Omega} f(\mathbf{x} \mid \theta)}$$

of the restricted (by H_0) and unrestricted (general model) maxima, and defines the critical region C as having the shape $C = \{\mathbf{x} : \Lambda(\mathbf{x}) \leq \lambda\}$. Evidently $0 \leq \Lambda(\mathbf{x}) \leq 1$, and the value of $\lambda (0 < \lambda \leq 1)$ is related to the significance level α by

$$\max_{\theta \in \Omega_0} \Pr(\mathbf{X} \in C \mid \theta) = \alpha$$

with the usual caveat concerning achievability of preassigned significance levels (see Section 4.3).

Note

1. It is not difficult to see that the LRT procedure is equivalent to the fundamental lemma when $\Omega = \{\theta_0, \theta_1\}$ and $\Omega_0 = \{\theta_0\}$, since, in this case,

$$\max_{\theta \in \Omega} f(\mathbf{x}|\theta) = f(\mathbf{x}|\theta_0) \quad \text{or} \quad f(\mathbf{x}|\theta_1)$$

and $\Lambda(\mathbf{x}) \leq \lambda \leq 1$ implies $f(\mathbf{x}|\theta_0) \leq \lambda f(\mathbf{x}|\theta_1)$.

2. Assuming $f(\mathbf{x}|\theta)$ is continuous as a function of θ, the maximum of $f(\mathbf{x}|\theta)$ for $\theta \in \Omega$ is just the value $f(\mathbf{x}|\hat{\theta})$ where $\hat{\theta}$ is the maximum likelihood estimate of θ. Similarly, maximizing $f(\mathbf{x}|\theta)$ over $\theta \in \Omega_0$ will give the value $f(\mathbf{x}|\hat{\theta}_0)$ where $\hat{\theta}_0$ is the restricted (by Ω_0) maximum likelihood estimate of θ.

The following examples illustrate the application of the LRT to the parameters of a normal distribution.

──────── **EXAMPLE 9.11   TESTING THE NORMAL MEAN** ────────

Suppose $X_1, X_2, \ldots, X_n$ is a random sample from the distribution $N(\mu, \sigma^2)$. We test the null hypothesis $H_0(\mu = \mu_0)$ against the alternative hypothesis $H_1(\mu \neq \mu_0)$ in the two cases, σ known and σ unknown:

1. $\sigma = \sigma_0$ known. The numerator in $\Lambda(\mathbf{x})$ requires no maximizing and is

$$f(\mathbf{x}|\mu_0, \sigma_0) = \frac{1}{(\sigma_0\sqrt{2\pi})^n} \exp\left[-\frac{1}{2\sigma_0^2}\sum_{i=1}^{n}(x_i - \mu_0)^2\right]$$

while the denominator, found by the method of Section 3.2, Example 3.5 (note that σ is constant), is

$$\max_{\mu} f(\mathbf{x}|\mu, \sigma_0) = \frac{1}{(\sigma_0\sqrt{2\pi})^n} \exp\left[-\frac{1}{2\sigma_0^2}\sum_{i=1}^{n}(x_i - \bar{x})^2\right]$$

Since

$$\sum_{i=1}^{n}(x_i - \bar{x})^2 = \sum_{i=1}^{n}(x_i - \mu_0)^2 - n(\bar{x} - \mu_0)^2$$

the ratio $\Lambda(\mathbf{x})$ is given by

$$\Lambda(\mathbf{x}) = \exp(-n(\bar{x} - \mu_0)^2/2\sigma_0^2)$$

and $\Lambda(\mathbf{x}) \leq \lambda$ is equivalent, by taking logarithms, to $n(\bar{x} - \mu_0)^2/2\sigma^2 \geq k$ or $\sqrt{n}|\bar{x} - \mu_0|/\sigma_0 \geq k_1$. Hence, the LRT provides the test suggested in Section 4.2, Example 4.15, and Section 9.2, Example 9.10.

2. σ unknown. Now we must write $\theta = (\mu, \sigma)$ and set $\Omega = \{(\mu, \sigma): -\infty < \mu < \infty, \sigma > 0\}$ and $\Omega_0 = \{(\mu_0, \sigma): \sigma > 0\}$, thereby expressing the null hypothesis in the form $H_0(\theta \in \Omega_0)$.

Using the MLE $\hat{\mu} = \bar{x}$ and

$$\hat{\sigma}^2 = \frac{1}{n} \sum_{i=1}^{n} (x_i - \bar{x})^2$$

calculated in Section 3.2, Example 3.5, the denominator in $\Lambda(\mathbf{x})$ is $f(\mathbf{x} \mid \hat{\mu}, \hat{\sigma})$ – that is,

$$\max_{\mu, \sigma} f(\mathbf{x} \mid \mu, \sigma) = \frac{1}{(\hat{\sigma}\sqrt{2\pi})^n} \exp\left[-\frac{1}{2\hat{\sigma}^2} \sum_{i=1}^{n} (x_i - \bar{x})^2 \right]$$

$$= (2\pi\hat{\sigma}^2 e)^{-n/2}$$

Similarly, with $\mu = \mu_0$ fixed, we maximize over σ to obtain $f(\mathbf{x} \mid \mu_0, \hat{\sigma}_0)$, where

$$\hat{\sigma}_0^2 = \frac{1}{n} \sum_{i=1}^{n} (x_i - \mu_0)^2$$

that is,

$$\max_{\sigma} f(\mathbf{x} \mid \mu_0, \sigma) = \frac{1}{(\hat{\sigma}_0\sqrt{2\pi})^n} \exp\left[-\frac{1}{2\hat{\sigma}_0^2} \sum_{i=1}^{n} (x_i - \mu_0)^2 \right]$$

$$= (2\pi\hat{\sigma}_0^2 e)^{-n/2}$$

Now

$$\Lambda(\mathbf{x}) = \left(\frac{\hat{\sigma}^2}{\hat{\sigma}_0^2} \right)^{n/2} \leq \lambda$$

is equivalent to

$$\frac{\sum_{i=1}^{n} (x_i - \bar{x})^2}{\sum_{i=1}^{n} (x_i - \mu_0)^2} \leq k$$

which, using the relation above, is, in turn, equivalent to

$$\frac{n(\bar{x} - \mu_0)^2}{\sum_{i=1}^{n} (x_i - \bar{x})^2} \geq k_1$$

Writing

$$s^2 = \frac{1}{n-1} \sum_{i=1}^{n} (x_i - \bar{x})^2$$

and $t = (\bar{x} - \mu_0)\sqrt{n}/s$, the last inequality simplifies to $|t| \geq c$. Thus the LRT procedure recovers the test structure used in TTEST (see Section 4.2).

The next example shows how the LRT produces the test statistic used in MINITAB's TWOSAMPLE command with the subcommand POOLED.

———— EXAMPLE 9.12 ————

Suppose $X_1, X_2, \ldots, X_n$ is a random sample from the distribution $N(\mu_1, \sigma^2)$, and $Y_1, Y_2, \ldots, Y_m$ a random sample from an independent distribution $N(\mu_2, \sigma^2)$. We test the composite null and alternative hypotheses $H_0(\mu_1 = \mu_2)$ and $H_1(\mu_1 \neq \mu_2)$.

Set $\theta = (\mu_1, \mu_2, \sigma)$, $\Omega = (\mu_1, \mu_2, \sigma)$: $-\infty < \mu_1 < \infty$, $-\infty < \mu_2 < \infty$, $\sigma > 0\}$ and $\Omega_0 = \{(\mu, \mu, \sigma): -\infty < \mu < \infty, \sigma > 0\}$ so that the null hypothesis is expressed in the form $H_0(\theta \in \Omega_0)$.

Now the joint p.d.f. of the independent sample variables $X_1, X_2, \ldots, X_n$, $Y_1, Y_2, \ldots, Y_m$ is

$$f(\mathbf{x}, \mathbf{y} | \mu_1, \mu_2, \sigma) = \frac{1}{(\sigma\sqrt{2\pi})^{n+m}} \exp\left(-\frac{1}{2\sigma^2}\left[\sum_{i=1}^{n}(x_i - \mu_1)^2 + \sum_{j=1}^{m}(y_j - \mu_2)^2\right]\right)$$

For the numerator of Λ – a function of $\mathbf{x}$ and $\mathbf{y}$ – we maximize $f(\mathbf{x}, \mathbf{y} | \mu, \mu, \sigma)$ by equating to zero the partial derivatives of log f with respect to μ and σ. Thus,

$$\sum_{i=1}^{n}(x_i - \mu) + \sum_{j=1}^{m}(y_j - \mu) = 0$$

and

$$-(n+m) + \frac{1}{\sigma^2}\left[\sum_{i=1}^{n}(x_i - \mu)^2 + \sum_{j=1}^{m}(y_j - \mu)^2\right] = 0$$

The restricted MLE, $\hat{\mu}_0$ and $\hat{\sigma}_0$, are therefore given by

$$(n+m)\hat{\mu}_0 = \sum_{i=1}^{n}x_i + \sum_{j=1}^{m}y_j = (n\bar{x} + m\bar{y})$$

and

$$(n+m)\hat{\sigma}_0^2 = \sum_{i=1}^{n}(x_i - \hat{\mu}_0)^2 + \sum_{j=1}^{m}(y_j - \hat{\mu}_0)^2$$

and the numerator of Λ is

$$f(\mathbf{x}, \mathbf{y} | \hat{\mu}_0, \hat{\mu}_0, \hat{\sigma}_0) = (2\pi\hat{\sigma}_0^2 e)^{-(n+m)/2}$$

In a similar way the denominator may be obtained by partial differentiation of log f with respect to the three variables μ_1, μ_2 and σ. Setting the partial derivatives equal to zero gives the three equations

$$\sum_{i=1}^{n} (x_i - \mu_1) = 0$$

$$\sum_{j=1}^{m} (y_j - \mu_2) = 0$$

and

$$-(n + m) + \frac{1}{\sigma^2} \left[\sum_{i=1}^{n} (x_i - \mu_1)^2 + \sum_{j=1}^{m} (y_j - \mu_2)^2 \right] = 0$$

with solutions $\hat{\mu}_1 = \bar{x}$, $\hat{\mu}_2 = \bar{y}$ and

$$(n + m)\hat{\sigma}^2 = \sum_{i=1}^{n} (x_i - \bar{x})^2 + \sum_{j=1}^{m} (y_j - \bar{y})^2$$

The denominator in Λ is then $f(\mathbf{x}, \mathbf{y} \mid \hat{\mu}_1, \hat{\mu}_2, \hat{\sigma}) = (2\pi\hat{\sigma}^2 e)^{-(n+m)/2}$, and thus we obtain

$$\Lambda(\mathbf{x}, \mathbf{y}) = \left(\frac{\hat{\sigma}^2}{\hat{\sigma}_0^2} \right)^{(n+m)/2}$$

Some simplifying algebra is needed to convert $\Lambda(\mathbf{x}, \mathbf{y})$ into the form used in TWOSAMPLE. Now

$$\sum_{i=1}^{n} (x_i - \hat{\mu}_0)^2 = \sum_{i=1}^{n} \left(x_i - \frac{n\bar{x} + m\bar{y})}{n + m} \right)^2$$

$$= \sum_{i=1}^{n} (x_i - \bar{x})^2 + n \left(\bar{x} - \frac{n\bar{x} + m\bar{y}}{n + m} \right)^2$$

$$= \sum_{i=1}^{n} (x_i - \bar{x})^2 + \frac{nm^2}{(n + m)^2} (\bar{x} - \bar{y})^2$$

and similarly,

$$\sum_{j=1}^{m} (y_j - \hat{\mu}_0)^2 = \sum_{j=1}^{m} (y_j - \bar{y})^2 + \frac{mn^2}{(m + n)^2} (\bar{x} - \bar{y})^2$$

Adding the last two equations we obtain

$$(n + m)\hat{\sigma}_0^2 = (n + m)\hat{\sigma}^2 + \frac{mn}{n + m} (\bar{x} - \bar{y})^2$$

The inequality $\Lambda(\mathbf{x}, \mathbf{y}) = (\hat{\sigma}^2/\hat{\sigma}_0^2)^{(n+m)/2} \leq \lambda$ is equivalent to $\hat{\sigma}_0^2/\hat{\sigma}^2 \geq \lambda^{-2/(n+m)} = k$ (say), and therefore to

$$1 + \frac{mn(\bar{x} - \bar{y})^2/(n+m)}{\sum_{i=1}^{n}(x_i - \bar{x})^2 + \sum_{j=1}^{m}(y_j - \bar{y})^2} \geq k$$

or

$$\frac{|\bar{x} - \bar{y}|\sqrt{mn(m+n-2)}}{\sqrt{(n+m)\left(\sum_{i=1}^{n}(x_i - \bar{x})^2 + \sum_{j=1}^{m}(y_j - \bar{y})^2\right)}} \geq c$$

Having introduced the constant term $n + m - 2$ in the last inequality, we see that the left-hand side is the observed value of the $t(n + m - 2)$ distributed statistic used by TWOSAMPLE with the subcommand POOLED.

The next two examples relate to two-tailed tests on the variance of a normal distribution.

──────── EXAMPLE 9.13 ────────────────────────────────

Suppose $X_1, X_2, \ldots, X_n$ is a random sample from the distribution $N(\mu, \sigma^2)$ where both μ and σ are unknown. We obtain the LRT critical region for a test of the null hypothesis $H_0(\sigma = \sigma_0)$ against the alternative hypothesis $H_1(\sigma \neq \sigma_0)$ at the level of significance α.

Set $\theta = (\mu, \sigma)$, $\Omega = \{(\mu, \sigma): -\infty < \mu < \infty, \sigma > 0\}$ and $\Omega_0 = \{(\mu, \sigma_0): -\infty < \mu < \infty\}$ to express the null hypothesis in the form $H_0(\theta \in \Omega_0)$.

Maximizing the joint p.d.f.,

$$f(\mathbf{x}|\mu, \sigma) = \frac{1}{(\sigma\sqrt{2\pi})^n} \exp\left[-\frac{1}{2\sigma^2}\sum_{i=1}^{n}(x_i - \mu)^2\right]$$

over μ and σ, we obtain the denominator of Λ as $f(\mathbf{x}|\hat{\mu}, \hat{\sigma}) = (2\pi\hat{\sigma}^2 e)^{-n/2}$, where $\hat{\mu} = \bar{x}$ and $n\hat{\sigma}^2 = \sum_{i=1}^{n}(x_i - \bar{x})^2$ (see Example 9.11).

Similarly, for the numerator of Λ, we obtain $\hat{\mu}_0 = \hat{\mu} = \bar{x}$ and the numerator is

$$f(\mathbf{x}|\hat{\mu}, \sigma_0) = \frac{1}{(2\pi\sigma_0^2)^{n/2}} \exp\left[-\frac{1}{2\sigma_0^2}\sum_{i=1}^{n}(x_i - \bar{x})^2\right]$$

$$= \frac{1}{(2\pi\sigma_0^2)^{n/2}} \exp\left[-\frac{n\hat{\sigma}^2}{2\sigma_0^2}\right]$$

and hence,

$$\Lambda(\mathbf{x}) = \frac{e^{n/2}}{n^{n/2}} \left(\frac{n\hat{\sigma}^2}{\sigma_0^2}\right)^{n/2} \exp\left(-\frac{n\hat{\sigma}^2}{2\sigma_0^2}\right)$$

Writing $y = n\hat{\sigma}^2/\sigma_0^2$, the inequality $\Lambda(\mathbf{x}) \le \lambda$ is equivalent to $y^{n/2} \exp(-y/2) \le k = \lambda n^{n/2} \exp(-n/2)$. A sketch of $h(y) = y^{n/2} \exp(-y/2)$ reveals that $y^{n/2} \exp(-y/2) \le k$ is equivalent to $y \le a$ or $y \ge b$ where

$$a^{n/2} \exp(-a/2) = b^{n/2} \exp(-b/2)$$

The LRT critical region is, therefore,

$$C = \{\mathbf{x}: (n-1)s^2/\sigma_0^2 \le a \text{ or } (n-1)s^2/\sigma_0^2 \ge b\}$$

where $(n-1)s^2 = \sum_{i=1}^n (x_i - \bar{x})^2 = n\hat{\sigma}^2$, and the values a and b, satisfying the equation above, are chosen so that $\Pr(\mathbf{X} \in C \mid \sigma = \sigma_0) = \alpha$.

The LRT procedure has recovered the $\chi^2(n-1)$ distributed statistic $W = (n-1)S^2/\sigma_0^2$ used in Section 4.2, Example 4.14, but does not specify the equi-tailed critical region used there. We now identify the LRT critical region as a type of equi-ordinate region.

Recall that the $\chi^2(m)$ distribution has p.d.f.

$$f_m(x) = \frac{1}{\Gamma\left(\frac{m}{2}\right) 2^{m/2}} x^{(m/2)-1} \exp(-x/2) \text{ for } x > 0$$

A critical region of the form $\{w : w \le a \text{ or } w \ge b\}$ is, therefore:

1. The equi-ordinate critical region based on the distribution W if $f_{n-1}(a) = f_{n-1}(b)$ or

$$a^{[(n-1)/2]-1} \exp(-a/2) = b^{[(n-1)/2]-1} \exp(-b/2)$$

2. The LRT critical region if

$$a^{n/2} \exp(-a/2) = b^{n/2} \exp(-b/2)$$

that is, $f_{n+2}(a) = f_{n+2}(b)$, so that the LRT critical region is equi-ordinate for the $\chi^2(n+2)$ distribution and

3. The UMPU critical region if

$$a^{(n-1)/2} \exp(-a/2) = b^{(n-1)/2} \exp(-b/2)$$

that is, $f_{n+1}(a) = f_{n+1}(b)$, so that the UMPU critical region is equi-ordinate for the $\chi^2(n+1)$ distribution (see reference 2, p. 194).

The next example uses tabled values for a and b, provided in Lindley *et al.* (see reference 3) to compare the UMPU critical region with the equi-tail region.

──────── **EXAMPLE 9.14** ────────

Let $W = (n-1)S^2/\sigma^2$. The values shown below are the equi-tail 5% values a_0 and b_0, obtained from tables such that

$$\Pr(W \le a_0) = \Pr(W \ge b_0) = 0.025$$

and the UMPU region values a and b taken from Lindley *et al.* (see reference 3):

n	Equi-tail (a_0, b_0)	UMPU (a, b)
6	(0.8312, 12.83)	(0.9892, 14.369)
10	(2.700, 19.02)	(2.9532, 20.305)
16	(6.262, 27.49)	(6.5908, 28.614)

We compare the performance of the two regions by calculating the probability of Type II error when $n = 10$ and $\sigma_0 = 1$, for two values, $\sigma = 2$ and $\sigma = \frac{1}{2}$, covered by the alternative hypothesis. In the one-dimensional form, with test statistic $W = (n-1)S^2$, since $\sigma_0 = 1$, the two critical regions are

$$R_1 = \{w: w \le 2.7 \text{ or } w \ge 19.02\}$$

and

$$R_2 = \{w: w \le 2.9532 \text{ or } w \ge 20.305\}$$

Under the alternative hypothesis, for a general $\sigma \ne 1$, the statistic W/σ^2 has the distribution $\chi^2(9)$, and the INVCDF command, with subcommand CHIS 9, was used to obtain the probabilities given below for Type II errors. Thus:

$$\begin{aligned}
\Pr(W \notin R_1 \mid \sigma = 2) &= \Pr(2.7/4 < W/4 < 19.02/4) \\
&= \Pr(0.675 < W/4 < 4.755) \\
&= 0.1449 - 0.0001 = 0.1448
\end{aligned}$$

$$\begin{aligned}
\Pr(W \notin R_2 \mid \sigma = 2) &= \Pr(2.9532/4 < W/4 < 20.305/4) \\
&= \Pr(0.7383 < W/4 < 5.0762) \\
&= 0.1724 - 0.0002 = 0.1722
\end{aligned}$$

$$\begin{aligned}
\Pr(W \notin R_1 \mid \sigma = 0.5) &= \Pr(2.7 \times 4 < 4W < 19.02 \times 4) \\
&= \Pr(10.8 < 4W < 76.08) \\
&= 1 - 0.7108 = 0.2892
\end{aligned}$$

$$\begin{aligned}
\Pr(W \notin R_2 \mid \sigma = 0.5) &= \Pr(2.9532 \times 4 < 4W < 20.305 \times 4) \\
&= \Pr(11.8128 < 4W < 81.22) \\
&= 1 - 0.7760 = 0.2240
\end{aligned}$$

These calculations show that neither the UMPU critical region nor the (biased) equi-tailed critical region achieves the lower probability of Type II error uniformly over the range $\sigma \ne 1$. This fact, together with the simplicity of application, leads to the common recommendation that the equi-tailed critical region be employed for this test on the normal variance.

──────────────────────────────

The final example in this section shows how the LRT defines the test statistic used in the ONEWAY test for the equality of several means.

──────── **EXAMPLE 9.15** ────────────────────────

Suppose that, for each integer i in the range $1 \leq i \leq m$, we have a random sample $X_{i1}, X_{i2}, \ldots, X_{in}$ from a $N(\mu_i, \sigma^2)$ distribution. We apply the LRT to the null hypothesis that all the means are equal against the alternative that at least one pair of means is distinct.

Setting $\theta = (\mu_1, \mu_2, \ldots, \mu_m, \sigma)$,

$$\Omega = \{ (\mu_1, \mu_2, \ldots, \mu_m, \sigma): -\infty < \mu_i < \infty \text{ for } 1 \leq i \leq m, \sigma > 0 \}$$

and $\Omega_0 = \{ (\mu, \mu, \ldots, \mu, \sigma): -\infty < \mu < \infty, \sigma > 0 \}$, then the null hypothesis takes the form $H_0(\theta \in \Omega_0)$.

Writing $\mathbf{x}_i = (x_{i1}, x_{i2}, \ldots, x_{in})$ and $\mathbf{x} = (x_1, x_2, \ldots, x_m)$, the joint p.d.f. of the sample variables is

$$f(\mathbf{x}|\mu_1, \mu_2, \ldots, \mu_m, \sigma) = \frac{1}{(\sigma\sqrt{2\pi})^{nm}} \exp\left[-\frac{1}{2\sigma^2} \sum_{i=1}^{n} \sum_{j=1}^{m} (x_{ij} - \mu_i)^2 \right]$$

Now maximizing f over $\theta \in \Omega$ for the denominator of Λ amounts to minimizing

$$Q = \sum_{i=1}^{m} \sum_{j=1}^{n} (x_{ij} - \mu_i)^2$$

as a function of the means μ_i, and maximizing

$$\frac{1}{(\sigma\sqrt{2\pi})^{nm}} \exp(-Q/2\sigma^2)$$

as a function of σ.

From ONEWAY in Section 4.2 we obtain the estimates

$$\hat{\mu}_i = \bar{x}_i = \frac{1}{n} \sum_{j=1}^{n} x_{ij}$$

with the minimum value $\hat{Q} = \sum_{i=1}^{m} \sum_{j=1}^{n} (x_{ij} - \bar{x}_i)^2$ and the usual calculation gives the estimate $\hat{\sigma}^2 = \hat{Q}/nm$, so that

$$\max_{\theta \in \Omega} f(\mathbf{x}|\mu_1, \mu_2 \ldots, \mu_m, \sigma) = \frac{1}{(\hat{\sigma}\sqrt{2\pi})^{nm}} \exp(-nm/2)$$

Similarly, for the numerator we obtain $\hat{Q}_0 = \sum_{i=1}^{m} \sum_{j=1}^{n} (x_{ij} - \bar{x})^2$, where $\bar{x} = (1/m) \sum_{j=1}^{m} \bar{x}_i$, and $\hat{\sigma}_0^2 = \hat{Q}_0/nm$, so that

$$\max_{\theta \in \Omega_0} f(\mathbf{x}|\mu, \mu, \ldots, \mu, \sigma) = \frac{1}{(\hat{\sigma}_0\sqrt{2\pi})^{nm}} \exp(-nm/2)$$

Now $\Lambda(\mathbf{x}) = (\hat{\sigma}^2/\hat{\sigma}_0^2)^{nm/2}$, and the inequality $\Lambda(\mathbf{x}) \leqslant \lambda$ is equivalent to $\hat{\sigma}_0^2/\hat{\sigma}^2 \geqslant k$, that is,

$$\frac{\sum_{i=1}^{m}\sum_{j=1}^{n}(x_{ij}-\bar{x})^2}{\sum_{i=1}^{m}\sum_{j=1}^{n}(x_{ij}-\bar{x}_i)^2} \geqslant k$$

However, in the notation of Section 4.2, the left-hand side is the observed value of the ratio SS_T/SS_e, and the LRT has therefore defined the test suggested on more informal grounds in the discussion of ONEWAY.

For a wide range of problems, including all the examples above, the LRT method defines an intuitively acceptable critical region. On the other hand, unfortunately, there are examples where the LRT defines a critical region having low power when other considerations suggest the use of a critical region with power close to one (see reference 2, p. 341, Problem 18).

9.3.1   Problems

9.6   The independent random variables $X_1, X_2, \ldots, X_n$ have the common p.d.f. $f(x \mid \theta) = \theta e^{-\theta x}$ for $x > 0$. Find the critical region for the likelihood ratio test of the null hypothesis $H_0(\theta = 1)$ against the alternative hypothesis $H_1(\theta = 1)$.

9.7   The random variables X_1 and X_2 have the joint p.d.f. $f(\mathbf{x} \mid \lambda, \tau) = \lambda^2 \tau \exp[-\lambda(x_1 + \tau x_2)]$. Writing $\theta = (\lambda, \tau)$, apply the likelihood ratio test to the null hypothesis $H_0(\tau = 1)$ against the alternative hypothesis $H_1(\tau = 1)$ and show that the critical region is equivalent to the one-dimensional region of the form $\{v : v < a \text{ or } v > b\}$ for the statistic $V = X_1/X_2$, where $a/(a+1)^2 = b/(b+1)^2$.

9.4   Solutions to problems

9.1   The inequality $f(\mathbf{x} \mid \theta_1) \geqslant k f(\mathbf{x} \mid \theta_0)$ defining the critical region reduces to $1 \geqslant k x_1 x_2$, and hence the critical region. Now

$$\alpha = \Pr(\mathbf{X} \in C \mid H_0) = \iint_C 4x_1 x_2 \, dx_1 \, dx_2$$

$$= \int_0^c \int_0^1 4x_1 x_2 \, dx_1 \, dx_2 + \int_c^1 4x_1 \left(\int_0^{c/x_1} x_2 \, dx_2 \right) dx_1$$

$$= c^2 - 2c^2 \log c$$

The probability of Type II error is given by

$$\Pr(\mathbf{X} \notin C | H_1) = \int_c^1 \left(\int_{c/x_1}^1 dx_2 \right) dx_1$$

$$= 1 - c + c \log c \approx 0.6849$$

and the probabilities of Type II error for the critical regions $\square$ and $\triangle$ are $1 - \sqrt{a} \approx 0.7764$ and $1 - \frac{1}{2}\sqrt{6a} \approx 0.7261$, respectively.

9.2 $M(t) = \int_0^\infty e^{2\theta xt} \theta e^{-\theta x} dx = 1/(1 - 2t)$ so that, by the independence of the random variables $X_1, X_2, \ldots, X_n$, $M_W(t) = (1 - 2t)^{-n}$. From Section 1.4, this is the moment generating function of the chi-squared distribution $\chi^2(2n)$.

9.3 The distribution function of $-2 \log X_1$ is given by

$$\Pr(-2 \log X_1 \leq x) = \Pr(X_1 \geq e^{-x/2}) = 1 - e^{-x/2}$$

so that $-2 \log X_1$ has the p.d.f. $\frac{1}{2}e^{-x/2}$. Hence, from Problem 9.2 with $\theta = \frac{1}{2}$, W has the $\chi^2(2n)$ distribution.

9.4 The inequality determining the shape of the critical region is

$$\frac{1}{(\sqrt{2\pi})^n \sigma_1^n} \exp\left(-\frac{1}{2\sigma_1^2} \sum_{i=1}^n x_i^2\right) \geq k \frac{1}{(\sqrt{2\pi})^n \sigma_0^n} \exp\left(-\frac{1}{2\sigma_0^2} \sum_{i=1}^n x_i^2\right)$$

or

$$\exp\left(\left(\frac{1}{2\sigma_0^2} - \frac{1}{2\sigma_1^2}\right) \sum_{i=1}^n x_i^2\right) \geq k_1$$

which, since $\sigma_0 < \sigma_1$, implies that $\sum_{i=1}^n x_i^2 \geq c$.

When $n = 16$, $\sigma_0 = 1$, $\sigma_1 = 2$ and $\alpha = 0.05$ then $\sum_{i=1}^{16} X_i^2$ has the $\chi^2(16)$ distribution under H_0, so that $\Pr(\sum_{i=1}^{16} X_i^2 \geq c | H_0) = 0.05$ gives, from tables, $c = 26.3$. The power of the test is given by $\Pr(\sum_{i=1}^{16} X_i^2 \geq 26.3 | H_1)$, and, under H_1, $\frac{1}{4} \sum_{i=1}^{16} X_i^2$ has the $\chi^2(16)$ distribution. Hence if W has the $\chi^2(16)$ distribution we require $\Pr(\frac{1}{4} \sum_{i=1}^n X_i^2 \geq 6.575 | H_1) = \Pr(W \geq 6.575) = 0.9806$.

9.5 Following Example 9.5, the critical region is $C = \{x: \sum_{i=1}^n x_i \geq c\}$. Writing $S_n = \sum_{i=1}^n X_i$, we require $\Pr(S_n \geq c | p = \frac{1}{2}) = 0.01$ and $\Pr(S_n \geq c | p = \frac{3}{4}) = 0.99$. Applying the normal approximation $Z = (S_n - np)/\sqrt{np(1 - p)}$, we require

$$\Pr(Z \geq 2(c - n/2)/\sqrt{n}) = 0.01 \quad \text{or} \quad (2c - n)/\sqrt{n} \approx 2.326$$

and

$$\Pr(Z \geq 4(c - 3n/4)/\sqrt{3n}) = 0.99 \quad \text{or} \quad (4c - 3n)/\sqrt{3n} \approx -2.326$$

giving $\sqrt{n} \approx 2.326(2 + \sqrt{3})$ or $n \approx 75.35$

9.6 The joint p.d.f. is $f(\mathbf{x} | \theta) = \theta^n \exp(-\theta t)$, where $t = \sum x_i$. The numerator of the likelihood ratio is $\exp(-t)$ – the value when $\theta = 1$ – and the denominator may be obtained by differentiating log f, as the value when $\theta = n/t$ – that is, $(n/t)^n e^{-n}$. Thus, the ratio is $\Lambda(\mathbf{x}) = t^n e^{-t}/n^n e^{-n}$, and the inequality $\Lambda(\mathbf{x}) < \lambda$ is equivalent to $t^n \exp(-t) < k$. The n-dimensional critical region is $\{\mathbf{x}: t < a$ or $t > b\}$, with a corresponding one-dimensional region $\{t: t < a$ or $t > b\}$, where $a^n e^{-a} = b^n e^{-b}$.

9.7 For the numerator of the likelihood ratio statistic we have to maximize $f = \lambda^2 \exp[-\lambda(x_1 + x_2)]$ with respect to λ, and obtain (as in Solution 9.6) the maximum $\{2/(x_1 + x_2)\}^2 e^{-2}$. For the denominator we have to maximize the joint p.d.f. with respect to both parameters, and accomplish this by partial differentiation of $\log f$. This gives the equations

$$-(x_1 + \tau x_2) + 2/\lambda = 0 \quad \text{and} \quad -\lambda x_2 + 1/\tau = 0$$

so that the maximum is obtained when a $\lambda \tau x_2 = 1 = \lambda x_1$, or when $\lambda^2 \tau x_1 x_2 = 1$ and $\lambda(x_1 + \tau x_2) = 2$. Thus, the denominator is $1/(x_1 x_2 e^2)$ and hence the ratio is $4 x_1 x_2/(x_1 + x_2)^2$. Now, with $v = x_1/x_2$, the inequality $4 x_1 x_2/(x_1 + x_2)^2 < \lambda$ is equivalent to $v/(v + 1)^2 < k$, and this produces the one-dimensional critical region $\{v: v < a$ or $v > b\}$ where $a/(a + 1)^2 = b/(b + 1)^2$.

References

1. Beaumont, G. P., *Intermediate Mathematical Statistics*, Chapman & Hall, London, 1980.
2. Lehman, E. H., *Testing Statistical Hypotheses*, 2nd edition, John Wiley, Chichester, 1986.
3. Lindley, D. V., East, D. A. and Hamilton, P. A., Tables for making inferences about the variance of a normal distribution., *Biometrika*, **47**, 1960, 433–7.

The sampling distributions

Density of the chi-squared distribution

We show, by induction, that if $X_1, X_2, \ldots, X_n$ is a random sample of size n from the distribution $N(0, 1)$, then the p.d.f. of $U_n = \sum_{i=1}^{n} X_i^2$ is

$$\frac{1}{2^{(n/2)}\Gamma(n/2)} \, x^{(1/2)n-1} e^{-(1/2)x} \quad 0 \leqslant x < \infty \tag{A1.1}$$

For $n = 1$, if Z is distributed $N(0, 1)$, then $\Pr(Z^2 \leqslant x) = \Pr(|Z| \leqslant \sqrt{x}) = 2\Pr(0 \leqslant Z \leqslant \sqrt{x})$

$$= 2 \int_0^{\sqrt{x}} \frac{1}{\sqrt{2\pi}} \, e^{-(1/2)t^2} \, \mathrm{d}t = \int_0^x \frac{1}{\sqrt{2\pi}} \, t^{-(1/2)} e^{-(1/2)t} \, \mathrm{d}t$$

Hence equation (A1.1) holds for $n = 1$. Assume equation (A1.1) is true for the integer n and consider $U_{n+1} = \sum_{i=1}^{n+1} X_i^2 = \sum_{i=1}^{n} X_i^2 + X_{n+1}^2$ where X_{n+1} is distributed $N(0, 1)$. Hence the p.d.f. of U_{n+1} from Section 1.1, equation (1.1) is

$$\int_{-\infty}^{\infty} f_{U_n}(z - y) f_{X_{n+1}^2}(y) \, \mathrm{d}y$$

$$= \frac{1}{\sqrt{2\pi}2^{(n/2)}\Gamma(n/2)} \int_0^z (z - y)^{(1/2)n-1} e^{-(1/2)(z-y)} y^{-1/2} e^{-(1/2)y} \, \mathrm{d}y$$

substituting $y = wz$

$$= \frac{z^{(1/2)(n+1)-1} e^{-(1/2)z}}{\sqrt{2\pi}2^{(n/2)}\Gamma(\frac{1}{2})} \int_0^1 (1 - w)^{(1/2)n-1} w^{-1/2} \, \mathrm{d}w$$

$$= \frac{1}{2^{(n+1)/2}\Gamma[(n + 1)/2]} \, z^{(1/2)(n+1)-1} e^{-(1/2)z}$$

using the result for the beta distribution (Section 1.1). Hence equation (A1.1) holds for a random sample of size $n + 1$ and the result follows by induction.

Lemma

1. $\int_0^{kb} x^{(n/2)-1} \exp(-ax/2)\, dx = b^{n/2} \int_0^k y^{(n/2)-1} \exp(-aby/2)\, dy$

2. $\int_0^\infty y^{(n/2)-1} \exp(-cy/2)\, dy = \left(\dfrac{2}{c}\right)^{n/2} \Gamma(n/2)$

The first result is obtained by substituting $y = x/b$, the second by substituting $z = cy/2$.

Density of the F(n, m) distribution

In Section 1.4, the random variable W is defined to have the $F(n, m)$ distribution if W may be written $W = mU/nV$, where U and V are independent random variables, with $\chi^2(n)$ and $\chi^2(m)$ distributions respectively. We now show that W has p.d.f.

$$\frac{\Gamma[(m+n)/2]}{\Gamma(n/2)\Gamma(m/2)} \left(\frac{n}{m}\right)^{n/2} \frac{x^{(n/2)-1}}{(1+nx/m)^{(m+n)/2}} \quad x > 0 \tag{A1.2}$$

From equation (A1.1), the joint p.d.f. of U, V is

$$\frac{(\frac{1}{2})^{n/2} u^{(n/2)-1} e^{-(1/2)u}}{\Gamma(n/2)} \cdot \frac{(\frac{1}{2})^{m/2} v^{(m/2)-1} e^{-(1/2)v}}{\Gamma(m/2)}$$

Now the distribution function of W is given by

$$\Pr(mU/nV < x) = \Pr(U < nxV/m)$$

which can be written as the double integral

$$\frac{(\frac{1}{2})^{(m+n)/2}}{\Gamma(m/2)\Gamma(n/2)} \int_0^\infty v^{(m/2)-1} e^{-(1/2)v} \left(\int_0^{nxv/m} u^{(n/2)-1} e^{-(1/2)u}\, du\right) dv$$

which, using lemma (1) and changing the order of integration, is

$$\frac{(\frac{1}{2})^{(m+n)/2}}{\Gamma(m/2)\Gamma(n/2)} \left(\frac{n}{m}\right)^{n/2} \int_0^x w^{(n/2)-1} \left\{\int_0^\infty v^{(m+n)/2-1} \exp[-\tfrac{1}{2}v(1+nw/m)]\, dv\right\} dw$$

and by (2) of the lemma is equal to

$$\frac{\Gamma[(m+n)/2]}{\Gamma(m/2)\Gamma(n/2)} \left(\frac{n}{m}\right)^{n/2} \int_0^x \frac{w^{(n/2)-1}}{(1+nw/m)^{(m+n)/2}}\, dw$$

But this is the distribution function of W and the density is obtained by differentiating with respect to x.

Density of the *t*-distribution

If Z has the distribution $N(0,1)$ and if V has the $\chi^2(m)$ distribution and is independent of Z, then we shall deduce that $T = Z/\sqrt{V/m}$ has p.d.f.

$$\frac{1}{\sqrt{m\pi}} \frac{\Gamma[(m+1)/2]}{\Gamma(m/2)} \frac{1}{(1+t^2/m)^{(m+1)/2}} \quad -\infty < t < \infty$$

In Problem 1.25 of Section 1.4 we show that if T has the $t(m)$ distribution, then T^2 has the $F(1,m)$ distribution. Now

$$\Pr(-t \leqslant T \leqslant t) = Pr(T^2 \leqslant t^2)$$

That is,

$$2\Pr(T \leqslant t) - 1 = Pr(W \leqslant t^2)$$

where $\Pr(T \leqslant t)$ is the distribution function of T and $\Pr(W \leqslant t^2)$ is the distribution function of the $F(1,m)$ distribution. Differentiating we see that the density of T is $t\mathrm{f}(t^2)$ where $\mathrm{f}(.)$ is the density of the $F(1,m)$ distribution. Substituting in equation (A1.2) gives the density of T as stated.

Sampling finite populations

Consider a finite population of N elements, to each of which is attached a numerical value. Let these values be $x_1, x_2, ..., x_N$, with mean μ and variance $\sigma^2 = \sum_{i=1}^{N} (x_i - \mu)^2/N$. If a single element is drawn at random from this population, then the value on the element is a random variable, X, such that $\Pr(X = x_i) = 1/N$. Hence $E(X) = \mu$, $V(X) = \sigma^2$.

Now suppose a simple random sample of n elements is drawn, one at a time *without replacement*, from the population. Let $X_1, X_2, ..., X_n$ be the random variables for the members of the sample and let $\bar{X} = \sum_{i=1}^{n} X_i/n$. The X_i each have the same distribution, so that $E(X_i) = \mu$, $V(X_i) = \sigma^2$. Then $E(\bar{X}) = E(\sum_{i=1}^{n} X_i/n) = \mu$. But the X_i are not independent, so

$$V(\bar{X}) = V\left(\sum_{i=1}^{n} X_i \Big/ n\right) = V\left(\sum_{i=1}^{n} X_i\right) \Big/ n^2$$

$$= \left[\sum_{i=1}^{n} V(X_i) + \sum_{j \neq k}^{n} \sum^{n} \text{Cov}(X_j, X_k)\right] \Big/ n^2$$

Since all pairs X_j, X_k, $j \neq k$, have the same joint distribution, $\text{Cov}(X_j, X_k) = \rho\sigma^2$, where ρ is the common correlation coefficient. Hence

$$V(\bar{X}) = [n\sigma^2 + n(n-1)\rho\sigma^2]/n^2$$

But this result holds for all n such that $1 \leq n \leq N$. If $n = N$, the sample draws the whole population, $\bar{X}$ is constant, and $V(\bar{X}) = 0$. Hence $0 = N\sigma^2 + N(N-1)\rho\sigma^2$ or $\rho = -1/(N-1)$. Substituting for ρ,

$$V(\bar{X}) = \frac{\sigma^2}{n} \left(\frac{N-n}{N-1}\right)$$

------- **EXAMPLE A2.1** ---

Suppose Np of the population values $x_1, x_2, \ldots, x_N$ have the value 1 and the remainder have value zero. Then

$$\mu = \frac{1}{N}[(Np)1 + N(1-p)0] = p$$

$$\sigma^2 = \frac{1}{N}[(Np)1^2 + N(1-p)0^2] - p^2 = pq, \text{ where } q = 1-p$$

Hence

$$V(\bar{X}) = \frac{pq}{n}\left(\frac{N-n}{N-1}\right)$$

Since the *number in the sample* having value 1 is $n\bar{X}$, we deduce the mean and variance of the hypergeometric distribution (see Section 1.3).

------- **EXAMPLE A2.2** ---

Let $x_i = i$ for $i = 1, 2, \ldots, N$. Then

$$\mu = \frac{1}{N}(1 + 2 + \cdots + N) = \frac{(N+1)}{2}$$

$$\sigma^2 = \frac{\sum\limits_{i=1}^{N} i^2}{N} - \frac{(N+1)^2}{4} = \frac{N(N+1)(2N+1)}{6N} - \frac{(N+1)^2}{4}$$

$$= \frac{N^2 - 1}{12}$$

Hence

$$V(\bar{X}) = \frac{N^2 - 1}{12n}\left(\frac{N-n}{N-1}\right) = \frac{(N+1)(N-n)}{12n}$$

After noting that the *sum* of the sample values is $n\bar{X}$, by identifying this as a sum of ranks when there are no ties, we may obtain the variance of the Mann–Whitney Statistic (see Section 6.3).

------- **EXAMPLE A2.3** ---

We extend Example A2.2 to the case when not all the integers in the population are distinct. Suppose that the l successive integers $j, j+1, \ldots, j+l-1$ are replaced by their average, $j + (l-1)/2$. We say that these integers have been shared and clearly the overall population mean, $\mu = (N+1)/2$, is undisturbed. The variance of the population value is, however,

$$\frac{1}{N}\left[\sum_{r=1}^{j-1} r^2 + l\left(j + \frac{l-1}{2}\right)^2 + \sum_{r=j+l}^{N} r^2\right] - \mu^2$$

$$= \frac{1}{N}\left\{\sum_{r=1}^{N} r^2 - N\mu^2 - \left[\sum_{r=j}^{j+l-1} r^2 - l\left(j + \frac{l-1}{2}\right)^2\right]\right\}$$

$$= \frac{N^2 - 1}{12} - \frac{1}{N}\sum_{s=1}^{l}\left[s - \left(\frac{l+1}{2}\right)\right]^2$$

$$= \frac{N^2 - 1}{12} - \frac{l^3 - l}{12N}$$

EXAMPLE A2.4

We derive the variance of the permutation correlation coefficient. That is, we have n pairs (x_i, Y_i) in which the x_i, are fixed and $Y_1, Y_2, \ldots, Y_n$ are a random permutation of $y_1, y_2, \ldots, y_n$. Hence $\sum_{i=1}^{n} x_i$, $\sum_{i=1}^{n} Y_i = \sum_{j=1}^{n} y_j$ are fixed. The required correlation coefficient is

$$\frac{\sum_{i=1}^{n} x_i Y_i - n\bar{x}\bar{y}}{\sqrt{\sum_{i=1}^{n}(x_i - \bar{x})^2 \sum_{i=1}^{n}(y_i - \bar{y})^2}}$$

with variance

$$\frac{V\left(\sum_{i=1}^{n} x_i Y_i\right)}{n\sigma_X^2 n\sigma_Y^2}$$

Now

$$V\left(\sum_{i=1}^{n} x_i Y_i\right) = \sum_{i=1}^{n} x_i^2 V(Y_i) + \sum\sum_{j \neq k} x_j x_k \operatorname{Cov}(Y_j, Y_k)$$

$$= \sum_{i=1}^{n} x_i^2 \sigma_Y^2 + c\left[\left(\sum_{i=1}^{n} x_i\right)^2 - \sum_{i=1}^{n} x_i^2\right]$$

where c is $\text{Cov}(Y_j, Y_k)$, $j \neq k$. But $c = -\sigma_Y^2/(n-1)$, hence

$$\text{V}\left(\sum_{i=1}^{n} x_i Y_i\right) = \sigma_Y^2 \sum_{i=1}^{n} x_i^2 - \frac{\sigma_Y^2}{n-1} \left[\left(\sum_{i=1}^{n} x_i\right)^2 - \sum_{i=1}^{n} x_i^2\right]$$

$$= \frac{n^2 \sigma_Y^2 \sigma_X^2}{n-1}$$

Hence the variance of the correlation coefficient is $1/(n-1)$.

─────── **EXAMPLE A2.5** ───────

Suppose $x_i = s_i$, $Y_i = R_i$, where $s_i = \text{rank}(x_i)$, $r_i = \text{rank}(y_i)$, and tied ranks are shared, $i = 1, 2, \ldots, n$. Then Spearman's S is defined in Section 7.2 as

$$\sum_{i=1}^{n} (s_i - R_i)^2 = \sum_{i=1}^{n} s_i^2 + \sum_{i=1}^{n} r_i^2 - 2 \sum_{i=1}^{n} s_i R_i$$

$$\text{E}(S) = \sum_{i=1}^{n} s_i^2 + \sum_{i=1}^{n} r_i^2 - 2n\bar{s}\bar{r}$$

and from Example A2.4,

$$\text{V}(S) = 4\text{V}\left(\sum_{i=1}^{n} s_i R_i\right) = 4\left(\sum_{i=1}^{n} s_i^2 - n\bar{s}^2\right)\left(\sum_{i=1}^{n} r_i^2 - n\bar{r}^2\right) \bigg/ (n-1)$$

Tables of distribution functions

Table A3.1   The Wilcoxon signed-rank distribution (Section 5.2)
This table gives the left-tail probabilities $\Pr(W_+ \leq i)$ of the signed-rank statistic W_+ for sample sizes $n = 1, 2, ..., 20$. By symmetry, the right-tail probabilities are given by

$$\Pr(W_+ \geq j) = \Pr(W_+ \leq n(n+1)/2 - j)$$

i \ n	1	2	3	4	5	6	7
0	0.50000	0.25000	0.12500	0.06250	0.03125	0.01563	0.00781
1	1.00000	0.50000	0.25000	0.12500	0.06250	0.03125	0.01563
2			0.37500	0.18750	0.09375	0.04688	0.02344
3			0.62500	0.31250	0.15625	0.07813	0.03906
4				0.43750	0.21875	0.10938	0.05469
5				0.56250	0.31250	0.15625	0.07813
6					0.40625	0.21875	0.10938
7					0.50000	0.28125	0.14844
8						0.34375	0.18750
9						0.42188	0.23438
10						0.50000	0.28906
11							0.34375
12							0.40625
13							0.46875
14							0.53125

(continued)

Table A3.1   Continued

i \ n	8	9	10	11	12	13	14
0	0.00391	0.00195	0.00098	0.00049	0.00024	0.00012	0.00006
1	0.00781	0.00391	0.00195	0.00098	0.00049	0.00024	0.00012
2	0.01172	0.00586	0.00293	0.00146	0.00073	0.00037	0.00018
3	0.01953	0.00977	0.00488	0.00244	0.00122	0.00061	0.00031
4	0.02734	0.01367	0.00684	0.00342	0.00171	0.00085	0.00043
5	0.03906	0.01953	0.00977	0.00488	0.00244	0.00122	0.00061
6	0.05469	0.02734	0.01367	0.00684	0.00342	0.00171	0.00085
7	0.07422	0.03711	0.01855	0.00928	0.00464	0.00232	0.00116
8	0.09766	0.04883	0.02441	0.01221	0.00610	0.00305	0.00153
9	0.12500	0.06445	0.03223	0.01611	0.00806	0.00403	0.00201
10	0.15625	0.08203	0.04199	0.02100	0.01050	0.00525	0.00262
11	0.19141	0.10156	0.05273	0.02686	0.01343	0.00671	0.00336
12	0.23047	0.12500	0.06543	0.03369	0.01709	0.00854	0.00427
13	0.27344	0.15039	0.08008	0.04150	0.02124	0.01074	0.00537
14	0.32031	0.17969	0.09668	0.05078	0.02612	0.01331	0.00671
15	0.37109	0.21289	0.11621	0.06152	0.03198	0.01636	0.00830
16	0.42188	0.24805	0.13770	0.07373	0.03857	0.01990	0.01013
17	0.47266	0.28516	0.16113	0.08740	0.04614	0.02393	0.01227
18	0.52734	0.32617	0.18750	0.10303	0.05493	0.02869	0.01477
19		0.36719	0.21582	0.12012	0.06470	0.03406	0.01764
20		0.41016	0.24609	0.13916	0.07568	0.04016	0.02094
21		0.45508	0.27832	0.16016	0.08813	0.04712	0.02472
22		0.50000	0.31250	0.18262	0.10181	0.05493	0.02899
23			0.34766	0.20654	0.11670	0.06360	0.03381
24			0.38477	0.23242	0.13306	0.07324	0.03925
25			0.42285	0.25977	0.15063	0.08386	0.04529
26			0.46094	0.28857	0.16968	0.09546	0.05200
27			0.50000	0.31885	0.19019	0.10815	0.05945
28				0.35010	0.21191	0.12195	0.06763
29				0.38232	0.23486	0.13672	0.07654
30				0.41553	0.25928	0.15271	0.08630
31				0.44922	0.28467	0.16980	0.09686
32				0.48291	0.31104	0.18787	0.10828
33				0.51709	0.33862	0.20715	0.12061
34					0.36670	0.22742	0.13379
35					0.39551	0.24866	0.14789
36					0.42505	0.27087	0.16290
37					0.45483	0.29395	0.17877
38					0.48486	0.31775	0.19550
39					0.51514	0.34241	0.21313
40						0.36768	0.23157
41						0.39343	0.25079
42						0.41968	0.27081
43						0.44629	0.29150
44						0.47302	0.31287
45						0.50000	0.33490
46							0.35742
47							0.38043
48							0.40387
49							0.42761
50							0.45160
51							0.47577
52							0.50000

Table A3.1   Continued

i \ n	15	16	17	18	19	20
0	0.00003	0.00002	0.00001	0.00000	0.00000	0.00000
1	0.00006	0.00003	0.00002	0.00001	0.00000	0.00000
2	0.00009	0.00005	0.00002	0.00001	0.00001	0.00000
3	0.00015	0.00008	0.00004	0.00002	0.00001	0.00000
4	0.00021	0.00011	0.00005	0.00003	0.00001	0.00001
5	0.00031	0.00015	0.00008	0.00004	0.00002	0.00001
6	0.00043	0.00021	0.00011	0.00005	0.00003	0.00001
7	0.00058	0.00029	0.00014	0.00007	0.00004	0.00002
8	0.00076	0.00038	0.00019	0.00010	0.00005	0.00002
9	0.00101	0.00050	0.00025	0.00013	0.00006	0.00003
10	0.00131	0.00066	0.00033	0.00016	0.00008	0.00004
11	0.00168	0.00084	0.00042	0.00021	0.00010	0.00005
12	0.00214	0.00107	0.00053	0.00027	0.00013	0.00007
13	0.00269	0.00134	0.00067	0.00034	0.00017	0.00008
14	0.00336	0.00168	0.00084	0.00042	0.00021	0.00010
15	0.00418	0.00209	0.00105	0.00052	0.00026	0.00013
16	0.00513	0.00258	0.00129	0.00064	0.00032	0.00016
17	0.00623	0.00314	0.00158	0.00079	0.00039	0.00020
18	0.00754	0.00381	0.00192	0.00097	0.00048	0.00024
19	0.00903	0.00459	0.00232	0.00117	0.00059	0.00029
20	0.01077	0.00549	0.00278	0.00140	0.00071	0.00035
21	0.01279	0.00655	0.00333	0.00168	0.00085	0.00043
22	0.01508	0.00775	0.00395	0.00200	0.00101	0.00051
23	0.01767	0.00912	0.00467	0.00237	0.00120	0.00060
24	0.02063	0.01070	0.00549	0.00280	0.00142	0.00072
25	0.02396	0.01248	0.00643	0.00329	0.00167	0.00084
26	0.02768	0.01450	0.00750	0.00385	0.00196	0.00099
27	0.03186	0.01677	0.00871	0.00448	0.00229	0.00116
28	0.03650	0.01932	0.01008	0.00520	0.00266	0.00136
29	0.04163	0.02216	0.01161	0.00602	0.00309	0.00158
30	0.04730	0.02533	0.01334	0.00694	0.00357	0.00183
31	0.05350	0.02884	0.01526	0.00797	0.00412	0.00211
32	0.06027	0.03270	0.01740	0.00912	0.00473	0.00243
33	0.06769	0.03696	0.01977	0.01041	0.00541	0.00279
34	0.07571	0.04163	0.02238	0.01184	0.00618	0.00319
35	0.08441	0.04672	0.02527	0.01342	0.00703	0.00365
36	0.09381	0.05229	0.02844	0.01518	0.00799	0.00415
37	0.10388	0.05833	0.03191	0.01712	0.00904	0.00472
38	0.11465	0.06487	0.03571	0.01925	0.01021	0.00534
39	0.12619	0.07193	0.03984	0.02158	0.01149	0.00604
40	0.13843	0.07953	0.04433	0.02414	0.01291	0.00681
41	0.15140	0.08768	0.04919	0.02693	0.01447	0.00766
42	0.16513	0.09641	0.05444	0.02997	0.01617	0.00859
43	0.17957	0.10571	0.06010	0.03327	0.01803	0.00962
44	0.19470	0.11560	0.06618	0.03684	0.02007	0.01074
45	0.21060	0.12611	0.07272	0.04071	0.02228	0.01198
46	0.22714	0.13722	0.07969	0.04488	0.02468	0.01332
47	0.24435	0.14893	0.08713	0.04937	0.02729	0.01479
48	0.26224	0.16125	0.09505	0.05419	0.03010	0.01638
49	0.28070	0.17419	0.10345	0.05935	0.03314	0.01812
50	0.29974	0.18773	0.11234	0.06487	0.03642	0.01999
51	0.31934	0.20187	0.12175	0.07076	0.03994	0.02203
52	0.33939	0.21660	0.13166	0.07702	0.04371	0.02422
53	0.35986	0.23187	0.14208	0.08368	0.04776	0.02658

(continued)

Table A3.1 Continued

i	n 15	16	17	18	19	20
54	0.38077	0.24771	0.15302	0.09073	0.05208	0.02913
55	0.40198	0.26408	0.16447	0.09819	0.05669	0.03186
56	0.42346	0.28094	0.17644	0.10607	0.06159	0.03479
57	0.44519	0.29829	0.18891	0.11438	0.06681	0.03793
58	0.46704	0.31609	0.20188	0.12310	0.07234	0.04128
59	0.48898	0.33427	0.21534	0.13226	0.07820	0.04485
60	0.51102	0.35286	0.22929	0.14186	0.08440	0.04865
61		0.37178	0.24369	0.15190	0.09093	0.05270
62		0.39098	0.25854	0.16236	0.09782	0.05699
63		0.41045	0.27383	0.17327	0.10506	0.06155
64		0.43013	0.28953	0.18461	0.11266	0.06636
65		0.44997	0.30561	0.19637	0.12063	0.07145
66		0.46994	0.32207	0.20856	0.12896	0.07682
67		0.48997	0.33885	0.22115	0.13767	0.08248
68		0.51003	0.35595	0.23415	0.14675	0.08843
69			0.37333	0.24754	0.15620	0.09467
70			0.39095	0.26131	0.16603	0.10122
71			0.40879	0.27544	0.17623	0.10808
72			0.42682	0.28992	0.18680	0.11526
73			0.44498	0.30473	0.19773	0.12274
74			0.46326	0.31985	0.20902	0.13055
75			0.48161	0.33526	0.22067	0.13868
76			0.50000	0.35094	0.23266	0.14713
77				0.36686	0.24498	0.15590
78				0.38301	0.25764	0.16499
79				0.39935	0.27061	0.17441
80				0.41586	0.28388	0.18414
81				0.43252	0.29744	0.19419
82				0.44929	0.31128	0.20455
83				0.46614	0.32537	0.21522
84				0.48306	0.33971	0.22619
85				0.50000	0.35428	0.23745
86					0.36905	0.24900
87					0.38400	0.26084
88					0.39913	0.27294
89					0.41439	0.28530
90					0.42979	0.29791
91					0.44528	0.31076
92					0.46086	0.32383
93					0.47649	0.33711
94					0.49216	0.35059
95					0.50784	0.36425
96						0.37808
97						0.39206
98						0.40618
99						0.42041
100						0.43474
101						0.44916
102						0.46364
103						0.47816
104						0.49272
105						0.50728

Table A3.2   The runs distribution (Section 6.2)

This table gives the distribution function $\Pr(R \leq r)$ of the runs statistic R for sample sizes n and m such that $4 \leq n \leq m \leq 15$.

$m = 4$

r \ n	4	5	6	7	8	9	10	11
2	0.02857	0.01587	0.00952	0.00606	0.00404	0.00280	0.00200	0.00147
3	0.11429	0.07143	0.04762	0.03333	0.02424	0.01818	0.01399	0.01099
4	0.37143	0.26190	0.19048	0.14242	0.10909	0.08531	0.06793	0 05495
5	0.62857	0.50000	0.40476	0.33333	0.27879	0.23636	0.20280	0.17582
6	0.88571	0.78571	0.69048	0.60606	0.53333	0.47133	0.41858	0.37363
7	0.97143	0.92857	0.88095	0.83333	0.78788	0.74545	0.70629	0.67033
8	1.00000	0.99206	0.97619	0.95455	0.92929	0.90210	0.87413	0.84615
9	1.00000	1.00000	1.00000	1.00000	1.00000	1.00000	1.00000	1.00000

$m = 4$

r \ n	12	13	14	15
2	0.00110	0.00084	0.00065	0.00052
3	0.00879	0.00714	0.00588	0.00490
4	0.04505	0.03739	0.03137	0.02657
5	0.15385	0.13571	0.12059	0.10784
6	0.33516	0.30210	0.27353	0.24871
7	0.63736	0.60714	0.57941	0.55392
8	0.81868	0.79202	0.76634	0.74174
9	1.00000	1.00000	1.00000	1.00000

$m = 5$

r \ n	5	6	7	8	9	10	11	12
2	0.00794	0.00433	0.00253	0.00155	0.00100	0.00067	0.00046	0.00032
3	0.03968	0.02381	0.01515	0.01010	0.00699	0.00500	0.00366	0.00275
4	0.16667	0.11039	0.07576	0.05361	0.03896	0.02897	0.02198	0.01697
5	0.35714	0.26190	0.19697	0.15152	0.11888	0.09491	0.07692	0.06319
6	0.64286	0.52165	0.42424	0.34732	0.28671	0.23876	0.20055	0.16984
7	0.83333	0.73810	0.65152	0.57576	0.51049	0.45455	0.40659	0.36538
8	0.96032	0.91126	0.85354	0.79332	0.73427	0.67832	0.62637	0.57870
9	0.99206	0.97619	0.95455	0.92929	0.90210	0.87413	0.84615	0.81868
10	1.00000	0.99784	0.99242	0.98368	0.97203	0.95804	0.94231	0.92534
11	1.00000	1.00000	1.00000	1.00000	1.00000	1.00000	1.00000	1.00000

$m = 5$

r \ n	13	14	15
2	0.00023	0.00017	0.00013
3	0.00210	0.00163	0.00129
4	0.01331	0.01058	0.00851
5	0.05252	0.04412	0.03741
6	0.14496	0.12461	0.10784
7	0.32983	0.29902	0.27219
8	0.53525	0.49579	0.46001
9	0.79202	0.76634	0.74174
10	0.90756	0.88932	0.87087
11	1.00000	1.00000	1.00000

m = 6

r \ n	6	7	8	9	10	11	12	13
2	0.00216	0.00117	0.00067	0.00040	0.00025	0.00016	0.00011	0.00007
3	0.01299	0.00758	0.00466	0.00300	0.00200	0.00137	0.00097	0.00070
4	0.06710	0.04254	0.02797	0.01898	0.01324	0.00945	0.00690	0.00512
5	0.17532	0.12121	0.08625	0.06294	0.04695	0.03571	0.02763	0.02171
6	0.39177	0.29604	0.22611	0.17483	0.13686	0.10844	0.08689	0.07036
7	0.60823	0.50000	0.41259	0.34266	0.28671	0.24176	0.20540	0.17577
8	0.82468	0.73310	0.64569	0.56643	0.49650	0.43568	0.38316	0.33794
9	0.93290	0.87879	0.82051	0.76224	0.70629	0.65385	0.60537	0.56092
10	0.98701	0.96620	0.93706	0.90210	0.86364	0.82353	0.78313	0.74337
11	0.99784	0.99242	0.98368	0.97203	0.95804	0.94231	0.92534	0.90756
12	1.00000	0.99942	0.99767	0.99441	0.98951	0.98303	0.97511	0.96594
13	1.00000	1.00000	1.00000	1.00000	1.00000	1.00000	1.00000	1.00000

m = 6

r \ n	14	15
2	0.00005	0.00004
3	0.00052	0.00039
4	0.00387	0.00297
5	0.01729	0.01393
6	0.05753	0.04747
7	0.15144	0.13132
8	0.29902	0.26548
9	0.52038	0.48349
10	0.70485	0.66796
11	0.88932	0.87087
12	0.95573	0.94466
13	1.00000	1.00000

$m = 7$

r \ n	7	8	9	10	11	12	13	14	15
2	0.00058	0.00031	0.00017	0.00010	0.00006	0.00004	0.00003	0.00002	0.00001
3	0.00408	0.00233	0.00140	0.00087	0.00057	0.00038	0.00026	0.00018	0.00013
4	0.02506	0.01538	0.00979	0.00643	0.00434	0.00300	0.00212	0.00152	0.00111
5	0.07751	0.05128	0.03497	0.02448	0.01753	0.01282	0.00955	0.00722	0.00555
6	0.20862	0.14918	0.10839	0.08001	0.05995	0.04557	0.03509	0.02735	0.02155
7	0.38345	0.29604	0.23077	0.18182	0.14480	0.11652	0.09469	0.07766	0.06424
8	0.61655	0.51360	0.42657	0.35459	0.29563	0.24750	0.20820	0.17604	0.14962
9	0.79138	0.70396	0.62238	0.54895	0.48416	0.42760	0.37848	0.33591	0.29902
10	0.92249	0.86713	0.80594	0.74332	0.68213	0.62408	0.57005	0.52038	0.47510
11	0.97494	0.94872	0.91608	0.87937	0.84050	0.80090	0.76161	0.72330	0.68640
12	0.99592	0.98788	0.97483	0.95712	0.93552	0.91093	0.88421	0.85611	0.82727
13	0.99942	0.99767	0.99441	0.98951	0.98303	0.97511	0.96594	0.95573	0.94466
14	1.00000	0.99984	0.99930	0.99815	0.99623	0.99345	0.98978	0.98524	0.97988
15	1.00000	1.00000	1.00000	1.00000	1.00000	1.00000	1.00000	1.00000	1.00000

m = 8

r \ n	8	9	10	11	12	13	14	15
2	0.00016	0.00008	0.00005	0.00003	0.00002	0.00001	0.00001	0.00000
3	0.00124	0.00070	0.00041	0.00025	0.00016	0.00010	0.00007	0.00005
4	0.00886	0.00531	0.00329	0.00210	0.00138	0.00093	0.00064	0.00045
5	0.03170	0.02028	0.01337	0.00905	0.00627	0.00444	0.00320	0.00235
6	0.10023	0.06865	0.04792	0.03406	0.02461	0.01806	0.01344	0.01014
7	0.21445	0.15734	0.11703	0.08824	0.06740	0.05212	0.04076	0.03223
8	0.40482	0.31859	0.25141	0.19937	0.15909	0.12779	0.10337	0.08419
9	0.59518	0.50000	0.41937	0.35219	0.29662	0.25077	0.21293	0.18163
10	0.78555	0.70156	0.62094	0.54668	0.48000	0.42105	0.36945	0.32454
11	0.89977	0.84266	0.78219	0.72172	0.66337	0.60836	0.55728	0.51032
12	0.96830	0.93941	0.90313	0.86175	0.81741	0.77183	0.72632	0.68181
13	0.99114	0.97972	0.96360	0.94344	0.92010	0.89443	0.86718	0.83901
14	0.99876	0.99585	0.99047	0.98234	0.97145	0.95800	0.94231	0.92475
15	0.99984	0.99930	0.99815	0.99623	0.99345	0.98978	0.98524	0.97988
16	1.00000	0.99996	0.99979	0.99940	0.99869	0.99757	0.99598	0.99388
17	1.00000	1.00000	1.00000	1.00000	1.00000	1.00000	1.00000	1.00000

m = 9

r \ n	9	10	11	12	13	14	15
2	0.00004	0.00002	0.00001	0.00001	0.00000	0.00000	0.00000
3	0.00037	0.00021	0.00012	0.00007	0.00004	0.00003	0.00002
4	0.00300	0.00176	0.00107	0.00067	0.00043	0.00028	0.00019
5	0.01222	0.00761	0.00488	0.00322	0.00217	0.00149	0.00105
6	0.04447	0.02943	0.01989	0.01369	0.00960	0.00684	0.00494
7	0.10897	0.07672	0.05489	0.03989	0.02941	0.02198	0.01664
8	0.23797	0.17856	0.13491	0.10276	0.07895	0.06118	0.04782
9	0.39922	0.31859	0.25494	0.20493	0.16563	0.13467	0.11018
10	0.60078	0.50955	0.42998	0.36211	0.30495	0.25717	0.21736
11	0.76203	0.68141	0.60503	0.53501	0.47214	0.41641	0.36741
12	0.89103	0.83417	0.77307	0.71105	0.65046	0.59280	0.53890
13	0.95553	0.92328	0.88509	0.84308	0.79907	0.75449	0.71039
14	0.98778	0.97420	0.95511	0.93110	0.90310	0.87208	0.83901
15	0.99700	0.99239	0.98512	0.97511	0.96254	0.94768	0.93088
16	0.99963	0.99863	0.99655	0.99308	0.98801	0.98128	0.97288
17	0.99996	0.99979	0.99940	0.99869	0.99757	0.99598	0.99388
18	1.00000	0.99999	0.99994	0.99981	0.99956	0.99913	0.99847
19	1.00000	1.00000	1.00000	1.00000	1.00000	1.00000	1.00000

$m = 10$

r \ n	10	11	12	13	14	15
2	0.00001	0.00001	0.00000	0.00000	0.00000	0.00000
3	0.00011	0.00006	0.00003	0.00002	0.00001	0.00001
4	0.00099	0.00057	0.00034	0.00021	0.00013	0.00008
5	0.00449	0.00274	0.00172	0.00111	0.00073	0.00049
6	0.01852	0.01192	0.00784	0.00526	0.00359	0.00249
7	0.05126	0.03489	0.02417	0.01703	0.01218	0.00884
8	0.12764	0.09205	0.06704	0.04933	0.03668	0.02755
9	0.24221	0.18492	0.14206	0.10991	0.08568	0.06730
10	0.41407	0.33496	0.27066	0.21894	0.17755	0.14447
11	0.58593	0.50000	0.42498	0.36068	0.30617	0.26023
12	0.75779	0.68004	0.60503	0.53513	0.47153	0.41457
13	0.87236	0.81508	0.75506	0.69505	0.63690	0.58177
14	0.94874	0.91510	0.87509	0.83073	0.78389	0.73612
15	0.98148	0.96511	0.94368	0.91796	0.88888	0.85738
16	0.99551	0.98961	0.98042	0.96780	0.95188	0.93298
17	0.99901	0.99726	0.99420	0.98961	0.98338	0.97550
18	0.99989	0.99956	0.99879	0.99739	0.99519	0.99204
19	0.99999	0.99994	0.99981	0.99956	0.99913	0.99847
20	1.00000	1.00000	0.99998	0.99994	0.99985	0.99969
21	1.00000	1.00000	1.00000	1.00000	1.00000	1.00000

$m = 11$

r \ n	11	12	13	14	15
2	0.00000	0.00000	0.00000	0.00000	0.00000
3	0.00003	0.00002	0.00001	0.00001	0.00000
4	0.00031	0.00018	0.00011	0.00006	0.00004
5	0.00159	0.00095	0.00059	0.00037	0.00024
6	0.00733	0.00461	0.00297	0.00195	0.00130
7	0.02264	0.01499	0.01011	0.00693	0.00483
8	0.06347	0.04427	0.03126	0.02233	0.01614
9	0.13491	0.09919	0.07356	0.05505	0.04158
10	0.25994	0.20170	0.15685	0.12243	0.09600
11	0.40998	0.33496	0.27346	0.22348	0.18306
12	0.59002	0.50718	0.43337	0.36900	0.31366
13	0.74006	0.66504	0.59328	0.52665	0.46602
14	0.86509	0.80855	0.74875	0.68834	0.62926
15	0.93653	0.90081	0.85981	0.81539	0.76919
16	0.97736	0.95939	0.93596	0.90778	0.87580
17	0.99267	0.98501	0.97403	0.95975	0.94243
18	0.99841	0.99600	0.99188	0.98574	0.97741
19	0.99969	0.99905	0.99783	0.99584	0.99296
20	0.99997	0.99986	0.99959	0.99905	0.99814
21	1.00000	0.99998	0.99994	0.99985	0.99969
22	1.00000	1.00000	1.00000	0.99998	0.99995
23	1.00000	1.00000	1.00000	1.00000	1.00000

$m = 12$

r \ n	12	13	14	15
2	0.00000	0.00000	0.00000	0.00000
3	0.00001	0.00000	0.00000	0.00000
4	0.00010	0.00006	0.00003	0.00002
5	0.00055	0.00032	0.00020	0.00012
6	0.00278	0.00172	0.00108	0.00070
7	0.00950	0.00614	0.00404	0.00271
8	0.02963	0.02010	0.01382	0.00962
9	0.06990	0.04977	0.03581	0.02603
10	0.15044	0.11259	0.08467	0.06404
11	0.26320	0.20682	0.16285	0.12864
12	0.42107	0.34755	0.28598	0.23506
13	0.57893	0.50000	0.42964	0.36807
14	0.73680	0.66418	0.59382	0.52769
15	0.84956	0.79318	0.73454	0.67591
16	0.93010	0.89369	0.85181	0.80621
17	0.97037	0.95023	0.92510	0.89579
18	0.99050	0.98165	0.96908	0.95279
19	0.99722	0.99386	0.98863	0.98130
20	0.99945	0.99851	0.99677	0.99396
21	0.99990	0.99968	0.99921	0.99840
22	0.99999	0.99996	0.99986	0.99966
23	1.00000	1.00000	0.99998	0.99995
24	1.00000	1.00000	1.00000	0.99999
25	1.00000	1.00000	1.00000	1.00000

$m = 13$

r \ n	13	14	15
2	0.00000	0.00000	0.00000
3	0.00000	0.00000	0.00000
4	0.00003	0.00002	0.00001
5	0.00018	0.00011	0.00006
6	0.00102	0.00062	0.00038
7	0.00381	0.00242	0.00156
8	0.01312	0.00869	0.00584
9	0.03406	0.02359	0.01653
10	0.08118	0.05888	0.04300
11	0.15657	0.11887	0.09064
12	0.27719	0.22051	0.17534
13	0.41791	0.34755	0.28826
14	0.58209	0.50565	0.43648
15	0.72281	0.65245	0.58470
16	0.84343	0.78796	0.72989
17	0.91882	0.88113	0.83878
18	0.96594	0.94465	0.91818
19	0.98688	0.97641	0.96229
20	0.99619	0.99209	0.98582
21	0.99898	0.99758	0.99523
22	0.99982	0.99947	0.99876
23	0.99997	0.99989	0.99972
24	1.00000	0.99999	0.99996
25	1.00000	1.00000	0.99999
26	1.00000	1.00000	1.00000
27	1.00000	1.00000	1.00000

m = 14		
		n
r	14	15
2	0.00000	0.00000
3	0.00000	0.00000
4	0.00001	0.00001
5	0.00006	0.00003
6	0.00036	0.00022
7	0.00148	0.00092
8	0.00555	0.00360
9	0.01575	0.01065
10	0.04123	0.02911
11	0.08711	0.06417
12	0.16969	0.13061
13	0.27979	0.22474
14	0.42660	0.35762
15	0.57340	0.50000
16	0.72021	0.65187
17	0.83031	0.77526
18	0.91289	0.87492
19	0.95877	0.93583
20	0.98425	0.97274
21	0.99445	0.98935
22	0.99852	0.99673
23	0.99964	0.99908
24	0.99994	0.99981
25	0.99999	0.99997
26	1.00000	1.00000
27	1.00000	1.00000
28	1.00000	1.00000
29	1.00000	1.00000

m = 15	
	n
r	15
2	0.00000
3	0.00000
4	0.00000
5	0.00002
6	0.00013
7	0.00055
8	0.00226
9	0.00696
10	0.01988
11	0.04572
12	0.09739
13	0.17491
14	0.29118
15	0.42407
16	0.57593
17	0.70882
18	0.82509
19	0.90261
20	0.95428
21	0.98012
22	0.99304
23	0.99774
24	0.99945
25	0.99987
26	0.99998
27	1.00000
28	1.00000
29	1.00000
30	1.00000

Table A3.3   The Mann–Whitney distribution (Section 6.3)

This table gives the left-tail probabilities $\Pr(U_t \leqslant k)$ of the Mann–Whitney statistic U_t for sample sizes n and m such that $2 \leqslant n \leqslant m \leqslant 15$. By symmetry, the right-tail probabilities are given by

$$\Pr(U_t \geqslant k) = \Pr(U_t \leqslant nm - k)$$

$m = 2$

k \ n	2	3	4	5	6	7	8	9
0	0.16667	0.10000	0.06667	0.04762	0.03571	0.02778	0.02222	0.01818
1	0.33333	0.20000	0.13333	0.09524	0.07143	0.05556	0.04444	0.03636
2	0.66667	0.40000	0.26667	0.19048	0.14286	0.11111	0.08889	0.07273
3		0.60000	0.40000	0.28571	0.21429	0.16667	0.13333	0.10909
4			0.60000	0.42857	0.32143	0.25000	0.20000	0.16364
5				0.57143	0.42857	0.33333	0.26667	0.21818
6					0.57143	0.44444	0.35556	0.29091
7						0.55556	0.44444	0.36364
8							0.55556	0.45455
9								0.54545
10								
11								
12								
13								
14								
15								

$m = 2$

k \ n	10	11	12	13	14	15
0	0.01515	0.01282	0.01099	0.00952	0.00833	0.00735
1	0.03030	0.02564	0.02198	0.01905	0.01667	0.01471
2	0.06061	0.05128	0.04396	0.03810	0.03333	0.02941
3	0.09091	0.07692	0.06593	0.05714	0.05000	0.04412
4	0.13636	0.11538	0.09890	0.08571	0.07500	0.06618
5	0.18182	0.15385	0.13187	0.11429	0.10000	0.08824
6	0.24242	0.20513	0.17582	0.15238	0.13333	0.11765
7	0.30303	0.25641	0.21978	0.19048	0.16667	0.14706
8	0.37879	0.32051	0.27473	0.23810	0.20833	0.18382
9	0.45455	0.38462	0.32967	0.28571	0.25000	0.22059
10	0.54545	0.46154	0.39560	0.34286	0.30000	0.26471
11		0.53846	0.46154	0.40000	0.35000	0.30882
12			0.53846	0.46667	0.40833	0.36029
13				0.53333	0.46667	0.41176
14					0.53333	0.47059
15						0.52941

m = 3

k \ n	3	4	5	6	7	8	9	10
0	0.05000	0.02857	0.01786	0.01190	0.00833	0.00606	0.00455	0.00350
1	0.10000	0.05714	0.03571	0.02381	0.01667	0.01212	0.00909	0.00699
2	0.20000	0.11429	0.07143	0.04762	0.03333	0.02424	0.01818	0.01399
3	0.35000	0.20000	0.12500	0.08333	0.05833	0.04242	0.03182	0.02448
4	0.50000	0.31429	0.19643	0.13095	0.09167	0.06667	0.05000	0.03846
5		0.42857	0.28571	0.19048	0.13333	0.09697	0.07273	0.05594
6		0.57143	0.39286	0.27381	0.19167	0.13939	0.10455	0.08042
7			0.50000	0.35714	0.25833	0.18788	0.14091	0.10839
8				0.45238	0.33333	0.24848	0.18636	0.14336
9				0.54762	0.41667	0.31515	0.24091	0.18531
10					0.50000	0.38788	0.30000	0.23427
11						0.46061	0.36364	0.28671
12						0.53939	0.43182	0.34615
13							0.50000	0.40559
14								0.46853
15								0.53147
16								
17								
18								
19								
20								
21								
22								

m = 3

k \ n	11	12	13	14	15
0	0.00275	0.00220	0.00179	0.00147	0.00123
1	0.00549	0.00440	0.00357	0.00294	0.00245
2	0.01099	0.00879	0.00714	0.00588	0.00490
3	0.01923	0.01538	0.01250	0.01029	0.00858
4	0.03022	0.02418	0.01964	0.01618	0.01348
5	0.04396	0.03516	0.02857	0.02353	0.01961
6	0.06319	0.05055	0.04107	0.03382	0.02819
7	0.08516	0.06813	0.05536	0.04559	0.03799
8	0.11264	0.09011	0.07321	0.06029	0.05025
9	0.14560	0.11648	0.09464	0.07794	0.06495
10	0.18407	0.14725	0.11964	0.09853	0.08211
11	0.22802	0.18242	0.14821	0.12206	0.10172
12	0.27747	0.22418	0.18214	0.15000	0.12500
13	0.32967	0.26813	0.21964	0.18088	0.15074
14	0.38462	0.31648	0.26071	0.21618	0.18015
15	0.44231	0.36703	0.30536	0.25441	0.21324
16	0.50000	0.41978	0.35179	0.29559	0.24877
17		0.47253	0.40000	0.33824	0.28676
18		0.52747	0.45000	0.38382	0.32721
19			0.50000	0.42941	0.36887
20				0.47647	0.41176
21				0.52353	0.45588
22					0.50000

$m = 4$

k \ n	4	5	6	7	8	9	10	11	12	13	14	15
0	0.01429	0.00794	0.00476	0.00303	0.00202	0.00140	0.00100	0.00073	0.00055	0.00042	0.00033	0.00026
1	0.02857	0.01587	0.00952	0.00606	0.00404	0.00280	0.00200	0.00147	0.00110	0.00084	0.00065	0.00052
2	0.05714	0.03175	0.01905	0.01212	0.00808	0.00559	0.00400	0.00293	0.00220	0.00168	0.00131	0.00103
3	0.10000	0.05556	0.03333	0.02121	0.01414	0.00979	0.00699	0.00513	0.00385	0.00294	0.00229	0.00181
4	0.17143	0.09524	0.05714	0.03636	0.02424	0.01678	0.01199	0.00879	0.00659	0.00504	0.00392	0.00310
5	0.24286	0.14286	0.08571	0.05455	0.03636	0.02517	0.01798	0.01319	0.00989	0.00756	0.00588	0.00464
6	0.34286	0.20635	0.12857	0.08182	0.05455	0.03776	0.02697	0.01978	0.01484	0.01134	0.00882	0.00697
7	0.44286	0.27778	0.17619	0.11515	0.07677	0.05315	0.03796	0.02784	0.02088	0.01597	0.01242	0.00980
8	0.55714	0.36508	0.23810	0.15758	0.10707	0.07413	0.05295	0.03883	0.02912	0.02227	0.01732	0.01367
9		0.45238	0.30476	0.20606	0.14141	0.09930	0.07093	0.05201	0.03901	0.02983	0.02320	0.01832
10		0.54762	0.38095	0.26364	0.18384	0.13007	0.09391	0.06886	0.05165	0.03950	0.03072	0.02425
11			0.45714	0.32424	0.23030	0.16503	0.11988	0.08864	0.06648	0.05084	0.03954	0.03122
12			0.54286	0.39394	0.28485	0.20699	0.15185	0.11282	0.08516	0.06513	0.05065	0.03999
13				0.46364	0.34141	0.25175	0.18681	0.13993	0.10604	0.08151	0.06340	0.05005
14				0.53636	0.40404	0.30210	0.22677	0.17143	0.13077	0.10084	0.07876	0.06218
15					0.46667	0.35524	0.26973	0.20586	0.15824	0.12269	0.09608	0.07611
16					0.53333	0.41259	0.31768	0.24469	0.18956	0.14790	0.11634	0.09236
17						0.46993	0.36663	0.28571	0.22308	0.17521	0.13856	0.11042
18						0.53007	0.41958	0.33040	0.26044	0.20588	0.16373	0.13106
19							0.47253	0.37656	0.29945	0.23866	0.19085	0.15351
20							0.52747	0.42564	0.34176	0.27437	0.22092	0.17853
21								0.47473	0.38516	0.31176	0.25261	0.20537
22								0.52527	0.43077	0.35168	0.28693	0.23452
23									0.47637	0.39244	0.32255	0.26522
24									0.52363	0.43529	0.36046	0.29825
25										0.47815	0.39902	0.33230
26										0.52185	0.43922	0.36816
27											0.47941	0.40480
28											0.52059	0.44272
29												0.48065
30												0.51935

m = 5

k	n=5	6	7	8	9	10	11	12	13	14	15
0	0.00397	0.00216	0.00126	0.00078	0.00050	0.00033	0.00023	0.00016	0.00012	0.00009	0.00006
1	0.00794	0.00433	0.00253	0.00155	0.00100	0.00067	0.00046	0.00032	0.00023	0.00017	0.00013
2	0.01587	0.00866	0.00505	0.00311	0.00200	0.00133	0.00092	0.00065	0.00047	0.00034	0.00026
3	0.02778	0.01515	0.00884	0.00544	0.00350	0.00233	0.00160	0.00113	0.00082	0.00060	0.00045
4	0.04762	0.02597	0.01515	0.00932	0.00599	0.00400	0.00275	0.00194	0.00140	0.00103	0.00077
5	0.07540	0.04113	0.02399	0.01476	0.00949	0.00633	0.00435	0.00307	0.00222	0.00163	0.00123
6	0.11111	0.06277	0.03662	0.02253	0.01449	0.00966	0.00664	0.00469	0.00338	0.00249	0.00187
7	0.15476	0.08874	0.05303	0.03263	0.02098	0.01399	0.00962	0.00679	0.00490	0.00361	0.00271
8	0.21032	0.12338	0.07449	0.04662	0.02997	0.01998	0.01374	0.00970	0.00700	0.00516	0.00387
9	0.27381	0.16450	0.10101	0.06371	0.04146	0.02764	0.01900	0.01341	0.00969	0.00714	0.00535
10	0.34524	0.21429	0.13384	0.08547	0.05594	0.03763	0.02587	0.01826	0.01319	0.00972	0.00729
11	0.42063	0.26840	0.17172	0.11111	0.07343	0.04962	0.03434	0.02424	0.01751	0.01290	0.00967
12	0.50000	0.33117	0.21591	0.14219	0.09491	0.06460	0.04487	0.03184	0.02299	0.01694	0.01271
13		0.39610	0.26515	0.17716	0.11988	0.08225	0.05746	0.04089	0.02965	0.02184	0.01638
14		0.46537	0.31944	0.21756	0.14885	0.10323	0.07257	0.05187	0.03770	0.02786	0.02090
15		0.53463	0.37753	0.26185	0.18182	0.12721	0.09020	0.06480	0.04727	0.03500	0.02632
16			0.43813	0.31080	0.21878	0.15485	0.11058	0.07999	0.05859	0.04352	0.03277
17			0.50000	0.36208	0.25924	0.18548	0.13370	0.09729	0.07166	0.05341	0.04031
18				0.41647	0.30320	0.21978	0.15980	0.11716	0.08672	0.06493	0.04915
19				0.47164	0.34965	0.25674	0.18864	0.13930	0.10376	0.07800	0.05928
20				0.52836	0.39860	0.29704	0.22047	0.16419	0.12302	0.09297	0.07088
21					0.44905	0.33933	0.25481	0.19134	0.14437	0.10965	0.08398
22					0.50000	0.38395	0.29167	0.22107	0.16795	0.12831	0.09868
23						0.42957	0.33059	0.25291	0.19363	0.14878	0.11500
24						0.47652	0.37134	0.28717	0.22152	0.17131	0.13306
25						0.52348	0.41346	0.32304	0.25140	0.19565	0.15280
26							0.45650	0.36070	0.28315	0.22196	0.17428
27							0.50000	0.39948	0.31653	0.24991	0.19743
28								0.43940	0.35142	0.27967	0.22227
29								0.47964	0.38749	0.31080	0.24865
30								0.52036	0.42449	0.34340	0.27657
31									0.46207	0.37693	0.30579
32									0.50000	0.41151	0.33624
33										0.44659	0.36771
34										0.48220	0.40003
35										0.51780	0.43299
36											0.46640
37											0.50000

$m = 6$

k \ n	6	7	8	9	10	11	12	13
0	0.00108	0.00058	0.00033	0.00020	0.00012	0.00008	0.00005	0.00004
1	0.00216	0.00117	0.00067	0.00040	0.00025	0.00016	0.00011	0.00007
2	0.00433	0.00233	0.00133	0.00080	0.00050	0.00032	0.00022	0.00015
3	0.00758	0.00408	0.00233	0.00140	0.00087	0.00057	0.00038	0.00026
4	0.01299	0.00699	0.00400	0.00240	0.00150	0.00097	0.00065	0.00044
5	0.02056	0.01107	0.00633	0.00380	0.00237	0.00154	0.00102	0.00070
6	0.03247	0.01748	0.00999	0.00599	0.00375	0.00242	0.00162	0.00111
7	0.04654	0.02564	0.01465	0.00879	0.00549	0.00356	0.00237	0.00162
8	0.06602	0.03671	0.02131	0.01279	0.00799	0.00517	0.00345	0.00236
9	0.08983	0.05070	0.02964	0.01798	0.01124	0.00727	0.00485	0.00332
10	0.12013	0.06876	0.04063	0.02478	0.01561	0.01010	0.00673	0.00461
11	0.15476	0.09033	0.05395	0.03317	0.02098	0.01366	0.00910	0.00623
12	0.19697	0.11713	0.07093	0.04396	0.02797	0.01826	0.01223	0.00837
13	0.24242	0.14744	0.09058	0.05674	0.03634	0.02384	0.01600	0.01098
14	0.29437	0.18298	0.11422	0.07233	0.04670	0.03079	0.02074	0.01426
15	0.34957	0.22261	0.14119	0.09051	0.05894	0.03911	0.02645	0.01824
16	0.40909	0.26690	0.17249	0.11189	0.07355	0.04913	0.03340	0.02311
17	0.46861	0.31410	0.20679	0.13606	0.09028	0.06076	0.04153	0.02886
18	0.53139	0.36538	0.24542	0.16384	0.10989	0.07450	0.05123	0.03575
19		0.41783	0.28638	0.19421	0.13174	0.09009	0.06232	0.04371
20		0.47261	0.33100	0.22797	0.15659	0.10803	0.07525	0.05304
21		0.52739	0.37729	0.26434	0.18382	0.12807	0.08985	0.06369
22			0.42591	0.30350	0.21391	0.15053	0.10644	0.07589
23			0.47486	0.34446	0.24613	0.17510	0.12481	0.08956
24			0.52514	0.38781	0.28109	0.20217	0.14539	0.10501
25				0.43197	0.31768	0.23117	0.16774	0.12203
26				0.47732	0.35639	0.26244	0.19225	0.14090
27				0.52268	0.39623	0.29541	0.21854	0.16143
28					0.43744	0.33024	0.24682	0.18381
29					0.47890	0.36627	0.27661	0.20776
30					0.52110	0.40377	0.30823	0.23352
31						0.44182	0.34098	0.26069
32						0.48061	0.37513	0.28944
33						0.51939	0.41004	0.31940
34							0.44581	0.35062
35							0.48179	0.38268
36							0.51821	0.41571
37								0.44914
38								0.48305
39								0.51695
40								
41								
42								
43								
44								
45								

m = 6 Continued

k \ n	14	15
0	0.00003	0.00002
1	0.00005	0.00004
2	0.00010	0.00007
3	0.00018	0.00013
4	0.00031	0.00022
5	0.00049	0.00035
6	0.00077	0.00055
7	0.00114	0.00081
8	0.00165	0.00118
9	0.00232	0.00166
10	0.00322	0.00230
11	0.00436	0.00311
12	0.00586	0.00418
13	0.00769	0.00549
14	0.01001	0.00715
15	0.01282	0.00918
16	0.01628	0.01167
17	0.02038	0.01463
18	0.02534	0.01823
19	0.03109	0.02243
20	0.03787	0.02740
21	0.04567	0.03315
22	0.05467	0.03982
23	0.06483	0.04742
24	0.07642	0.05611
25	0.08929	0.06586
26	0.10372	0.07685
27	0.11956	0.08903
28	0.13702	0.10255
29	0.15593	0.11735
30	0.17652	0.13361
31	0.19850	0.15115
32	0.22208	0.17015
33	0.24698	0.19046
34	0.27332	0.21217
35	0.30077	0.23509
36	0.32949	0.25934
37	0.35903	0.28465
38	0.38953	0.31109
39	0.42056	0.33842
40	0.45219	0.36665
41	0.48398	0.39549
42	0.51602	0.42500
43		0.45480
44		0.48493
45		0.51507

m = 7

k \ n	7	8	9	10	11	12	13	14
0	0.00029	0.00016	0.00009	0.00005	0.00003	0.00002	0.00001	0.00001
1	0.00058	0.00031	0.00017	0.00010	0.00006	0.00004	0.00003	0.00002
2	0.00117	0.00062	0.00035	0.00021	0.00013	0.00008	0.00005	0.00003
3	0.00204	0.00109	0.00061	0.00036	0.00022	0.00014	0.00009	0.00006
4	0.00350	0.00186	0.00105	0.00062	0.00038	0.00024	0.00015	0.00010
5	0.00554	0.00295	0.00166	0.00098	0.00060	0.00038	0.00025	0.00016
6	0.00874	0.00466	0.00262	0.00154	0.00094	0.00060	0.00039	0.00026
7	0.01311	0.00699	0.00393	0.00231	0.00141	0.00089	0.00058	0.00039
8	0.01894	0.01026	0.00577	0.00339	0.00207	0.00131	0.00085	0.00057
9	0.02652	0.01445	0.00822	0.00483	0.00295	0.00187	0.00121	0.00081
10	0.03642	0.02005	0.01145	0.00679	0.00415	0.00262	0.00170	0.00114
11	0.04866	0.02704	0.01556	0.00926	0.00569	0.00359	0.00233	0.00156
12	0.06410	0.03605	0.02089	0.01249	0.00770	0.00488	0.00317	0.00212
13	0.08246	0.04693	0.02745	0.01651	0.01021	0.00649	0.00423	0.00282
14	0.10431	0.06030	0.03558	0.02154	0.01339	0.00853	0.00557	0.00372
15	0.12966	0.07599	0.04537	0.02766	0.01728	0.01105	0.00724	0.00484
16	0.15880	0.09464	0.05708	0.03512	0.02206	0.01417	0.00930	0.00623
17	0.19143	0.11593	0.07080	0.04391	0.02778	0.01792	0.01180	0.00793
18	0.22786	0.14048	0.08689	0.05440	0.03463	0.02247	0.01485	0.01000
19	0.26748	0.16783	0.10524	0.06654	0.04267	0.02782	0.01847	0.01248
20	0.31002	0.19845	0.12614	0.08063	0.05210	0.03417	0.02278	0.01545
21	0.35519	0.23170	0.14956	0.09662	0.06297	0.04156	0.02784	0.01894
22	0.40239	0.26791	0.17552	0.11477	0.07545	0.05013	0.03375	0.02305
23	0.45076	0.30629	0.20393	0.13492	0.08956	0.05992	0.04056	0.02781
24	0.50000	0.34716	0.23488	0.15739	0.10546	0.07111	0.04840	0.03334
25		0.38943	0.26801	0.18192	0.12315	0.08367	0.05731	0.03966
26		0.43326	0.30323	0.20866	0.14272	0.09778	0.06740	0.04689
27		0.47754	0.34030	0.23735	0.16415	0.11342	0.07872	0.05504
28		0.52246	0.37885	0.26810	0.18747	0.13071	0.09134	0.06423
29			0.41853	0.30044	0.21258	0.14956	0.10530	0.07448
30			0.45909	0.33453	0.23947	0.17012	0.12068	0.08588
31			0.50000	0.36981	0.26801	0.19223	0.13746	0.09843
32				0.40626	0.29808	0.21598	0.15569	0.11224
33				0.44339	0.32950	0.24121	0.17535	0.12726
34				0.48113	0.36212	0.26792	0.19644	0.14358
35				0.51887	0.39571	0.29590	0.21890	0.16115
36					0.43005	0.32518	0.24271	0.18003
37					0.46490	0.35542	0.26778	0.20013
38					0.50000	0.38662	0.29401	0.22148
39						0.41847	0.32132	0.24398
40						0.45090	0.34959	0.26763
41						0.48357	0.37865	0.29229
42						0.51643	0.40840	0.31794
43							0.43865	0.34441
44							0.46923	0.37166
45							0.50000	0.39950
46								0.42787
47								0.45656
48								0.48551
49								0.51449
50								
51								
52								

m = 7 Continued

k \ n	15
0	0.00001
1	0.00001
2	0.00002
3	0.00004
4	0.00007
5	0.00011
6	0.00018
7	0.00026
8	0.00039
9	0.00055
10	0.00077
11	0.00106
12	0.00144
13	0.00192
14	0.00254
15	0.00331
16	0.00426
17	0.00543
18	0.00686
19	0.00858
20	0.01064
21	0.01309
22	0.01597
23	0.01934
24	0.02326
25	0.02778
26	0.03296
27	0.03886
28	0.04554
29	0.05305
30	0.06147
31	0.07083
32	0.08118
33	0.09257
34	0.10503
35	0.11860
36	0.13330
37	0.14912
38	0.16610
39	0.18420
40	0.20343
41	0.22374
42	0.24510
43	0.26746
44	0.29075
45	0.31490
46	0.33982
47	0.36543
48	0.39163
49	0.41830
50	0.44533
51	0.47261
52	0.50000

m = 8

k \ n	8	9	10	11	12	13	14	15
0	0.00008	0.00004	0.00002	0.00001	0.00001	0.00000	0.00000	0.00000
1	0.00016	0.00008	0.00005	0.00003	0.00002	0.00001	0.00001	0.00000
2	0.00031	0.00016	0.00009	0.00005	0.00003	0.00002	0.00001	0.00001
3	0.00054	0.00029	0.00016	0.00009	0.00006	0.00003	0.00002	0.00001
4	0.00093	0.00049	0.00027	0.00016	0.00010	0.00006	0.00004	0.00002
5	0.00148	0.00078	0.00043	0.00025	0.00015	0.00009	0.00006	0.00004
6	0.00233	0.00123	0.00069	0.00040	0.00024	0.00015	0.00009	0.00006
7	0.00350	0.00185	0.00103	0.00060	0.00036	0.00022	0.00014	0.00009
8	0.00521	0.00276	0.00153	0.00089	0.00053	0.00033	0.00021	0.00014
9	0.00738	0.00395	0.00219	0.00127	0.00076	0.00047	0.00030	0.00020
10	0.01033	0.00555	0.00311	0.00180	0.00108	0.00067	0.00043	0.00028
11	0.01406	0.00761	0.00427	0.00249	0.00149	0.00092	0.00059	0.00038
12	0.01896	0.01032	0.00583	0.00340	0.00205	0.00127	0.00081	0.00053
13	0.02494	0.01370	0.00777	0.00455	0.00275	0.00171	0.00109	0.00071
14	0.03248	0.01798	0.01026	0.00603	0.00365	0.00227	0.00145	0.00094
15	0.04149	0.02320	0.01332	0.00787	0.00478	0.00298	0.00190	0.00124
16	0.05245	0.02962	0.01714	0.01017	0.00620	0.00387	0.00248	0.00162
17	0.06519	0.03723	0.02171	0.01297	0.00793	0.00497	0.00318	0.00208
18	0.08026	0.04636	0.02726	0.01638	0.01007	0.00632	0.00406	0.00266
19	0.09744	0.05697	0.03380	0.02044	0.01262	0.00796	0.00513	0.00337
20	0.11725	0.06940	0.04157	0.02531	0.01571	0.00995	0.00642	0.00423
21	0.13932	0.08359	0.05055	0.03101	0.01935	0.01231	0.00797	0.00526
22	0.16410	0.09979	0.06099	0.03771	0.02367	0.01512	0.00983	0.00650
23	0.19114	0.11794	0.07286	0.04542	0.02869	0.01841	0.01201	0.00797
24	0.22090	0.13830	0.08641	0.05433	0.03455	0.02228	0.01459	0.00971
25	0.25268	0.16063	0.10154	0.06442	0.04125	0.02674	0.01758	0.01174
26	0.28687	0.18519	0.11849	0.07588	0.04894	0.03191	0.02107	0.01411
27	0.32269	0.21172	0.13714	0.08869	0.05763	0.03780	0.02508	0.01686
28	0.36045	0.24035	0.15771	0.10300	0.06747	0.04452	0.02969	0.02003
29	0.39922	0.27071	0.17997	0.11877	0.07843	0.05210	0.03491	0.02365
30	0.43924	0.30292	0.20412	0.13614	0.09067	0.06063	0.04085	0.02779
31	0.47956	0.33649	0.22988	0.15502	0.10414	0.07013	0.04751	0.03247
32	0.52044	0.37149	0.25739	0.17553	0.11899	0.08070	0.05499	0.03775
33		0.40740	0.28630	0.19752	0.13514	0.09233	0.06330	0.04367
34		0.44418	0.31672	0.22107	0.15270	0.10513	0.07252	0.05028
35		0.48128	0.34823	0.24601	0.17157	0.11907	0.08266	0.05761
36		0.51872	0.38091	0.27238	0.19187	0.13423	0.09380	0.06573
37			0.41428	0.29994	0.21343	0.15056	0.10593	0.07464
38			0.44838	0.32870	0.23633	0.16814	0.11912	0.08441
39			0.48270	0.35839	0.26040	0.18688	0.13335	0.09506
40			0.51730	0.38899	0.28568	0.20684	0.14867	0.10662
41				0.42019	0.31194	0.22790	0.16504	0.11910
42				0.45195	0.33922	0.25011	0.18252	0.13255
43				0.48392	0.36725	0.27332	0.20101	0.14693
44				0.51608	0.39604	0.29753	0.22057	0.16230
45					0.42532	0.32260	0.24109	0.17861
46					0.45505	0.34850	0.26259	0.19589
47					0.48496	0.37505	0.28495	0.21407
48					0.51504	0.40222	0.30817	0.23318
49						0.42981	0.33212	0.25313
50						0.45777	0.35677	0.27393
51						0.48589	0.38197	0.29548
52						0.51411	0.40770	0.31776
53							0.43378	0.34067

m = 8 Continued

k \ n	8	9	10	11	12	13	14	15
54							0.46017	0.36418
55							0.48669	0.38817
56							0.51331	0.41259
57								0.43732
58								0.46230
59								0.48741
60								0.51259

m = 9

k \ n	9	10	11	12	13	14	15
0	0.00002	0.00001	0.00001	0.00000	0.00000	0.00000	0.00000
1	0.00004	0.00002	0.00001	0.00001	0.00000	0.00000	0.00000
2	0.00008	0.00004	0.00002	0.00001	0.00001	0.00000	0.00000
3	0.00014	0.00008	0.00004	0.00002	0.00001	0.00001	0.00001
4	0.00025	0.00013	0.00007	0.00004	0.00002	0.00001	0.00001
5	0,00039	0.00021	0.00011	0.00006	0.00004	0.00002	0.00001
6	0.00062	0.00032	0.00018	0.00010	0.00006	0.00004	0.00002
7	0.00093	0.00049	0.00027	0.00015	0.00009	0.00006	0.00003
8	0.00138	0.00073	0.00040	0.00023	0.00013	0.00008	0.00005
9	0.00200	0.00105	0.00058	0.00033	0.00020	0.00012	0.00007
10	0.00282	0.00149	0.00082	0.00047	0.00028	0.00017	0.00011
11	0.00389	0.00207	0.00114	0.00065	0.00039	0.00023	0.00015
12	0.00531	0.00284	0.00157	0.00090	0.00053	0.00032	0.00020
13	0.00710	0.00381	0.00212	0.00122	0.00072	0.00044	0.00027
14	0.00938	0.00507	0.00283	0.00163	0.00097	0.00059	0.00037
15	0.01222	0.00664	0.00372	0.00215	0.00128	0.00078	0.00049
16	0.01573	0.00861	0.00485	0.00281	0.00167	0.00102	0.00064
17	0.01999	0.01101	0.00623	0.00363	0.00217	0.00133	0.00083
18	0.02515	0.01396	0.00795	0.00464	0.00278	0.00171	0.00107
19	0.03126	0.01749	0.01002	0.00588	0.00353	0.00217	0.00137
20	0.03850	0.02174	0.01253	0.00739	0.00445	0.00275	0.00173
21	0.04696	0.02674	0.01552	0.00919	0.00557	0.00344	0.00218
22	0.05675	0.03263	0.01906	0.01136	0.00690	0.00428	0.00271
23	0.06796	0.03945	0.02323	0.01391	0.00849	0.00529	0.00336
24	0.08075	0.04736	0.02810	0.01693	0.01038	0.00649	0.00413
25	0.09513	0.05638	0.03372	0.02045	0.01260	0.00790	0.00504
26	0.11121	0.06665	0.04020	0.02454	0.01520	0.00957	0.00612
27	0.12904	0.07820	0.04759	0.02924	0.01821	0.01151	0.00739
28	0.14866	0.09116	0.05597	0.03464	0.02169	0.01377	0.00887
29	0.17005	0.10551	0.06540	0.04077	0.02568	0.01637	0.01059
30	0.19325	0.12140	0.07597	0.04773	0.03023	0.01937	0.01257
31	0.21814	0.13876	0.08771	0.05553	0.03540	0.02279	0.01485
32	0.24471	0.15769	0.10068	0.06427	0.04123	0.02669	0.01746
33	0.27285	0.17812	0.11493	0.07397	0.04778	0.03109	0.02043
34	0.30241	0.20009	0.13049	0.08471	0.05509	0.03605	0.02379
35	0.33324	0.22348	0.14736	0.09650	0.06321	0.04159	0.02758
36	0.36522	0.24835	0.16558	0.10942	0.07219	0.04779	0.03184
37	0.39809	0.27448	0.18511	0.12345	0.08206	0.05465	0.03659
38	0.43165	0.30189	0.20595	0.13866	0.09288	0.06224	0.04189
39	0.46571	0.33036	0.22805	0.15501	0.10465	0.07058	0.04775

m = 9 Continued

k \ n	9	10	11	12	13	14	15
40	0.50000	0.35985	0.25136	0.17255	0.11743	0.07972	0.05423
41		0.39009	0.27582	0.19122	0.13121	0.08968	0.06134
42		0.42105	0.30134	0.21105	0.14603	0.10050	0.06914
43		0.45241	0.32782	0.23196	0.16187	0.11219	0.07763
44		0.48412	0.35515	0.25394	0.17874	0.12478	0.08686
45		0.51588	0.38322	0.27690	0.19663	0.13828	0.09685
46			0.41188	0.30080	0.21552	0.15270	0.10761
47			0.44099	0.32553	0.23536	0.16803	0.11918
48			0.47042	0.35104	0.25614	0.18429	0.13156
49			0.50000	0.37719	0.27779	0.20144	0.14475
50				0.40390	0.30026	0.21947	0.15877
51				0.43103	0.32349	0.23835	0.17362
52				0.45850	0.34738	0.25806	0.18928
53				0.48614	0.37188	0.27854	0.20574
54				0.51386	0.39688	0.29976	0.22299
55					0.42229	0.32163	0.24100
56					0.44802	0.34413	0.25974
57					0.47396	0.36715	0.27918
58					0.50000	0.39066	0.29926
59						0.41454	0.31995
60						0.43874	0.34119
61						0.46315	0.36293
62						0.48771	0.38509
63						0.51229	0.40763
64							0.43046
65							0.45351
66							0.47672
67							0.50000

m = 10

k \ n	10	11	12	13	14	15
0	0.00001	0.00000	0.00000	0.00000	0.00000	0.00000
1	0.00001	0.00001	0.00000	0.00000	0.00000	0.00000
2	0.00002	0.00001	0.00001	0.00000	0.00000	0.00000
3	0.00004	0.00002	0.00001	0.00001	0.00000	0.00000
4	0.00006	0.00003	0.00002	0.00001	0.00001	0.00000
5	0.00010	0.00005	0.00003	0.00002	0.00001	0.00001
6	0.00016	0.00009	0.00005	0.00003	0.00002	0.00001
7	0.00024	0.00013	0.00007	0.00004	0.00002	0.00001
8	0.00036	0.00019	0.00010	0.00006	0.00003	0.00002
9	0.00053	0.00028	0.00015	0.00008	0.00005	0.00003
10	0.00075	0.00039	0.00021	0.00012	0.00007	0.00004
11	0.00104	0.00055	0.00030	0.00017	0.00010	0.00006
12	0.00144	0.00076	0.00042	0.00024	0.00014	0.00008
13	0.00194	0.00103	0.00056	0.00032	0.00019	0.00011
14	0.00260	0.00138	0.00076	0.00043	0.00025	0.00015
15	0.00342	0.00183	0.00101	0.00057	0.00033	0.00020
16	0.00447	0.00239	0.00132	0.00075	0.00044	0.00027
17	0.00575	0.00310	0.00172	0.00098	0.00058	0.00035

m = 10 *Continued*

k \ n	10	11	12	13	14	15
18	0.00734	0.00397	0.00221	0.00127	0.00075	0.00045
19	0.00927	0.00505	0.00282	0.00162	0.00096	0.00058
20	0.01162	0.00636	0.00357	0.00206	0.00122	0.00074
21	0.01440	0.00794	0.00448	0.00259	0.00153	0.00093
22	0.01773	0.00983	0.00558	0.00324	0.00192	0.00117
23	0.02163	0.01208	0.00688	0.00401	0.00239	0.00145
24	0.02621	0.01474	0.00845	0.00494	0.00295	0.00180
25	0.03151	0.01785	0.01029	0.00605	0.00363	0.00222
26	0.03763	0.02148	0.01246	0.00736	0.00443	0.00271
27	0.04460	0.02567	0.01498	0.00889	0.00537	0.00330
28	0.05256	0.03050	0.01791	0.01068	0.00648	0.00400
29	0.06150	0.03600	0.02129	0.01276	0.00777	0.00481
30	0.07157	0.04226	0.02516	0.01516	0.00927	0.00576
31	0.08275	0.04931	0.02957	0.01792	0.01101	0.00686
32	0.09516	0.05723	0.03458	0.02108	0.01301	0.00814
33	0.10878	0.06606	0.04021	0.02466	0.01529	0.00961
34	0.12373	0.07587	0.04654	0.02873	0.01790	0.01129
35	0.13993	0.08668	0.05360	0.03330	0.02086	0.01321
36	0.15750	0.09856	0.06146	0.03843	0.02420	0.01539
37	0.17634	0.11151	0.07012	0.04415	0.02796	0.01786
38	0.19652	0.12560	0.07966	0.05051	0.03216	0.02064
39	0.21794	0.14081	0.09010	0.05753	0.03685	0.02376
40	0.24063	0.15719	0.10148	0.06528	0.04206	0.02725
41	0.26442	0.17468	0.11381	0.07376	0.04782	0.03114
42	0.28937	0.19334	0.12715	0.08303	0.05417	0.03546
43	0.31526	0.21308	0.14147	0.09310	0.06113	0.04023
44	0.34211	0.23393	0.15681	0.10402	0.06875	0.04549
45	0.36968	0.25578	0.17314	0.11579	0.07704	0.05125
46	0.39797	0.27863	0.19049	0.12844	0.08604	0.05757
47	0.42671	0.30237	0.20879	0.14196	0.09577	0.06444
48	0.45590	0.32695	0.22808	0.15639	0.10625	0.07192
49	0.48526	0.35225	0.24825	0.17170	0.11749	0.08001
50	0.51474	0.37822	0.26933	0.18791	0.12953	0.08875
51		0.40470	0.29121	0.20498	0.14234	0.09814
52		0.43163	0.31387	0.22290	0.15596	0.10821
53		0.45885	0.33720	0.24165	0.17037	0.11898
54		0.48627	0.36117	0.26119	0.18558	0.13045
55		0.51373	0.38565	0.28147	0.20156	0.14262
56			0.41060	0.30246	0.21831	0.15552
57			0.43588	0.32408	0.23579	0.16913
58			0.46143	0.34630	0.25400	0.18345
59			0.48712	0.36903	0.27288	0.19847
60			0.51288	0.39222	0.29242	0.21419
61				0.41577	0.31255	0.23057
62				0.43963	0.33324	0.24761
63				0.46369	0.35442	0.26527
64				0.48789	0.37605	0.28353
65				0.51211	0.39805	0.30234
66					0.42038	0.32167
67					0.44294	0.34148
68					0.46570	0.36171
69					0.48855	0.38233
70					0.51145	0.40327

$m = 10$ *Continued*

k \ n	10	11	12	13	14	15
71						0.42448
72						0.44591
73						0.46748
74						0.48915
75						0.51085

$m = 11$

k \ n	11	12	13	14	15
0	0.00000	0.00000	0.00000	0.00000	0.00000
1	0.00000	0.00000	0.00000	0.00000	0.00000
2	0.00001	0.00000	0.00000	0.00000	0.00000
3	0.00001	0.00001	0.00000	0.00000	0.00000
4	0.00002	0.00001	0.00000	0.00000	0.00000
5	0.00003	0.00001	0.00001	0.00000	0.00000
6	0.00004	0.00002	0.00001	0.00001	0.00000
7	0.00006	0.00003	0.00002	0.00001	0.00001
8	0.00009	0.00005	0.00003	0.00002	0.00001
9	0.00014	0.00007	0.00004	0.00002	0.00001
10	0.00020	0.00010	0.00006	0.00003	0.00002
11	0.00028	0.00014	0.00008	0.00004	0.00003
12	0.00038	0.00020	0.00011	0.00006	0.00004
13	0.00052	0.00027	0.00015	0.00008	0.00005
14	0.00070	0.00037	0.00020	0.00011	0.00006
15	0.00093	0.00049	0.00027	0.00015	0.00009
16	0.00122	0.00065	0.00035	0.00020	0.00011
17	0.00159	0.00084	0.00046	0.00026	0.00015
18	0.00205	0.00109	0.00060	0.00034	0.00020
19	0.00262	0.00140	0.00077	0.00044	0.00025
20	0.00332	0.00178	0.00098	0.00056	0.00032
21	0.00416	0.00224	0.00124	0.00071	0.00041
22	0.00519	0.00281	0.00156	0.00089	0.00052
23	0.00642	0.00349	0.00195	0.00111	0.00065
24	0.00788	0.00431	0.00241	0.00138	0.00081
25	0.00962	0.00529	0.00297	0.00171	0.00100
26	0.01165	0.00644	0.00364	0.00210	0.00123
27	0.01403	0.00780	0.00442	0.00256	0.00151
28	0.01680	0.00940	0.00535	0.00311	0.00184
29	0.01999	0.01125	0.00644	0.00375	0.00223
30	0.02365	0.01340	0.00771	0.00451	0.00269
31	0.02783	0.01587	0.00918	0.00539	0.00322
32	0.03258	0.01871	0.01088	0.00642	0.00385
33	0.03795	0.02194	0.01283	0.00760	0.00458
34	0.04397	0.02561	0.01506	0.00897	0.00542
35	0.05071	0.02975	0.01760	0.01053	0.00639
36	0.05820	0.03441	0.02048	0.01231	0.00750
37	0.06650	0.03962	0.02372	0.01434	0.00877
38	0.07564	0.04542	0.02738	0.01663	0.01021
39	0.08566	0.05185	0.03146	0.01921	0.01185
40	0.09659	0.05896	0.03601	0.02211	0.01370
41	0.10847	0.06677	0.04106	0.02536	0.01578

m = 11 Continued

k \ n	11	12	13	14	15
42	0.12132	0.07533	0.04665	0.02898	0.01811
43	0.13516	0.08466	0.05281	0.03299	0.02072
44	0.14999	0.09479	0.05956	0.03743	0.02363
45	0.16582	0.10575	0.06694	0.04233	0.02686
46	0.18265	0.11756	0.07498	0.04771	0.03043
47	0.20046	0.13022	0.08371	0.05360	0.03437
48	0.21923	0.14377	0.09314	0.06004	0.03870
49	0.23893	0.15819	0.10330	0.06703	0.04345
50	0.25952	0.17350	0.11421	0.07461	0.04863
51	0.28094	0.18967	0.12589	0.08280	0.05428
52	0.30316	0.20671	0.13833	0.09162	0.06041
53	0.32609	0.22457	0.15156	0.10109	0.06705
54	0.34968	0.24325	0.16557	0.11123	0.07422
55	0.37383	0.26270	0.18035	0.12204	0.08194
56	0.39847	0.28289	0.19591	0.13354	0.09022
57	0.42350	0.30376	0.21221	0.14574	0.09908
58	0.44883	0.32527	0.22926	0.15864	0.10854
59	0.47436	0.34735	0.24702	0.17223	0.11861
60	0.50000	0.36995	0.26546	0.18651	0.12929
61		0.39297	0.28454	0.20148	0.14059
62		0.41638	0.30423	0.21712	0.15253
63		0.44006	0.32449	0.23341	0.16509
64		0.46395	0.34526	0.25033	0.17827
65		0.48797	0.36648	0.26786	0.19208
66		0.51203	0.38811	0.28596	0.20650
67			0.41007	0.30460	0.22151
68			0.43231	0.32375	0.23711
69			0.45476	0.34335	0.25327
70			0.47734	0.36337	0.26998
71			0.50000	0.38375	0.28719
72				0.40445	0.30489
73				0.42541	0.32304
74				0.44657	0.34161
75				0.46788	0.36055
76				0.48929	0.37983
77				0.51071	0.39940
78					0.41922
79					0.43924
80					0.45941
81					0.47968
82					0.50000

m = 12

k \ n	12	13	14	15
0	0.00000	0.00000	0.00000	0.00000
1	0.00000	0.00000	0.00000	0.00000
2	0.00000	0.00000	0.00000	0.00000
3	0.00000	0.00000	0.00000	0.00000
4	0.00000	0.00000	0.00000	0.00000
5	0.00001	0.00000	0.00000	0.00000

m = 12 *Continued*

k \ n	12	13	14	15
6	0.00001	0.00001	0.00000	0.00000
7	0.00002	0.00001	0.00000	0.00000
8	0.00002	0.00001	0.00001	0.00000
9	0.00004	0.00002	0.00001	0.00001
10	0.00005	0.00003	0.00001	0.00001
11	0.00007	0.00004	0.00002	0.00001
12	0.00010	0.00005	0.00003	0.00002
13	0.00014	0.00007	0.00004	0.00002
14	0.00019	0.00010	0.00005	0.00003
15	0.00025	0.00013	0.00007	0.00004
16	0.00033	0.00017	0.00009	0.00005
17	0.00043	0.00023	0.00012	0.00007
18	0.00056	0.00029	0.00016	0.00009
19	0.00072	0.00038	0.00021	0.00011
20	0.00091	0.00048	0.00026	0.00015
21	0.00116	0.00062	0.00034	0.00019
22	0.00146	0.00078	0.00042	0.00024
23	0.00182	0.00097	0.00053	0.00030
24	0.00226	0.00121	0.00067	0.00038
25	0.00278	0.00150	0.00083	0.00047
26	0.00341	0.00184	0.00102	0.00058
27	0.00415	0.00225	0.00125	0.00071
28	0.00502	0.00274	0.00153	0.00087
29	0.00605	0.00331	0.00185	0.00106
30	0.00725	0.00399	0.00224	0.00128
31	0.00864	0.00478	0.00269	0.00155
32	0.01024	0.00570	0.00322	0.00186
33	0.01209	0.00676	0.00384	0.00222
34	0.01421	0.00799	0.00456	0.00264
35	0.01662	0.00939	0.00538	0.00313
36	0.01936	0.01100	0.00633	0.00370
37	0.02245	0.01283	0.00742	0.00435
38	0.02593	0.01491	0.00867	0.00510
39	0.02983	0.01726	0.01008	0.00596
40	0.03418	0.01990	0.01168	0.00694
41	0.03901	0.02286	0.01349	0.00804
42	0.04437	0.02616	0.01552	0.00929
43	0.05027	0.02984	0.01780	0.01071
44	0.05675	0.03392	0.02034	0.01229
45	0.06384	0.03842	0.02318	0.01407
46	0.07158	0.04338	0.02632	0.01606
47	0.07999	0.04882	0.02980	0.01827
48	0.08909	0.05477	0.03363	0.02072
49	0.09890	0.06126	0.03785	0.02343
50	0.10946	0.06831	0.04246	0.02643
51	0.12076	0.07594	0.04751	0.02972
52	0.13283	0.08417	0.05300	0.03334
53	0.14567	0.09304	0.05896	0.03729
54	0.15929	0.10254	0.06542	0.04161
55	0.17368	0.11271	0.07239	0.04631
56	0.18884	0.12355	0.07990	0.05140
57	0.20476	0.13506	0.08795	0.05691
58	0.22142	0.14727	0.09658	0.06286
59	0.23879	0.16016	0.10578	0.06927

m = 12 Continued

k \ n	12	13	14	15
60	0.25686	0.17375	0.11558	0.07615
61	0.27558	0.18801	0.12598	0.08351
62	0.29494	0.20295	0.13699	0.09138
63	0.31486	0.21855	0.14862	0.09976
64	0.33533	0.23480	0.16086	0.10868
65	0.35627	0.25166	0.17372	0.11813
66	0.37764	0.26912	0.18720	0.12812
67	0.39937	0.28715	0.20127	0.13867
68	0.42142	0.30571	0.21595	0.14977
69	0.44369	0.32476	0.23119	0.16143
70	0.46615	0.34426	0.24701	0.17364
71	0.48870	0.36417	0.26336	0.18641
72	0.51130	0.38444	0.28022	0.19972
73		0.40502	0.29758	0.21356
74		0.42586	0.31539	0.22793
75		0.44690	0.33363	0.24281
76		0.46808	0.35225	0.25818
77		0.48935	0.37122	0.27402
78		0.51065	0.39050	0.29032
79			0.41005	0.30704
80			0.42981	0.32415
81			0.44974	0.34163
82			0.46979	0.35945
83			0.48992	0.37756
84			0.51008	0.39594
85				0.41455
86				0.43334
87				0.45228
88				0.47133
89				0.49043
90				0.50957

m = 13			

k \ n	13	14	15
0	0.00000	0.00000	0.00000
1	0.00000	0.00000	0.00000
2	0.00000	0.00000	0.00000
3	0.00000	0.00000	0.00000
4	0.00000	0.00000	0.00000
5	0.00000	0.00000	0.00000
6	0.00000	0.00000	0.00000
7	0.00000	0.00000	0.00000
8	0.00001	0.00000	0.00000
9	0.00001	0.00000	0.00000
10	0.00001	0.00001	0.00000
11	0.00002	0.00001	0.00001
12	0.00003	0.00001	0.00001
13	0.00004	0.00002	0.00001
14	0.00005	0.00003	0.00001
15	0.00007	0.00003	0.00002
16	0.00009	0.00005	0.00002
17	0.00011	0.00006	0.00003
18	0.00015	0.00008	0.00004
19	0.00019	0.00010	0.00005
20	0.00025	0.00013	0.00007
21	0.00031	0.00017	0.00009
22	0.00040	0.00021	0.00011
23	0.00050	0.00026	0.00014
24	0.00062	0.00033	0.00018
25	0.00077	0.00041	0.00022
26	0.00096	0.00051	0.00028
27	0.00117	0.00063	0.00034
28	0.00143	0.00077	0.00042
29	0.00174	0.00094	0.00052
30	0.00211	0.00114	0.00063
31	0.00254	0.00137	0.00076
32	0.00304	0.00165	0.00092
33	0.00362	0.00198	0.00110
34	0.00430	0.00236	0.00132
35	0.00508	0.00280	0.00157
36	0.00599	0.00331	0.00186
37	0.00702	0.00390	0.00220
38	0.00820	0.00457	0.00259
39	0.00955	0.00535	0.00304
40	0.01107	0.00623	0.00356
41	0.01280	0.00724	0.00415
42	0.01474	0.00838	0.00482
43	0.01691	0.00966	0.00558
44	0.01935	0.01111	0.00644
45	0.02206	0.01273	0.00742
46	0.02507	0.01455	0.00851
47	0.02840	0.01658	0.00974
48	0.03208	0.01883	0.01112

m = 13 Continued			

k \ n	13	14	15
49	0.03613	0.02133	0.01265
50	0.04057	0.02409	0.01436
51	0.04542	0.02713	0.01625
52	0.05072	0.03047	0.01835
53	0.05647	0.03413	0.02065
54	0.06270	0.03814	0.02320
55	0.06943	0.04250	0.02599
56	0.07669	0.04724	0.02904
57	0.08449	0.05238	0.03238
58	0.09284	0.05794	0.03601
59	0.10177	0.06393	0.03996
60	0.11128	0.07038	0.04423
61	0.12139	0.07729	0.04886
62	0.13211	0.08469	0.05385
63	0.14343	0.09259	0.05922
64	0.15537	0.10101	0.06499
65	0.16792	0.10994	0.07117
66	0.18108	0.11941	0.07777
67	0.19484	0.12942	0.08481
68	0.20921	0.13998	0.09230
69	0.22415	0.15108	0.10026
70	0.23966	0.16274	0.10868
71	0.25572	0.17494	0.11758
72	0.27230	0.18769	0.12697
73	0.28938	0.20098	0.13685
74	0.30693	0.21480	0.14723
75	0.32492	0.22913	0.15810
76	0.34332	0.24397	0.16947
77	0.36207	0.25930	0.18134
78	0.38116	0.27509	0.19369
79	0.40052	0.29132	0.20652
80	0.42013	0.30798	0.21983
81	0.43992	0.32502	0.23360
82	0.45987	0.34242	0.24782
83	0.47991	0.36016	0.26248
84	0.50000	0.37819	0.27755
85		0.39648	0.29302
86		0.41499	0.30886
87		0.43369	0.32506
88		0.45253	0.34158
89		0.47148	0.35840
90		0.49049	0.37549
91		0.50951	0.39282
92			0.41036
93			0.42808
94			0.44593
95			0.46390
96			0.48193
97			0.50000

m = 14

k	14	15
0	0.00000	0.00000
1	0.00000	0.00000
2	0.00000	0.00000
3	0.00000	0.00000
4	0.00000	0.00000
5	0.00000	0.00000
6	0.00000	0.00000
7	0.00000	0.00000
8	0.00000	0.00000
9	0.00000	0.00000
10	0.00000	0.00000
11	0.00000	0.00000
12	0.00001	0.00000
13	0.00001	0.00000
14	0.00001	0.00001
15	0.00002	0.00001
16	0.00002	0.00001
17	0.00003	0.00002
18	0.00004	0.00002
19	0.00005	0.00003
20	0.00007	0.00003
21	0.00008	0.00004
22	0.00011	0.00006
23	0.00013	0.00007
24	0.00017	0.00009
25	0.00021	0.00011
26	0.00026	0.00014
27	0.00032	0.00017
28	0.00040	0.00021
29	0.00049	0.00026
30	0.00059	0.00031
31	0.00072	0.00038
32	0.00086	0.00046
33	0.00104	0.00056
34	0.00124	0.00067
35	0.00148	0.00080
36	0.00176	0.00095
37	0.00208	0.00113
38	0.00245	0.00134
39	0.00288	0.00158
40	0.00337	0.00185
41	0.00393	0.00217
42	0.00457	0.00253
43	0.00530	0.00295
44	0.00612	0.00342
45	0.00705	0.00395
46	0.00810	0.00456
47	0.00928	0.00524
48	0.01059	0.00601
49	0.01206	0.00687
50	0.01370	0.00784
51	0.01551	0.00892
52	0.01752	0.01012
53	0.01974	0.01146

m = 14 Continued

k	14	15
54	0.02218	0.01294
55	0.02487	0.01458
56	0.02781	0.01639
57	0.03102	0.01837
58	0.03452	0.02055
59	0.03833	0.02294
60	0.04246	0.02554
61	0.04693	0.02839
62	0.05176	0.03148
63	0.05696	0.03483
64	0.06255	0.03846
65	0.06853	0.04238
66	0.07494	0.04661
67	0.08178	0.05115
68	0.08906	0.05603
69	0.09679	0.06126
70	0.10499	0.06685
71	0.11366	0.07281
72	0.12282	0.07915
73	0.13246	0.05859
74	0.14259	0.09304
75	0.15321	0.10060
76	0.16433	0.10859
77	0.17594	0.11700
78	0.18805	0.12586
79	0.20063	0.13515
80	0.21369	0.14489
81	0.22722	0.15507
82	0.24120	0.16570
83	0.25562	0.17678
84	0.27046	0.18829
85	0.28571	0.20025
86	0.30134	0.21263
87	0.31733	0.22544
88	0.33367	0.23865
89	0.35031	0.25227
90	0.36723	0.26627
91	0.38441	0.28065
92	0.40181	0.29537
93	0.41940	0.31043
94	0.43714	0.32581
95	0.45501	0.34148
96	0.47297	0.35741
97	0.49098	0.37360
98	0.50902	0.39000
99		0.40659
100		0.42335
101		0.44024
102		0.45724
103		0.47431
104		0.49143
105		0.50857

$m = 15$			$m = 15$ *Continued*			$m = 15$ *Continued*	
k \ n	15		k \ n	15		k \ n	15
0	0.00000		53	0.00640		106	0.40317
1	0.00000		54	0.00726		107	0.41907
2	0.00000		55	0.00822		108	0.43510
3	0.00000		56	0.00928		109	0.45123
4	0.00000		57	0.01045		110	0.46744
5	0.00000		58	0.01175		111	0.48371
6	0.00000		59	0.01318		112	0.50000
7	0.00000		60	0.01475			
8	0.00000		61	0.01647			
9	0.00000		62	0.01836			
10	0.00000		63	0.02042			
11	0.00000		64	0.02267			
12	0.00000		65	0.02511			
13	0.00000		66	0.02776			
14	0.00000		67	0.03064			
15	0.00000		68	0.03375			
16	0.00001		69	0.03710			
17	0.00001		70	0.04071			
18	0.00001		71	0.04460			
19	0.00001		72	0.04876			
20	0.00002		73	0.05322			
21	0.00002		74	0.05799			
22	0.00003		75	0.06307			
23	0.00004		76	0.06849			
24	0.00004		77	0.07424			
25	0.00006		78	0.08034			
26	0.00007		79	0.08680			
27	0.00009		80	0.09363			
28	0.00011		81	0.10084			
29	0.00013		82	0.10843			
30	0.00016		83	0.11641			
31	0.00020		84	0.12478			
32	0.00024		85	0.13355			
33	0.00029		86	0.14272			
34	0.00035		87	0.15230			
35	0.00042		88	0.16227			
36	0.00050		89	0.17266			
37	0.00059		90	0.18344			
38	0.00070		91	0.19462			
39	0.00083		92	0.20619			
40	0.00098		93	0.21814			
41	0.00115		94	0.23048			
42	0.00135		95	0.24318			
43	0.00158		96	0.25624			
44	0.00184		97	0.26965			
45	0.00213		98	0.28339			
46	0.00247		99	0.29744			
47	0.00285		100	0.31180			
48	0.00328		101	0.32644			
49	0.00377		102	0.34134			
50	0.00432		103	0.35648			
51	0.00494		104	0.37185			
52	0.00563		105	0.38742			

Table A3.4 Spearman's S distribution (Section 7.2)
This table gives the left-tail probabilities $\Pr(S \leq k)$ of Spearman's S statistic for sample sizes
$n = 4, 5, ..., 10$. By symmetry, the right-tail probabilities are given by

$$\Pr(S \geq k) = \Pr(S \leq n(n^2 - 1)/3 - k)$$

n = 4		n = 5		n = 6		n = 7		n = 8	
S	P	S	P	S	P	S	P	S	P
0	0.04167	0	0.00833	0	0.00139	0	0.00020	0	0.00002
2	0.16667	2	0.04167	2	0.00833	2	0.00139	2	0.00020
4	0.20833	4	0.06667	4	0.01667	4	0.00337	4	0.00057
6	0.37500	6	0.11667	6	0.02917	6	0.00615	6	0.00112
8	0.45833	8	0.17500	8	0.05139	8	0.01190	8	0.00228
10	0.54167	10	0.22500	10	0.06806	10	0.01706	10	0.00362
		12	0.25833	12	0.08750	12	0.02401	12	0.00536
		14	0.34167	14	0.12083	14	0.03313	14	0.00769
		16	0.39167	16	0.14861	16	0.04405	16	0.01089
		18	0.47500	18	0.17778	18	0.05476	18	0.01396
		20	0.52500	20	0.20972	20	0.06944	20	0.01838
				22	0.24861	22	0.08333	22	0.02292
				24	0.28194	24	0.10000	24	0.02879
				26	0.32917	26	0.11786	26	0.03470
				28	0.35694	28	0.13333	28	0.04154
				30	0.40139	30	0.15119	30	0.04809
				32	0.45972	32	0.17679	32	0.05749
				34	0.50000	34	0.19782	34	0.06615
						36	0.22222	36	0.07557
						38	0.24881	38	0.08549
						40	0.27798	40	0.09831
						42	0.29742	42	0.10809
						44	0.33075	44	0.12153
						46	0.35655	46	0.13373
						48	0.39127	48	0.14960
						50	0.41984	50	0.16342
						52	0.45317	52	0.17994
						54	0.48175	54	0.19469
						56	0.51825	56	0.21394
								58	0.23090
								60	0.25040
								62	0.26820
								64	0.29107
								66	0.30955
								68	0.33229
								70	0.35166
								72	0.37602
								74	0.39650
								76	0.42006
								78	0.44100
								80	0.46744
								82	0.48839
								84	0.51161

(*continued*)

Table A3.4 Continued

n = 9				n = 10			
S	P	S	P	S	P	S	P
0	0.00000	82	0.20504	0	0.00000	82	0.07201
2	0.00002	84	0.21831	2	0.00000	84	0.07741
4	0.00008	86	0.23150	4	0.00001	86	0.08314
6	0.00018	88	0.24667	6	0.00002	88	0.08931
8	0.00037	90	0.26032	8	0.00005	90	0.09562
10	0.00066	92	0.27585	10	0.00010	92	0.10222
12	0.00101	94	0.29047	12	0.00017	94	0.10910
14	0.00154	96	0.30670	14	0.00027	96	0.11627
16	0.00225	98	0.32182	16	0.00040	98	0.12374
18	0.00304	100	0.33887	18	0.00057	100	0.13158
20	0.00413	102	0.35403	20	0.00080	102	0.13941
22	0.00538	104	0.37177	22	0.00109	104	0.14783
24	0.00691	106	0.38781	24	0.00143	106	0.15645
26	0.00861	108	0.40499	26	0.00185	108	0.16520
28	0.01069	110	0.42159	28	0.00236	110	0.17437
30	0.01275	112	0.44005	30	0.00291	112	0.18395
32	0.01556	114	0.45581	32	0.00361	114	0.19345
34	0.01843	116	0.47420	34	0.00439	116	0.20346
36	0.02163	118	0.49078	36	0.00527	118	0.21354
38	0.02516	120	0.50922	38	0.00628	120	0.22414
40	0.02944			40	0.00746	122	0.23487
42	0.03328			42	0.00870	124	0.24588
44	0.03802			44	0.01012	126	0.25674
46	0.04286			46	0.01168	128	0.26833
48	0.04840			48	0.01338	130	0.28014
50	0.05400			50	0.01528	132	0.29183
52	0.06029			52	0.01733	134	0.30366
54	0.06639			54	0.01948	136	0.31607
56	0.07376			56	0.02190	138	0.32836
58	0.08090			58	0.02449	140	0.34104
60	0.08883			60	0.02722	142	0.35358
62	0.09690			62	0.03016	144	0.36650
64	0.10626			64	0.03337	146	0.37947
66	0.11491			66	0.03671	148	0.39251
68	0.12496			68	0.04030	150	0.40564
70	0.13479			70	0.04414	152	0.41911
72	0.14560			72	0.04814	154	0.43257
74	0.15625			74	0.05244	156	0.44581
76	0.16813			76	0.05693	158	0.45919
78	0.17929			78	0.06155	160	0.47299
80	0.19266			80	0.06670	162	0.48648
						164	0.50000

Table A3.5   Kendall's rank correlation coefficient distribution (Section 7.3)
This table gives the right-tail probabilities $\Pr(R_K \geq x)$ of Kendall's rank correlation coefficient R_K for sample sizes $n = 4, 5, ..., 20$. By symmetry, the left-tail probabilities are given by

$$\Pr(R_K \leq -x) = \Pr(R_K \geq x)$$

	x	$\Pr(R_K \geq x)$		x	$\Pr(R_K \geq x)$		x	$\Pr(R_K \geq x)$
$n=4$	0.00000	0.62500	$n=9$	0.00000	0.54027	$n=11$	0.01818	0.50000
	0.33333	0.37500		0.05556	0.45973		0.05455	0.43963
	0.66667	0.16667		0.11111	0.38071		0.09091	0.38058
	1.00000	0.04167		0.16667	0.30610		0.12727	0.32405
				0.22222	0.23835		0.16364	0.27113
$n=5$	0.00000	0.59167		0.27778	0.17924		0.20000	0.22269
	0.20000	0.40833		0.33333	0.12976		0.23636	0.17936
	0.40000	0.24167		0.38889	0.09009		0.27273	0.14148
	0.60000	0.11667		0.44444	0.05972		0.30909	0.10917
	0.80000	0.04167		0.50000	0.03759		0.34545	0.08229
	1.00000	0.00833		0.55556	0.02231		0.38182	0.06049
				0.61111	0.01237		0.41818	0.04328
$n=6$	0.06667	0.50000		0.66667	0.00633		0.45455	0.03009
	0.20000	0.35972		0.72222	0.00294		0.49091	0.02027
	0.33333	0.23472		0.77778	0.00121		0.52727	0.01319
	0.46667	0.13611		0.83333	0.00043		0.56364	0.00827
	0.60000	0.06806		0.88889	0.00012		0.60000	0.00497
	0.73333	0.02778		0.94444	0.00002		0.63636	0.00285
	0.86667	0.00833					0.67273	0.00155
	1.00000	0.00139	$n=10$	0.02222	0.50000		0.70909	0.00080
				0.06667	0.43090		0.74545	0.00038
$n=7$	0.04762	0.50000		0.11111	0.36374		0.78182	0.00017
	0.14286	0.38631		0.15556	0.30033		0.81818	0.00007
	0.23810	0.28095		0.20000	0.24216		0.85455	0.00002
	0.33333	0.19067		0.24444	0.19036		0.89091	0.00001
	0.42857	0.11944		0.28889	0.14562			
	0.52381	0.06806		0.33333	0.10819	$n=12$	0.00000	0.52672
	0.61905	0.03452		0.37778	0.07787		0.03030	0.47328
	0.71429	0.01508		0.42222	0.05416		0.06061	0.42029
	0.80952	0.00536		0.46667	0.03628		0.09091	0.36865
	0.90476	0.00139		0.51111	0.02331		0.12121	0.31918
	1.00000	0.00020		0.55556	0.01430		0.15152	0.27260
				0.60000	0.00833		0.18182	0.22951
$n=8$	0.00000	0.54757		0.64444	0.00457		0.21212	0.19035
	0.07143	0.45243		0.68889	0.00234		0.24242	0.15541
	0.14286	0.35977		0.73333	0.00111		0.27273	0.12479
	0.21429	0.27421		0.77778	0.00047		0.30303	0.09847
	0.28571	0.19938		0.82222	0.00018		0.33333	0.07630
	0.35714	0.13755		0.86667	0.00006		0.36364	0.05798
	0.42857	0.08943		0.91111	0.00001		0.39394	0.04316
	0.50000	0.05434					0.42424	0.03143
	0.57143	0.03051					0.45455	0.02237
	0.64286	0.01558					0.48485	0.01553
	0.71429	0.00707					0.51515	0.01049
	0.78571	0.00275					0.54545	0.00689
	0.85714	0.00087					0.57576	0.00438
	0.92857	0.00020					0.60606	0.00269
	1.00000	0.00002					0.63636	0.00159
							0.66667	0.00090
							0.69697	0.00049
							0.72727	0.00025
							0.75758	0.00012
							0.78788	0.00005
							0.81818	0.00002
							0.84848	0.00001

Table A3.5   Continued

	x	$Pr(R_K \geq x)$		x	$Pr(R_K \geq x)$		x	$Pr(R_K \geq x)$
n = 13	0.00000	0.52382	n = 14	0.01099	0.50000	n = 15	0.00952	0.50000
	0.02564	0.47618		0.03297	0.45724		0.02857	0.46129
	0.05128	0.42887		0.05495	0.41496		0.04762	0.42292
	0.07692	0.38251		0.07692	0.37359		0.06667	0.38525
	0.10256	0.33772		0.09890	0.33359		0.08571	0.34859
	0.12821	0.29502		0.12088	0.29532		0.10476	0.31325
	0.15385	0.25489		0.14286	0.25913		0.12381	0.27948
	0.17949	0.21769		0.16484	0.22528		0.14286	0.24753
	0.20513	0.18370		0.18681	0.19399		0.16190	0.21756
	0.23077	0.15309		0.20879	0.16541		0.18095	0.18973
	0.25641	0.12593		0.23077	0.13959		0.20000	0.16412
	0.28205	0.10218		0.25275	0.11656		0.21905	0.14078
	0.30769	0.08174		0.27473	0.09625		0.23810	0.11973
	0.33333	0.06443		0.29670	0.07858		0.25714	0.10092
	0.35897	0.04999		0.31868	0.06339		0.27619	0.08429
	0.38462	0.03816		0.34066	0.05051		0.29524	0.06973
	0.41026	0.02863		0.36264	0.03973		0.31429	0.05712
	0.43590	0.02109		0.38462	0.03083		0.33333	0.04632
	0.46154	0.01524		0.40659	0.02359		0.35238	0.03717
	0.48718	0.01079		0.42857	0.01778		0.37143	0.02950
	0.51282	0.00748		0.45055	0.01320		0.39048	0.02315
	0.53846	0.00506		0.47253	0.00964		0.40952	0.01795
	0.56410	0.00334		0.49451	0.00692		0.42857	0.01375
	0.58974	0.00214		0.51648	0.00488		0.44762	0.01040
	0.61538	0.00134		0.53846	0.00337		0.46667	0.00776
	0.64103	0.00081		0.56044	0.00228		0.48571	0.00570
	0.66667	0.00047		0.58242	0.00151		0.50476	0.00413
	0.69231	0.00026		0.60440	0.00098		0.52381	0.00295
	0.71795	0.00014		0.62637	0.00062		0.54286	0.00207
	0.74359	0.00007		0.64835	0.00038		0.56190	0.00143
	0.76923	0.00004		0.67033	0.00023		0.58095	0.00097
	0.79487	0.00002		0.69231	0.00013		0.60000	0.00064
	0.82051	0.00001		0.71429	0.00007		0.61905	0.00042
				0.73626	0.00004		0.63810	0.00027
				0.75824	0.00002		0.65714	0.00017
				0.78022	0.00001		0.67619	0.00010
							0.69524	0.00006
							0.71429	0.00003
							0.73333	0.00002
							0.75238	0.00001
							0.77143	0.00001

(continued)

Table A3.5   Continued

	x	$Pr(R_K \geq x)$		x	$Pr(R_K \geq x)$		x	$Pr(R_K \geq x)$
$n = 16$	0.00000	0.51765	$n = 17$	0.00000	0.51617	$n = 18$	0.00654	0.50000
	0.01667	0.48235		0.01471	0.48383		0.01961	0.47026
	0.03333	0.44718		0.02941	0.45160		0.03268	0.44068
	0.05000	0.41240		0.04412	0.41967		0.04575	0.41141
	0.06667	0.37828		0.05882	0.38825		0.05882	0.38262
	0.08333	0.34505		0.07353	0.35752		0.07190	0.35444
	0.10000	0.31293		0.08824	0.32765		0.08497	0.32701
	0.11667	0.28213		0.10294	0.29882		0.09804	0.30047
	0.13333	0.25282		0.11765	0.27116		0.11111	0.27491
	0.15000	0.22514		0.13235	0.24480		0.12418	0.25043
	0.16667	0.19921		0.14706	0.21984		0.13725	0.22713
	0.18333	0.17510		0.16176	0.19636		0.15033	0.20507
	0.20000	0.15286		0.17647	0.17442		0.16340	0.18429
	0.21667	0.13252		0.19118	0.15406		0.17647	0.16484
	0.23333	0.11407		0.20588	0.13528		0.18954	0.14673
	0.25000	0.09746		0.22059	0.11809		0.20261	0.12996
	0.26667	0.08264		0.23529	0.10245		0.21569	0.11453
	0.28333	0.06952		0.25000	0.08833		0.22876	0.10041
	0.30000	0.05802		0.26471	0.07567		0.24183	0.08757
	0.31667	0.04802		0.27941	0.06439		0.25490	0.07596
	0.33333	0.03941		0.29412	0.05443		0.26797	0.06553
	0.35000	0.03206		0.30882	0.04569		0.28105	0.05621
	0.36667	0.02584		0.32353	0.03808		0.29412	0.04794
	0.38333	0.02063		0.33824	0.03151		0.30719	0.04064
	0.40000	0.01631		0.35294	0.02588		0.32026	0.03425
	0.41667	0.01277		0.36765	0.02109		0.33333	0.02868
	0.43333	0.00989		0.38235	0.01704		0.34641	0.02387
	0.45000	0.00758		0.39706	0.01366		0.35948	0.01973
	0.46667	0.00574		0.41176	0.01086		0.37255	0.01620
	0.48333	0.00430		0.42647	0.00856		0.38562	0.01322
	0.50000	0.00318		0.44118	0.00668		0.39869	0.01070
	0.51667	0.00232		0.45588	0.00517		0.41176	0.00860
	0.53333	0.00167		0.47059	0.00396		0.42484	0.00687
	0.55000	0.00119		0.48529	0.00300		0.43791	0.00544
	0.56667	0.00083		0.50000	0.00225		0.45098	0.00427
	0.58333	0.00057		0.51471	0.00167		0.46405	0.00333
	0.60000	0.00039		0.52941	0.00123		0.47712	0.00257
	0.61667	0.00026		0.54412	0.00089		0.49020	0.00197
	0.63333	0.00017		0.55882	0.00064		0.50327	0.00149
	0.65000	0.00011		0.57353	0.00045		0.51634	0.00112
	0.66667	0.00007		0.58824	0.00031		0.52941	0.00084
	0.68333	0.00004		0.60294	0.00022		0.54248	0.00062
	0.70000	0.00002		0.61765	0.00015		0.55556	0.00045
	0.71667	0.00001		0.63235	0.00010		0.56863	0.00032
	0.73333	0.00001		0.64706	0.00006		0.58170	0.00023
				0.66176	0.00004		0.59477	0.00016
				0.67647	0.00003		0.60784	0.00011
				0.69118	0.00002		0.62092	0.00008
				0.70588	0.00001		0.63399	0.00005
				0.72059	0.00001		0.64706	0.00003
							0.66013	0.00002
							0.67320	0.00001
							0.68627	0.00001
							0.69935	0.00001

Table A3.5 Continued

	x	$Pr(R_K \geq x)$		x	$Pr(R_K \geq x)$	x	$Pr(R_K \geq x)$
$n = 19$	0.00585	0.50000	$n = 20$	0.00000	0.51277	0.62105	0.00003
	0.01754	0.47250		0.01053	0.48723	0.63158	0.00002
	0.02924	0.44513		0.02105	0.46175	0.64211	0.00001
	0.04094	0.41802		0.03158	0.43642	0.65263	0.00001
	0.05263	0.39127		0.04211	0.41134	0.66316	0.00001
	0.06433	0.36502		0.05263	0.38661		
	0.07602	0.33936		0.06316	0.36232		
	0.08772	0.31441		0.07368	0.33855		
	0.09942	0.29026		0.08421	0.31540		
	0.11111	0.26699		0.09474	0.29293		
	0.12281	0.24467		0.10526	0.27121		
	0.13450	0.22337		0.11579	0.25030		
	0.14620	0.20313		0.12632	0.23026		
	0.15789	0.18400		0.13684	0.21113		
	0.16959	0.16600		0.14737	0.19293		
	0.18129	0.14915		0.15789	0.17570		
	0.19298	0.13345		0.16842	0.15945		
	0.20468	0.11889		0.17895	0.14419		
	0.21637	0.10546		0.18947	0.12992		
	0.22807	0.09313		0.20000	0.11663		
	0.23977	0.08187		0.21053	0.10431		
	0.25146	0.07165		0.22105	0.09294		
	0.26316	0.06240		0.23158	0.08249		
	0.27485	0.05408		0.24211	0.07292		
	0.28655	0.04665		0.25263	0.06421		
	0.29825	0.04003		0.26316	0.05630		
	0.30994	0.03418		0.27368	0.04917		
	0.32164	0.02903		0.28421	0.04275		
	0.33333	0.02452		0.29474	0.03701		
	0.34503	0.02061		0.30526	0.03190		
	0.35673	0.01722		0.31579	0.02738		
	0.36842	0.01430		0.32632	0.02338		
	0.38012	0.01181		0.33684	0.01988		
	0.39181	0.00969		0.34737	0.01682		
	0.40351	0.00791		0.35789	0.01416		
	0.41520	0.00641		0.36842	0.01187		
	0.42690	0.00516		0.37895	0.00989		
	0.43860	0.00413		0.38947	0.00820		
	0.45029	0.00328		0.40000	0.00677		
	0.46199	0.00259		0.41053	0.00555		
	0.47368	0.00203		0.42105	0.00453		
	0.48538	0.00158		0.43158	0.00367		
	0.49708	0.00122		0.44211	0.00296		
	0.50877	0.00093		0.45263	0.00237		
	0.52047	0.00071		0.46316	0.00189		
	0.53216	0.00053		0.47368	0.00150		
	0.54386	0.00040		0.48421	0.00118		
	0.55556	0.00029		0.49474	0.00092		
	0.56725	0.00021		0.50526	0.00071		
	0.57895	0.00016		0.51579	0.00055		
	0.59064	0.00011		0.52632	0.00042		
	0.60234	0.00008		0.53684	0.00032		
	0.61404	0.00006		0.54737	0.00024		
	0.62573	0.00004		0.55789	0.00018		
	0.63743	0.00003		0.56842	0.00013		
	0.64912	0.00002		0.57895	0.00010		
	0.66082	0.00001		0.58947	0.00007		
	0.67251	0.00001		0.60000	0.00005		
	0.68421	0.00001		0.61053	0.00004		

Index